The
Chemical Physics
of Surfaces

The Chemical Physics of Surfaces

S. Roy Morrison

Stanford Research Institute
Menlo Park, California

PLENUM PRESS · NEW YORK AND LONDON

Library of Congress Cataloging in Publication Data

Morrison, Stanley Roy.
  The chemical physics of surfaces.

  Includes bibliographical references and index.
  1. Surface chemistry. I. Title.
QD506.M67   541'.3453   76-54152
ISBN 0-306-30960-2

First Printing — March 1977
Second Printing — October 1978

© 1977 Plenum Press, New York
A Division of Plenum Publishing Corporation
227 West 17th Street, New York, N.Y. 10011

Printed in the United States of America

✳ | Preface

At the present stage of development of surface science, there has seemed to be a need for a book-length review spanning the disciplines of surface physics and surface chemistry—a review to summarize and show the connection between the observations from each discipline. The various results and theories, derived on the one hand from studies of the physical, electronic, and optical properties of surfaces and on the other hand from studies of the chemical activity of surfaces, supplement each other in the search for a realistic model of the surface. The improved understanding possible with such an interdisciplinary approach has been confirmed by recent developments which cannot be classified as either surface chemistry or surface physics. Specifically, recent new experimental techniques and quantum mechanical models have provided a much more accurate picture of the nature of the electronic energy levels (bonding orbitals) present at a solid surface. With this more accurate picture we are now able to reconcile the various chemical and physical models that appeared in the early literature on surfaces.

The objective of this work has therefore been to describe the results and current models of surface science spanning a broad gray area between surface physics and surface chemistry with some overlap into each of these disciplines. Relevant aspects of surface chemistry are discussed; we cover chemical interactions where bonding and electronic properties dominate, but stop short of specialized topics such as surfactants or liquid/liquid interfaces. Relevant aspects of surface physics are discussed; we cover physical effects at the solid/gas and solid/liquid interfaces, but stop short of specialized topics such as solid/solid interfaces or grain boundaries. Throughout the book the objective has been to present the subject as much as possible as a unified science, "chemical physics," to identify the conditions where each of the various classical models becomes the most useful representation.

In preparing this book two groups were always in the forefront of consideration: the mature surface chemist who wants to become more familiar with recent developments in surface physics, and the mature surface physicist who wants to become more familiar with the chemistry occurring on his sample surfaces. However, the student or scientist interested in entering the field should also find the book of value as a general introduction to surface science.

An extensive list of references, far more than normal for a book of this type, has been included. An effort has been made to make the book self-sufficient, and to discuss the points adequately so it is not necessary to delve often into the literature. Ample references were, however, included to help the reader initiate a deeper study of those areas of particular concern to him. The liberal use of references has an additional advantage in the fast-moving field of surface science. Many stimulating new concepts which are not completely proved can be included as the opinion proffered in certain references, and these new concepts add greatly to our understanding of the surface. Thus, in the discussions of recent developments, more references will be noted than in the discussions of classical surface science.

I would like to acknowledge my indebtedness to several of my coworkers in the Materials Group at Stanford Research Institute for valuable comments on the text. These include Jon McCarty, Morris Landstrass, Paul Wentrcek, and Don MacMillan. In particular I should like to thank Dan Cubicciotti and Michihiro Nakamura for their extensive reviews and comments. In addition I am pleased to acknowledge my indebtedness to my wife, Phyllis Morrison, for literary criticism, assistance in proofreading, and general support through the long period when she was a book widow.

S. Roy Morrison

✳ Contents

6. Nonvolatile Foreign Additives on the Solid Surface

10. Surface Sites in Heterogeneous Catalysis

✳ Notation*

Roman Letters

a . . . . . Ionic radius

A . . . . . Area, electron affinity of a free atom, symbol for an atom (as in an AB crystal)

b . . . . . Constant in Langmuir isotherm [Equation (7.3)]

B . . . . Symbol for an atom (as in an AB crystal)

$\bar{c}$ . . . . . Thermal velocity of electrons

c_i . . . . Concentration of ion i in solution

c_{ox} . . . . Concentration of oxidizing agent (energy level unoccupied) in solution

c_{red} . . . . Concentration of reducing agent (energy level occupied) in solution

C . . . . Capacity [Equation (3.7)]

C_H . . . . . Helmholtz capacity

C_{sc} . . . . Space charge layer capacity

D . . . . Dispersion (defined in Section 6.2.1)

e . . . . . Electron, electronic charge

e^- . . . . Electron

E . . . . Energy

E^o . . . . Standard oxidation potential (redox potential) for a redox couple in solution

E^o_{redox} . . . An energy parameter in solution resembling a Fermi energy [Equation (8.19)]

---

* See Table 3.1 for the abbreviations used for measurement techniques such as UPS, AES, etc.

E_A . . . . Activation energy of adsorption

E_c . . . . Energy of conduction band edge

E_{cs} . . . . Conduction band edge at the surface

E_D . . . . Activation energy of desorption

E_e . . . . Energy of a free electron at infinity

E_F . . . . Fermi energy

$E_F(\text{ref})$ . . Fermi energy of the metal contact to a reference electrode in solution, $E_F(H_2)$, to a hydrogen reference electrode

E_g . . . . Forbidden energy gap of a semiconductor

E_m . . . . Effective surface state energy [Equation (5.17)]

E_{ox} . . . . Mean energy level of an oxidizing agent (energy level unoccupied) in solution

E_{O_2}, E_O . . Energy level on an oxygen molecule, atom

E_p . . . . Energy in polarized dielectric

E_{red} . . . Mean energy level of a reducing agent (energy level occupied) in solution

E_t, E_1, E_2 . Energy level of surface states

E_v . . . . Energy of the valence band edge

E_{vs} . . . . Valence band edge at the surface

f . . . . . Fermi distribution function [Equation (2.14)]

G . . . . Conductance of a sample

ΔG . . . . Free energy change in a reaction

h . . . . . Planck's constant

h^+ . . . . Hole

H_0 . . . . Acid strength, acidity of a site

H . . . . Hamiltonian

J . . . . . Current density

J_a . . . . Anodic current density at an electrode

J_0 . . . . Exchange current density at an electrode, the current in each direction when the net current is zero

J_p, J_e . . . Current density due to holes, electrons

k . . . . . Boltzmann's constant

K_n . . . . Rate constant for electron capture at a surface state

K_p . . . . Rate constant for hole capture at a surface state

L . . . . Langmuir $= 10^{-6}$ Torr-sec

LDS . . . Local density of states

M . . . . Symbol for electropositive atom, as in the compound MX

n_1 . . . . Emission constant for electrons [Equation (2.37)]

n_b . . . . Bulk electron density in the conduction band

n_s . . . . Density of electrons in the conduction band at the surface (per unit volume)

n_t . . . . Density of electrons in a surface state (per unit area)

$N(E)$ . . . Density of electronic energy levels as a function of energy

ΔN . . . . Change in electron density (per unit area) from flat band condition

N_A . . . . Acceptor density

N_c . . . . Effective density of states in the conduction band

N_D . . . . Donor density

N_{sc} . . . . Density of immobile charge in the space charge region

N_t . . . . Density of surface states

N_v . . . . Effective density of states in the valence band

O_L^{2-}, O_L^- . Surface lattice oxide ion, an oxide ion with an electron removed

Ox . . . . Symbol for an oxidizing agent (electron acceptor)

p . . . . . Pressure (p_{O_2} = oxygen partial pressure, etc.), density of holes

ΔP . . . . Change in hole density (per unit area) from flat band condition

p_1 . . . . Emission constant for holes [Equation (2.42)]

p_b . . . . Bulk hole density in the valence band

p_t . . . . Density of holes in a surface state (density of unoccupied states)

q . . . . . Electronic charge

Q . . . . Charge (coulombs), heat of adsorption

Q_c, Q_p . . Heat of chemisorption, physisorption

Red . . . Symbol for a reducing agent (electron donor)

s . . . . . Surface recombination velocity [Equation (9.19)], sticking coefficient in adsorption

S . . . . A site on the surface, the configuration of molecules in the neighborhood of the ion of interest

T . . . . Temperature, usually °K

U . . . . Steady state carrier capture rate at the surface [Equation (9.1)]

U_p, U_n . . Net rate of hole, electron capture at a surface state

U_s . . . . eV_s/kT

$W(E)$ . . . Probability that a fluctuating energy level will have the energy E

x . . . . . Distance from the surface, a distance parameter

x_0 . . . . Position where surface space charge layer begins

X . . . . Symbol for an electronegative atom as in the compound MX

z . . . . . Partial charge on an atom

z_1, z_A, z_B . . Parameters in the Hückel theory

Z . . . . Formal valence number of an ion, thermal velocity of ions in solution

Greek Letters

α . . . . . The Coulomb integral in the Hückel model [Equation (4.6)], the transfer coefficient in electrochemical theory, as defined following Equation (8.14)

$\bar{\alpha}$ . . . . . Mean value of α for an AB crystal

β . . . . . The resonance integral in the Hückel model [Equation (4.7)]

Γ_x . . . . Surface density of the adsorbate X

Γ_n . . . . Density of unoccupied states

Γ_t . . . . Surface density of adsorption sites

Δ . . . . A parameter measured in ellipsometry (Section 3.1.9)

ε_0 . . . . Permittivity of free space

η . . . . . Overvoltage at an electrode, voltage relative to the zero current voltage

θ . . . . . Fractional coverage of the surface (Γ/Γ_t)

$\varkappa, \varkappa_s$ . . . Dielectric constant

$\varkappa_{op}$ . . . . Optical dielectric constant

λ . . . . . Reorganization energy of a polar medium

μ . . . . $E_c - E_F$

μ_n, μ_p . . Mobility of electrons, holes

ν . . . . . Frequency

ϱ . . . . . Charge density

σ . . . . . Capture cross section, conductivity

$\Delta\sigma$ . . . . Surface conductivity [Equation (3.5)]

ϕ . . . . Potential, atomic orbital or hydrogenic wave function, work function of a solid

ϕ_b . . . . Potential in the bulk of the semiconductor

ϕ_s . . . . Potential at the surface

ψ . . . . A parameter measured in ellipsometry (Section 3.1.9), a wave function [Equation (4.1)]

Superscripts and Subscripts

ss . . . . Surface states

sc . . . . Space charge

0 . . . . .   Value at equilibrium, value before perturbation
t . . . . .   Value on a surface state
s . . . . .   Value at the surface
b . . . . .   Value in the bulk solid
v, μ . . . .   Integers, number assigned to an atom in the chain in Hückel
              theory
n, m . . . .   Small integers, as in chemical formulas
* . . . . .   Activated form or excited form
ox . . . .   Oxidized form (energy level of interest unoccupied)
red . . . .   Reduced form (energy level of interest occupied)
L . . . .   On a lattice site

1 | Introduction

1.1. Surface States and Surface Sites

1.1.1. The Chemical Versus Electronic Representation of the Surface

There have been two dominant models used in describing the chemical and electronic behavior of a surface. One is the atomistic model, or the "surface molecule" model; the other is the band model, often called for emphasis the "rigid band" model. The atomistic model has been preferred in general in discussions of the chemical processes at a solid surface; the rigid band model is often preferred in discussions of electron exchange between the solid and a surface group. The atomistic model describes the solid surface in terms of surface sites, atoms, or groups of atoms at the surface, essentially ignoring the band structure of the solid. The rigid band model describes the surface in terms of surface states, localized electronic energy levels available at the surface, to a great extent ignoring the microscopic details of atom/atom interaction between the surface species and its neighboring substrate atom.

The best example of this schizophrenia and how it is being resolved arises in theoretical and experimental work on adsorbed species. Up to a few years ago the two approaches appeared endlessly diverging. Many authors[1,2] have emphasized the rigid band model in the theory of the surface. Electrons are described with nonlocalized Bloch wave functions, with surface states occurring at the surface because the adsorbed atom has an electron affinity different from atoms of the substrate. The adsorbate provides surface state energy levels able to capture or donate electrons to the solid, but any local adsorbate/solid interaction is substantially neglected. The advantage of the model is that the mathematical description, in the first approximation, is independent of the chemical species at the surface and independent of the details of the local chemical interaction. Others[3-5] have emphasized the other alternative, ignoring the presence of the solid as a whole and considering a completely local bonding of the adsorbate atom to one particular surface atom or group of atoms on the solid. If one is primarily interested in the microscopic details of chemical bonding, this model is obviously superior. This schizophrenia has not been only among theorists. Most physicists and some chemists have weighted the band model much too heavily in trying to describe the observed behavior of adsorbed species. And others have ignored band effects completely in their interpretation of results.

With the advent of better approaches and better computers, modern theories[6-10] are finally combining the approaches. The surface atom plus a cluster of nearby atoms is analyzed without the neglect of the background bands of the total solid. The results of the calculations are coming close to experimental values. The present period is a very exciting one in the development of surface physics and chemistry, primarily because of the synthesis of the two approaches that is under way. However, to understand the synthesis, the two approaches must first be understood: the approach emphasizing surface sites and the approach emphasizing surface states to describe the behavior of the solid surface.

We will use the term "surface site" to describe a microscopic group of atoms at the surface which is in some way reactive. Sites can be associated with the uniform surface; more reactive sites can be found on heterogeneous regions of the surface. Examples of a site on a uniform clean surface would be a surface atom of the host lattice with a "dangling bond," an unoccupied bonding orbital of high electron affinity, or an occupied bonding orbital with a low ionization potential. A site can be one of the cations on the surface of an ionic crystal which, uncompensated by a full complement of anions in the way that bulk cations are compensated, exhibits a high electron affinity. Or it can be a location with a high electric field, such that polar molecules are attracted to the site. Such sites may be present in density the order of $10^{19}/m^2$, one for each surface atom. On the other hand, sites may be associated with imperfections on a heterogeneous surface. Examples of the latter would be sites at a crystallographic step or at the intersection of a dislocation with the surface. At a crystallographic step there will be sites where an adsorbing species can interact with several lattice atoms, and the total interaction can be very strong indeed. The probability of an oxygen molecule sticking after striking a silicon surface is 500 times higher when the surface is covered by steps.[11] Or the site can be an impurity at the surface. Mixed oxides often show a much stronger attraction for gaseous donors (bases) such as ammonia than does either oxide individually, as discussed in Section 4.3.3.

A surface state is a localized electronic energy level at the surface. We are particularly interested in energy levels which are active in exchanging or sharing electrons with the nonlocalized energy bands in the bulk of the solid. Thus when we discuss surface states we are often more interested in macroscopic properties than in the microscopic atomic orbitals that provide the surface states. Quantum mechanics shows that localized energy levels will arise at the surface of a solid simply because the periodicity is broken. Such surface states are called "intrinsic states," and originate in band model calculations in various ways (as will be discussed in Section 4.2). For example, there are Shockley states, which arise particularly on covalent materials where for the bulk atoms there is a large overlap of valence electron orbitals. The exchange energy stabilizes these bulk valence electrons. Some of the electrons on the surface atoms on the other hand are in orbitals directed toward the surface ("dangling bonds") and do not overlap a neighboring orbital, so they occupy different levels from bulk valence electrons. As another example there are Tamm states, which arise when the surface atom has an electron affinity different from bulk atoms; again the electrons on these surface atoms will occupy states of energy different from bulk electrons. More generally, localized surface state energy levels can arise from many causes. Impurities at the surface, or adsorbates, or even ions nearby in solution that are close enough to exchange electrons with the bands of the solid will give rise to localized surface energy levels, and therefore satisfy the definition (and more important, obey the mathematics) of surface states in a rigid band model.

As an illustration of the use of the two approaches, consider the case of ferrous and ferric ions as impurities present on the surface of a semiconductor. First let us assume that the energy levels associated with the iron ions are affected very little by the presence of the solid (as will be discussed later, this permits the rigid band model to give a useful description of the system). Then we can consider that the ferric and ferrous ions can exchange electrons with the semiconductor in accordance with a redox process such as

$$Fe^{2+} = Fe^{3+} + e \qquad (1.1)$$

In the rigid band model both the species Fe^{2+} and Fe^{3+} would be represented by the same surface state energy level. If the surface state is occupied by an electron, the species is in the form Fe^{2+}, Equation (1.1), and if the surface state is unoccupied, the species is Fe^{3+}. The surface state associated with Equation (1.1), the Fe^{2+}/Fe^{3+} surface state, would have an energy determined by the tendency of the reduced species Fe^{2+} to give up electrons. A stronger reducing agent, such as V^{2+}, with its greater driving force to give up electrons, would have a higher surface state energy level (V^{2+}/V^{3+}) than the Fe^{2+}/Fe^{3+} level. In other words, there is a direct relation between the chemical redox properties of a surface group and its expected surface state energy level.[12–14] For example, in Section 8.2.3 we explore the relation between the redox potential of a couple and its surface state energy level.

On the other hand, in a strictly surface site (surface molecule) model, interaction with the bands is ignored and local interaction with the neighboring substrate atom(s) is emphasized. The interaction of a substrate atom with an Fe^{3+} ion is expected to be quite different from its interaction with an Fe^{2+} ion, so correspondingly two quite different "surface molecules" will form. If the interactions are strong, the resulting electronic energy levels will be quite different for the two cases. The stronger the interactions, the more we diverge from the simple rigid band picture of a single surface state. In the surface molecule picture, the surface site picture, we look to chemical valence theory to describe the strong interactions at the surface.

1.1.2. The Surface State on the Band Diagram

In Figure 1.1 a band diagram of a semiconductor is introduced to illustrate the relation in the rigid band model between chemical properties and the resulting surface state energy level. The band picture of a solid that will normally be used for illustration in this book is of the form of

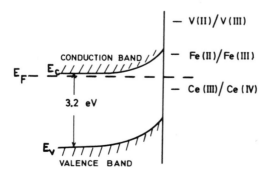

Figure 1.1. Surface states of chemical couples, to illustrate the expected energy level as the chemical properties of the species is varied. ZnO is used as the example semiconductor; the indicated position of the surface states relative to the bands is discussed further in the text.

Figure 1.1, a plot of the energy levels of an electron versus a distance parameter. In a solid, due to overlap of orbitals, the single electronic energy levels from each atom are combined and broadened into a band. In a metal, the uppermost occupied band overlaps the next higher band, or the uppermost occupied band is only partly occupied by electrons. For a semiconductor or an insulator, the bands of most interest are the band occupied by the valence electrons in the solid and the next excited state, an unoccupied (essentially) band. These are shown shaded in Figure 1.1. The band occupied by the valence electrons is termed the valence band, and its uppermost energy level, the valence band edge, is denoted by E_v. The next higher band, the excited state, is termed the conduction band (since electrons introduced into this band are capable of conducting electricity), and the lowermost energy level, the conduction band edge, is denoted by E_c. If there are donor impurities in the material donating excess electrons to the conduction band, the semiconductor is an n-type semiconductor. If there are acceptor impurities capturing electrons from the valence band (leaving unoccupied levels that are called holes), the material is p-type. Holes behave as unit positive charges in the conduction of electricity through the material.

Since we are using distance as the parameter on the abscissa in this diagram, we can terminate the band diagram with a surface, as shown. Surface states are indicated as localized energy levels in the region of this termination. For a surface impurity such as Fe^{2+}, lying beyond the last plane of the crystal, the energy level would be indicated beyond the termination.

A single energy level is shown for the redox couple Fe^{2+}/Fe^{3+}, which can exchange electrons with the solid according to the reaction (1.1). This represents the simple case of the rigid band model where the energy level is the same whether it is occupied or unoccupied. If the level is occupied the chemical species is Fe^{2+}; if the level is unoccupied, the chemical species is Fe^{3+}. This model is entirely analogous to the usual solid state model for a bulk impurity—an indium donor in a ZnO lattice is represented by a single donor impurity level; if the electron is on the donor, the species is In^{2+}; if the donor is ionized, the species is In^{3+}. With bulk impurities only electronic transitions are considered, so the chemical form of the ion is unimportant, and the description in terms of oxidation state is seldom mentioned. With a surface impurity which can undergo either electronic or chemical reactions, both representations are important, and the relation between the surface state and oxidation state representations must be accurately described.

As discussed in the preceding section, a simple interpretation of chemical activity suggests the surface state energy level for V^{2+}/V^{3+} will be higher than that for the iron couple, the energy level for Ce^{3+}/Ce^{4+} will be lower, and thus the levels are assigned as shown in Figure 1.1. The curvature of the bands as indicated reflects a surface double layer, discussed in Chapter 2.

1.1.3. The Fermi Energy in the Surface State Model

In addition, in Figure 1.1 is shown the Fermi energy or Fermi level E_F, which is equivalent to the electrochemical potential for electrons in the solid. At the Fermi energy the probability of an electron occupying a level is $\frac{1}{2}$, according to the Fermi distribution function. The probability that a higher available energy level is occupied by an electron is less than $\frac{1}{2}$, decreasing exponentially with energy. The probability that a level well below the Fermi energy (more than say $2\,kT$, 0.05 eV at room temperature) is occupied is close to unity. In Figure 1.1 then if this is the representation at equilibrium, the iron levels are essentially all unoccupied by electrons (they are the Fe^{3+} species) and the cerium levels are essentially all occupied (they are the species Ce^{3+}).

The Fermi energy is the energy associated with electrons transferring to or from the solid in equilibrium processes. This feature arises since the Fermi energy is the electrochemical potential of electrons in the solid. Electrons moving into the solid appear to terminate at the Fermi energy, and electrons emerging from the solid appear to originate on the average at the Fermi energy. Thus the mean energy of the electron e in Equation (1.1) is E_F.

As will be discussed in detail in Chapter 2, the position of the Fermi energy relative to the surface state is a dominant variable in the rigid band model. The band curvature, or band bending, shown in Figure 1.1 dramatically varies the position of the surface state relative to the Fermi energy and thus varies the surface state occupancy (the oxidation state of the adsorbate). This double-layer effect is hard to work into the surface molecule model.

1.1.4. Need for Both Surface Site and Surface State Models

In general the description of the surface in terms of surface states and the rigid band model has been most useful when dealing with semiconductors. However it will be seen in later discussions that it is desirable to introduce the features of the band model, to a greater or lesser extent, for discussions of the surface of all solids: metals, semiconductors, and insulators. Most of the interesting surface reactions cannot be simply relegated to the "microscopic" or to the "macroscopic" categories, but are in the broad gray area where both types of interactions must be simultaneously considered.

One of the primary objectives of this monograph is to show the parallelism and overlap of the two models, the surface molecule and the rigid band models, attempting as much as possible with modern theories to provide a combined picture. This goal is entirely consistent with the most recent theoretical descriptions of surface states where the energy level of the surface state is derived from a model based on a combination of a localized molecular orbital description and a nonlocalized band approximation. These new developments will be studied carefully, but their application is limited to simple configurations and processes. Most features of surface behavior can be understood only in a semiquantitative way; they are too complex for quantum models to be successful. The interpretation of these features can be made in terms of surface sites, as modified by the existence of bands in the solid, or in terms of surface states, as modified by local chemical processes. Seldom do cases occur where purely one or the other model provides a completely satisfactory description of the surface behavior.

1.2. Bonding of Foreign Species to the Solid Surface

1.2.1. Types of Interaction

The interaction between a foreign species and the surface of a solid will determine both the chemical and electrical properties of the surface. With respect to the chemical behavior this fact is so obvious as to be barely worth mentioning. Clearly adsorption, catalysis, corrosion, and the passivation techniques to prevent such processes, all depend upon the bonding of a reactant to the surface of the solid of interest. On the other hand the importance of the bonding mechanism in determining physical effects has been somewhat neglected to date. As will be discussed many times, the surface state energy level associated with a foreign species will depend on the chemical bond it forms with the surface. Now a valuable feature of the surface state approach is the ability to ignore the origin of the surface states and concentrate on the measurement. And such an approach has been fruitful if at times naive. It is of great interest, however, to attempt to derive the expected surface state energy from the chemical properties of the surface species and from the interaction of the surface species with the substrate.

In the present section we introduce a very qualitative discussion of the types of bonding, the various effects which influence the energy of bonding, and the relation between bonding and the energy levels and chemical characteristics of the surface group.

At the one extreme of adsorbate bonding is ionosorption, where a free electron from the conduction band or a free hole from the valence band of a solid becomes captured on or injected by the surface species and this "electron transfer" process is the only process releasing or absorbing energy. Ionosorption is a case of an ionic bond in the real sense. The foreign adsorbate species are ionized, and the counterions may be hundreds of angstroms from the adsorbed ions, so there is no sharing of electrons as found in the covalent bond discussed below. Figure 1.2a shows the formation of O_2^- by ionosorption. In the semiquantitative calculations of Green and Lee[15] an energy cycle is examined where an electron is removed from the solid, placed on the foreign atom, and then the resulting ion is permitted to approach the surface. Their calculation of the net energy change shows that for gas adsorption this pure case of ionosorption will seldom happen. By these calculations only oxygen molecules have sufficient electron affinity to capture an electron in this way from the conduction band of a typical semiconductor. The modification of Green and Lee's theory by including electrostatic (Madelung) effects will improve the situation somewhat.[16] Madelung effects are pure electrostatic effects occurring on an ionic solid (Sections 4.2.2 and 5.2.3.1). The charged adsorbate can locate itself at a region of attractive potential (the Madelung potential), increasing the energy released but not requiring a formal chemical bond to the surface atom of the solid.

Another class of foreign adsorbates where the bonding can be described best as ionosorption is that of ionized surface additives. An example, ferric/ferrous additives, introduced perhaps as chloride salts, were discussed in the preceding section. In this case, the surface state energy level can be in the conduction band region, the energy gap region (between the conduction band and valence band edges) or in the valence band region of the solid, depending on the relative chemical properties of the redox system (the adsorbate) and the solid.

In general, ionosorption involving only pure electrostatic effects (with no local chemical bond formation) will be rare.

$$
\begin{array}{ccc}
\underline{\quad O_2^- \quad} & \underline{\quad \begin{matrix} O & O & O & O \\ \| & \| & \| & \| \end{matrix} \quad} & \underline{\quad \begin{matrix} Ge & O & Ge & O \end{matrix} \quad} \\
-Ge-Ge-Ge-Ge- & -Ge \;\; Ge-Ge \;\; Ge- & O \;\; Ge \;\; O \;\; Ge \\
\;\;\;\;\;\;\;\;\;\; | & \;\; |\;\;\;\; |\;\;\;\; |\;\;\;\; | & \\
-Ge-Ge-Ge-Ge- & -Ge-Ge-Ge-Ge- & -Ge-Ge-Ge-Ge- \\
\;\;\;| \;\; \overset{+}{}\;\;\;| \;\;\;\; | & |\;\;\;| \;\;\;\; | \;\;\;\; | & |\;\;\;\;\; | \;\;\;\; | \;\;\;\; | \\
-Ge-P^+-Ge-Ge- & -Ge-\;P\;-Ge-Ge- & -Ge-\;P\;-Ge-Ge- \\
|\;\;\;\;\;|\;\;\;\;\;|\;\;\;\;\;| & |\;\;\;\;\;|\;\;\;\;\;|\;\;\;\;\;| & |\;\;\;\;\;|\;\;\;\;\;|\;\;\;\;\;| \\
\text{(a)} & \text{(b)} & \text{(c)}
\end{array}
$$

Figure 1.2. Various forms of bonding of oxygen to a solid: (a) ionosorption, with counter charge on phosphorus donor, (b) covalent bonding, and (c) incipient oxide formation.

At the other extreme is the case when the adsorbed foreign species is held with only localized forces, pure "chemical bonding;" when there is no electron transfer from the bands of the solid, but only chemical bonding between the adsorbate and a surface atom or a few surface atoms of the solid. Such adsorption may occur, for example on the surface of an insulator, or on a semiconductor with a "dangling bond" surface state in which a localized electron is trapped or a localized orbital is unoccupied and available at the surface. Then a covalent chemical bond can develop as described in the Section 1.2.2. In Figure 1.2b oxygen is shown adsorbed on a germanium surface, forming a local double bond with the surface atoms. Another form of local bonding that may occur on an ionic semiconductor or insulator is an acid/base covalent bond, with the solid attracting gaseous bases or acids as described in the next two sections. Sorption on metal surfaces is expected to depend dominantly on covalent chemical bonding, but it is expected that the bond strength will depend to some extent on the collective electron properties (the Fermi energy) and so will not be "pure" localized chemical bonding.

Local bonding between an atom and a surface species may involve dipole/dipole attraction, which results in strong forces if a polar molecule adsorbs on an ionic solid, but weak (van der Waals) forces if a nonpolar molecule adsorbs on a nonpolar solid. In the latter case there is a weak attractive interaction due to induced dipoles.

Local bonding may alternatively depend on what are called crystal field or ligand field effects. The crystal field theory originated in models of bonding in solids emphasizing the directional characteristics of the bonding. The theory is most important when analyzing bonding to the transition metal atoms where the d orbitals have interesting directional characteristics. Polar groups or charged groups becoming bonded to the transition metal or ion are termed its ligands. The energy of the system can be lowered if these ligands attach themselves with certain orientations related to the directional characteristics of the d orbitals. The insertion of the ligands raises the energy of some of the d electrons and lowers the energy of others. If there is an incomplete set of d electrons, so most can be located in the lowered energy levels, a net gain in energy obtains. This, together with some covalent interaction of the type described below, leads to a strong bond.

Geometry can be important in local bonding. In adsorption on transition metals or transition metal ions with dangling d orbitals, bonds are favored to species that have orbitals that match the directional characteristics of the d orbitals.[4] Bonds can be mixed, say part acid/base and part geometry-dependent crystal field bonding to d orbitals.[17] In the adsorption of large molecules,[18] the strength of the bond may depend strongly on the match between lattice spacing and bond length in the adsorbate.

A third class of foreign atom interaction at a solid surface, the possibility of which must always be considered, is the formation of another phase. A sketch of this possibility is shown in Figure 1.2c. The obvious example is the oxidation of a metal. The question of when adsorbed oxygen begins to behave like a surface oxide is difficult to define, particularly when relocation of the atoms of the solid must be considered as part of the process of adsorption as well as being considered part of the process of a second phase formation. This problem will be discussed further in Section 5.2.

1.2.2. The Chemical Bond

It is beyond the scope of the present work to review the theories of the covalent chemical bond between two molecules or atoms. A short review is presented here that is intended to provide a minimal background for discussions of theoretical models of local adsorbate/solid interaction. Supplementary reading is, however, recommended.

The valence bond theory and the molecular orbital theory each attempts to describe what is termed the covalent bond between two atoms. The covalent bond can vary between the homopolar covalent bond, a bond formed between two identical atoms, to the ionic bond where one of the two atoms has a comparatively high electron affinity and the bonding electrons are primarily in orbitals around that nucleus.

The valence bond theory describes a molecule in terms of wave functions representing an electron assigned to each atom, and the total wave function becomes a series of products of electron wave functions. The straightforward model is corrected by recognizing that in addition to the obvious locations, electron 2 could equally well be occupying the orbital around atom 1, and electron 1 be on atom 2. The possibility of thus exchanging the electrons brings another term into the expression which provides substantial lowering of the energy. The greater the overlap between the orbitals of the two bonding electrons, the greater the exchange term and the lower the electronic energy. The bonding electrons must be unpaired in the original lattice (or become unpaired during the bonding reaction) or the overlap is not possible due to the Pauli principle.

The atomic orbitals used to construct these wave functions can be hydrogen-like orbitals, or can be "hybrid" orbitals, built up of a linear combination of hydrogen-like orbitals. These hybrid orbitals are chosen to have special directional characteristics that will increase the overlap. For example p orbitals have dominant lobes at 90° angles, an sp orbital (a linear combination of one s and one p orbital) shows two linearly directed lobes making a 180° angle, an sp^2 orbital has three lobes in a plane making angles of 120°, and sp^3 orbitals (a hybrid of s and the three p orbitals) can be formed with tetrahedral symmetry, having four lobes making 109°20′ angles. Such directional characteristics become extremely important when discussing the bonding of adsorbates to the surface atom of a solid.

The molecular orbital theory is the one we shall refer to most often. Here a molecule is described in terms of a combination (sum) of atomic orbitals. In other words, one calculates the available energy levels in the molecule as a whole (summing atomic orbitals as the first representation) and then fills these molecular orbitals with the appropriate number of electrons, starting with the lowest energy orbital.

In cases where the original atomic orbitals show overlap, half of the molecular energy levels originating from two overlapping atomic orbitals are "bonding orbitals" lower in energy than the original atomic levels. The other half of the molecular energy levels are "antibonding orbitals" higher in energy than the original levels. Thus if the original atomic orbitals are half-occupied, so the electrons occupying the molecular orbitals can be accommodated in the lower energy bonding orbitals alone, there is a considerable gain in energy by the covalent bonding process, and a substantial lowering of the energy level of the electrons. On the other hand, if the original atomic orbitals are initially completely occupied, then both the bonding and the antibonding molecular orbitals will be completely occupied, so there will be no gain in energy by forming the bond. It is possible but unusual to have overlapping orbitals where the sign of the wave functions are such that there is no energy change, and these are called "nonbonding" orbitals.

A convenient qualitative way to view the available molecular orbitals associated with a pair of atoms is thus to consider three types, the bonding orbital, the nonbonding orbital, and the antibonding orbital. Comparing the energy level of each to the energy level of an unpaired electron on its isolated atomic orbital, we can say that the energy level of the bonding orbital will be lower, the energy of the nonbonding orbital will be about equal, and the energy of the antibonding orbital will be higher. This qualitative classification will be useful as a starting point in the estimation of surface state energy not only for "simple" bonding of adsorbed species to the surface, but also for complex configurations of surface atoms or for groups of atoms at the surface. To a first approximation the bonding orbital is a molecular orbital where a single electron occupies two atomic orbitals that overlap, with the overlap such that the sign of the wave function in the overlap region is the same for both atomic orbitals. The nonbonding molecular orbital is one with the overlap such that in half of the overlap region the sign of the wave function is the same but in the other half it is opposite. A nonbonding orbital can also be one in which there is no overlap. An antibonding orbital is a molecular orbital in which there is overlap, but the sign of the wave function in the overlap region is opposite for the two atomic orbitals.

A molecular orbital where the overlap of the electronic wave functions occurs on the axis, the line between the bonding atoms, is termed a σ bond. A molecular orbital with overlap in the off-axis region is termed a π bond. Two atoms can be bonded together with two or even three bonds, involving both σ electrons and π electrons, the orbitals extending in different directions from the axis between the two atoms. For example, Figure 1.3 shows a sketch of two atoms each with three p orbitals (nitrogen, for example). The p_x orbital is directed along the x axis with the wave function positive in one direction and negative in the other. Similarly for p_y and p_z orbitals. If these two atoms in Figure 1.3 were moved together, it is clear that there would be a substantial overlap of the p_y orbitals, leading to a strong covalent σ bond. In addition, there would be some overlap of the p_x and the p_z orbitals, forming two off-axis π bonds.

When a covalent bond forms between unlike atoms, there is a shift in the electron distribution leading to a "partial charge" on each of the atoms, a partial negative charge on the one, a partial positive charge on the other.[19] Two identical atoms each able to provide an unpaired electron will form a homopolar covalent bond where the partial charge is zero. Two unlike atoms will form a heteropolar covalent bond with the partial charge on each atom not zero, for one of the atoms will inevitably be

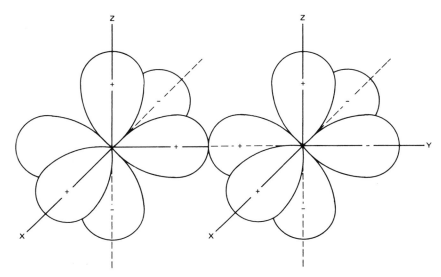

Figure 1.3. Triple bond formation using p_x, p_y, and p_z orbitals. As the separation of the atoms decreases, the p_y orbitals overlap to form a σ bond, the p_x and p_z orbitals overlap to form π bonds.

somewhat more "electronegative" than the other, and the shared electron pair will have a greater wave function amplitude in the neighborhood of the more electronegative species.[20] This partial charge concept is important not only on individual molecules, but also on solids where one moves[21] from a pure covalent crystal to a pure ionic crystal as the partial charge increases. The expected adsorbate bonding to the crystal surface depends strongly on the ionicity of the crystal, as will be discussed in the next section.

Another type of bond which becomes particularly important in the discussion of surfaces is the bond between two molecules *A* and *B*, one of which, *B*, has an electron pair which is not being shared by another atom, and the other of which, *A*, has an unoccupied orbital at a convenient energy (near or below the energy of the electron pair orbital on molecule *B*) and a convenient directional characteristic. Again in this case, the occupied orbitals of molecule *B* can overlap the unoccupied orbitals of molecule *A*, forming the usual bond with the usual energy advantage. The fact that both electrons come from *B* instead of one from *B* and the one from *A* has little effect on the molecular orbital theory, although obviously molecule *B* will normally acquire a partial positive charge, and molecule *A* will acquire a partial negative charge. The symbols *A* and *B* were chosen for these molecules as molecule A is defined as an acid, a Lewis acid, molecule B is defined as a base, a Lewis base. Such donation or acceptance of an electron pair is an acid/base reaction.

1.2.3. Acid and Basic Surface Sites on Solids

The concepts of covalent bonding of adsorbates to a solid are so analogous to the equivalent concepts in atom/atom bonding that the qualitative description of covalent bonding in the preceding section can be easily extended to the sorbate/solid case, pending a more detailed analysis in Chapter 5. The analogy is not so clear, however, for acid/base bonding at a surface. Thus it is of value to examine more closely the concepts of acid/base bonding as visualized at the solid surface.

As with the molecular Lewis acid, a Lewis acid site on a solid is a site which has an unoccupied orbital with a high electron affinity for an electron pair, so that a major decrease in energy is obtained when such a site shares an electron pair donated by an adsorbed base molecule. Lewis base sites on the surface can also arise which have electron pairs available at a high energy level, and a major decrease in energy is obtained if they share this electron pair with an adsorbed electron pair acceptor.

The acid/base concept on a solid is further complicated because there are two forms of acid sites, Lewis and Brønsted acid sites. The Lewis acid site has a high electron affinity as described in the preceding paragraph; the Brønsted acid site has a tendency to give up a proton. In many cases the two can be related by the presence of water:

$$L^+ + H_2O = (L{:}OH) + H_s{}^+ \tag{1.2}$$

where L^+, the Lewis site, a cation on the surface, shares an electron pair with the OH^- ion from the water molecule, and the remaining proton $H_s{}^+$ is adsorbed but easily removed in a chemical reaction. Thus the Lewis activity has been converted to a Brønsted activity.[22]

Most measurement techniques (Section 3.3.3) fail to distinguish between Lewis and Brønsted activity. The acid strength of an acid site is measured by its ability to convert an adsorbing acid molecule into its conjugate base. Lewis acid or basic strengths are thus closely related to electron affinities of the sites (a site with a higher electron affinity will have greater power to extract the electron donor groups from the molecular base and thus convert the molecular base into its conjugate acid). However, the acid strength of a site by this definition is not solely determined by its electron affinity, but includes other factors such as the geometry of the site and the orientation of the unoccupied orbitals. Unfortunately the acidity of the site also depends upon these same two factors on the adsorbing base molecule, and so a series of solids which have a given order of acidity with one test base molecule may not have the same order with another. Thus the measurement of acidity or basicity of surface sites must remain semiquantitative.

The acid strength, or acidity, of a Brønsted site on a solid is thus indicated by its ability to drive Equation (1.3) to the right:

$$H^+ + B = BH^+ \tag{1.3}$$

where B is the adsorbate in its basic form, BH^+ is the conjugate acid form, and H^+ is the proton provided by the Brønsted site. The equilibrium constant for the reaction is K:

$$[B] \cdot [H^+]/[BH^+] = K = 10^{-pK} \tag{1.4}$$

where the square brackets indicate concentrations,* and the form on the right is simply to introduce the term pK, which represents a characteristic of the adsorbate, not the solid, and describes the tendency of the adsorbed base to convert to its acid form.

The acidity of the acid sites on solids is related to the effective concentration of protons and defined as H_0, where

$$H_0 = -\log[H^+] \tag{1.5}$$

The ratio of the basic to the acid form of the adsorbate is thus given in terms of the acidity of the sites on the solid by

$$\log[B]/[BH^+] = H_0 - pK \tag{1.6}$$

Analogously, the definition of acidity H_0 for a Lewis acid site on a solid surface is

$$H_0 = -\log a_A \tag{1.7}$$

where a_A is the activity of the electron pair acceptor. Perhaps an easier form to visualize is the analog of Equation (1.6): the ratio of the basic to acid form of the adsorbate is given in terms of the Lewis acidity H_0 by the equation

$$\log[B]/[AB] = H_0 - pK \tag{1.8}$$

---

* We will assume that the chemical activity coefficients are unity. The more rigorous expressions for K and, below, for H_0 including activity coefficients will be found in Tanabe.[23]

where B is the base and AB is the form of the base when its available electron pair is shared. The point of Equation (1.6) and (1.8) arises in the use of acid/base indicators, where there is a color difference between B and BH^+, or between B and AB, so that whether H_0 is greater or less than the pK of the indicator is immediately obvious by the color of the adsorbed species.

It seems strange that a concept so similar to that of surface states, the electron affinity of surface orbitals, has led us so far afield, to a discussion of acid/base indicators. However, the tendency to electron sharing must be evaluated by the use of molecules that can share electrons, so such chemical techniques are valuable.

Methods other than acid/base indicators are used of course (as will be discussed in Chapter 3), and experimental work suggests[24] that there can be a one-to-one relation between acidity and the energy level of surface states at the same site. But to evaluate properly the tendency to bond by electron pair sharing, there seems no substitute at present to methods where an adsorbing molecule is provided with which the electron pair can be shared.

1.2.4. Adsorbate Bonding on Various Solid Types

An insulating ionic solid such as sodium chloride is expected to show little tendency to form covalent or even acid/base chemical bonds with foreign species at the surface. Its behavior can be described with reference to Figure 1.1. With a wide band gap insulator, the valence band electrons are paired and at a very low energy and therefore unlikely to be shared with foreign species (either acids or species with unpaired electrons) because unless the foreign species is highly electronegative, its unoccupied orbitals will be at a higher energy. The unoccupied orbitals of the solid (the conduction band) are at a very high energy, and most donor adsorbates (either bases or species with unpaired electrons) will have their bonding electrons at too low an energy level to have any interaction with the unoccupied conduction band levels. There is however, a possibility not easily analyzable. When one examines the available orbitals more closely, as will be done in Section 4.2.2 in the discussion of the Madelung approach to the description of the surface of ionic solids, one finds that the surface ions are more accommodating than the band model makes it appear. The unoccupied levels associated with surface sodium ions in sodium chloride are substantially lower in energy than the levels of the bulk sodium levels, the conduction band. The occupied levels associated with surface chloride ions are substantially higher energy than the bulk chloride levels, the valence band. However, even with due consideration that these surface states are more reactive, one expects that a solid with such an electropositive cation and electronegative anion will form acid/base or covalent bonds only with the most reactive adsorbates.

As the next class consider ionic solids in which the anion is less electronegative (than the halides), so that the occupied orbitals in the valence band may be at an energy suitable for electron sharing. Na_2O, for example, may show strongly basic properties with occupied oxygen orbitals sharing electron pairs with adsorbing gaseous acids such as phenol.[23] Alternatively, consider ionic solids in which the cation has a greater electron affinity (than the alkali metals). Then the surface states will become more reactive. For example, $ZnCl_2$ shows reasonable Lewis acid activity, with unoccupied zinc orbitals available which can share electrons with adsorbing gaseous bases, such as ammonia.[23]

As the electronegativity difference between the cation and anion diminishes the bonding of the cation to anion becomes more homopolar, and hybrid orbitals develop where the electrons tend to be located between the atoms. Then at the surface these hybrid orbitals appear as dangling bonds (unpaired electrons), and covalent bonds can be formed with adsorbing species. Transition metals offer the lobes of d orbitals for bonding. Qualitatively the interaction can be viewed as in Section 1.2.2. However the covalent bonding is not as simple as in the atom/atom bond described in that section, and band effects should be taken into consideration. With a metal, for example, the electrons in the solid have a wave function extending through the crystal. Somehow the band nature of these electronic wave functions and how they affect the bond must be considered. Substantial theoretical activity in this area is presently in progress, and will be described in Chapters 4 and 5.

1.2.5. Movement of Surface Atoms: Relaxation, Reconstruction, and Relocation

In recent years it has been determined that in many cases atoms in the surface layer of a crystal do not occupy the positions that would be expected from a simple continuation of the bulk lattice structure. The reason is obvious. Small shifts in atom position either normal to or parallel to the surface of the crystal can be made with relatively little expenditure of energy by a surface atom that is constricted on only three sides. Because the bonding configuration of the surface atom is quite different from that of the bulk atoms, a substantial increase in bonding energy may be obtained by a slight shift in position from the "normal" lattice site. Thus movement of surface atoms can result in a net lowering of the energy of the system.

"Relaxation" is the term applied to motion of the surface plane normal to the surface. Observed relaxation on a clean surface has been always toward the crystal from the "normal" lattice site, representing a stronger bond formed by the surface atom to the substrate atoms. "Reconstruction" is the term applied to motion parallel to the surface, particularly motion that results in a periodic spacing which can be observed by low-energy electron diffraction (LEED) discussed in Chapter 3. "Relocation" is a more general term meaning shifts in lattice position to a configuration not necessarily specified.

Consider, for example, the elemental and covalent material germanium. In principle the surface atoms are incompletely coordinated (do not have their complete set of nearest neighbors), and therefore, in consideration of the orbitals of their four tetrahedrally directed valence electrons, at least one orbital should be directed out from the surface. This is the dangling bond, which, in principle, is available to form a strong bond with an adsorbing species. In practice, however, with no foreign species present the surface layer of germanium atoms distorts, and the energy of the system is lowered by "reconstruction" of the surface layer[25] so that the dangling bonds can overlap somewhat with each other.

The ability and tendency of surface atoms or ions to move has an important consequence relative to Franck–Condon effects in electron transfer. The Franck–Condon principle states that electron transfer to or from an energy level (in our case a surface group) is rapid compared to the period of vibration of a molecule.[26] After such electron transfer, the surface group has a different charge and will often form a different chemical bond to the surface. If such a change in bonding requires a relocation of an atom, two ground state configurations must be allowed for, one when the surface state is occupied, one when unoccupied. The result is that on the average electron capture will occur primarily on groups that have one bonding configuration; the reverse process, electron injection, will occur when the groups have a different bonding configuration. As revealed by the Franck–Condon principle the configuration cannot change during the electron transfer process.

The net effect of such relocation associated with electron transfer is that the unoccupied energy levels will be at a higher energy than the occupied energy levels. This can be termed Franck–Condon splitting of the surface states, and its inclusion is the first step in introducing some "surface molecule" concepts into the "rigid band" picture of the surface. Such Franck–

Condon splitting, discussed further in Section 5.5, is to be expected with most surface states at the free surface. If such splitting is highly energetic, it will mean that the rigid band representation is inappropriate for the first approximation; one must prefer the surface molecule model, where the bonding of the two oxidation states of the species is considered individually.

1.2.6. The Electronic Energy Level (Surface State) of a Sorbate/Solid Complex

Even without consideration of relocation effects, the energy level of the valence electrons of an adsorbing species will be affected substantially by local bonding with the solid. Or, from the opposite point of view, the surface states on a clean surface will be affected substantially if they act as the site for the adsorption of a foreign species. In this section the relation between bonding and surface state energy is introduced. More quantitative considerations are presented in Section 5.5.

The energy of bonding of a foreign species to an intrinsic surface state (a Tamm or Shockley state) will control the surface state energy level of the adsorbed species. With no local bonding interaction between the adsorbate and such surface states, one can assume a relatively simple relation between the ionization potential of the sorbate and the resulting surface state energy. If a covalent bond forms between the surface and the impinging adsorbate, valence electrons will be lowered in energy when they occupy the bonding orbitals.[27] If other types of interaction occur, the electronic energy levels will similarly be lowered. On an ionic solid the interaction can be a simple matter of an attractive or repulsive electrostatic potential at the site. The potential at a surface site can in principle be calculated using the Madelung approach, which sums the potential due to all the separate ions in the crystal at the surface site. Then an ion which when free has a certain electronic energy level will have that electronic energy level lowered or raised in energy by the magnitude of the Madelung potential.[16] In addition there is the possibility of an acid/base interaction—if a gaseous base adsorbs on a surface site that has an unoccupied orbital at an energy near or below the energy of the unshared electron pair of the gaseous base, there will be electron sharing and the electrons will be stabilized. The energy level of the electrons will be lowered.

The conclusion from these qualitative arguments, which carries through the more quantitative considerations to be discussed later, is that even without relocation effects the nature of the bond formed between an adsorbed atom and the atom of the solid will have a strong effect on the surface state energy level associated with the resulting surface complex. The energy level of the complex will not be the same as the energy level of the dangling bond on the free surface, nor will it be the same as the energy level of the free molecule before adsorption. In order to estimate what energy level will be associated with a given surface atom, and to estimate whether the surface states will have a single level or a broad distribution of energies, one requires information on the type of bonding to be expected and the heterogeneity of the surface.

1.3. Surface Hydration on Ionic Solids

Highly polar adsorbates, in particular water, are an important feature of the surfaces of ionic solids. Water adsorbs strongly, forms OH groups at the surface, and plays a dominating role in the chemical and electronic properties of the material. Adsorbed water is not so dominating on most clean elemental solids, but the attraction for water begins upon the formation of the first layer of oxide.

The bonding of water is of various forms. It can adsorb in an acid/base reaction on a Lewis acid with the OH group sharing its electrons with the Lewis acid site, leaving the proton perhaps weakly bonded and available for reaction. Or it can adsorb on a Lewis base (a lattice oxygen ion or its equivalent), with the proton strongly bonded, sharing the electrons provided by the basic site, with a weakly bonded OH^- group left over providing residual basic character to the solid. A weaker bond can be formed by a simple polarization of the molecule in the electric field at the surface of the ionic solid. Polarization in the crystal field of a transition metal ion will often lead to bonding of this type.[17] Finally it can bond by "hydrogen bonding," where one of the protons is shared between the oxygen ion of the water and an anion of the solid. These various forms of bonding are just variations on a theme, and water will seldom be purely bonded by one or the other, but the bond can best be considered to have elements of each form. There is also the possibility[28] of more homopolar bonding on a covalent material, where the water dissociates into H and OH groups for development of unpaired electrons. However, this possibility will not be discussed in detail here, as it is less generally important.

The adsorption of water on an oxide or other multivalent ionic solid is energetically favorable because it terminates the crystal with a single-valent species in accordance with Pauling's valency rule.[29-32] In general, it is energetically sound to terminate ionic solids with a polar molecule which can be oriented to neutralize surface electric fields, rather than abruptly to terminate the crystal with a layer of ions in normal lattice positions, leaving intense fields directed into space.

The presence or absence of water on the surface of an ionic solid strongly affects the chemical activity of the surface. Details will be developed in later sections, but a qualitative introduction is of value early in the development of the subject of surfaces. We will discuss the chemical activity of oxides as a function of the amount of adsorbed water present. If several monolayers of water are adsorbed, resulting from exposure of a surface to liquid or a high relative humidity of water at room temperature, such a surface is readily attacked by appropriate corrosive agents which can chemically react with the solid. Electrochemical activity can even occur, for oxidants and reductants can become particularly active in an aqueous-like environment and surface heterogeneity will cause circulating currents. Moderate drying of the surface will remove the "physically adsorbed" water and leave the surface covered by adsorbed OH groups. This removal is easily reversed, and such a "hydroxylated" surface is easily rehydrated if immersed in water. Further drying will begin the desorption of these OH groups and will leave the underlying ions exposed.[33] Generally this signals the onset of the surface activity associated with acid and basic centers, for the partial dehydration will induce regions of high electrical potential of one sign or the other (Section 4.3). Outgassing at even higher temperature to remove all the residual water will of course eliminate Brønsted activity (involving protons). It may require higher temperatures to eliminate Lewis activity (involving centers which tend to share electrons with adsorbates). At high temperatures the surface ions can relocate to neutralize high electric fields. In fact, neutralization by this high-temperature mechanism will often be effective enough that the surface is no longer reactive, and even readsorption of water is extremely slow.

The electrical effects of water in electron transfer to a surface are understood relatively poorly. There is no question that they are significant —electron transfer to or from the bands of a semiconductor upon adsorption or desorption of water has been reported for most semiconductors. The mechanism involved has not been satisfactorily explained, perhaps because there are many possible effects which will occur simultaneously. Three models may be considered. First is the obvious direct removal of an electron from an adsorbing water molecule (water usually causes electron injection into a semiconductor). On most semiconductors this model is unsatisfactory, as it requires substantial energy to thus remove an electron from a water molecule, or even an OH^- ion, and the unoccupied energy levels in the semiconductor are not of low enough energy. A second model is associated with chemical bonding between the OH^- or H^+ and an acid or basic group on the surface, where the acid or basic group was acting as a surface state. The electron affinity of the surface group may change substantially when it forms a bond with an OH^- or an H^+ from a water molecule. The third model has to do with the influence of coadsorption of water on a surface where there is an active surface state present. As will be described in Section 2.3, the presence of a polar coadsorbate may have a substantial effect on the kinetics of electron transfer and on the energy of the surface state. Thus for example, in many typical observations of the electrical effect of water, it may be that the water is affecting the adsorption of oxygen at O_2^- through its effect on the kinetics of electron transfer to and from the adsorbed oxygen.

1.4. Surface Heterogeneity

In 1931 Taylor[34] focused the attention of surface chemists on the probable heterogeneity of the surface of powders used as catalysts. Such heterogeneity must have a strong influence on both the chemical and electrical properties of the surface. It means that one expects a broad spectrum of adsorption sites. With a given species being adsorbed, it means there will be a broad spectrum of activation energies (the kinetic factor) and of heats of adsorption (the equilibrium factor). With the more recent awareness of surface states and their dominance in the electrical and chemical properties of surfaces, it must be recognized that there also will be a broad spectrum of possible bonding configurations and surface state energy levels.

There are many forms and degrees of heterogeneity depending on the sample under study, varying from the extreme heterogeneity of a mixed oxide powder to the minimal heterogeneity of a carefully prepared single crystal. Some are illustrated in Figure 1.4. For the powdered catalysts such as those Taylor was interested in, there is heterogeneity arising because various crystal planes are exposed, each of which will offer different sites.[35] Also in those early days of catalytic research, when Taylor developed his concept, purification and outgassing (vacuum) techniques were not at the level of sophistication that is available today, so unidentified foreign adsorbates were probably present, even in the most careful work. There are probably fewer present in the most careful of the modern studies.

Even on single crystals, there will be a spectrum of sites arising due to heterogeneity. Crystallographic steps or "terraces" are almost inevitable.

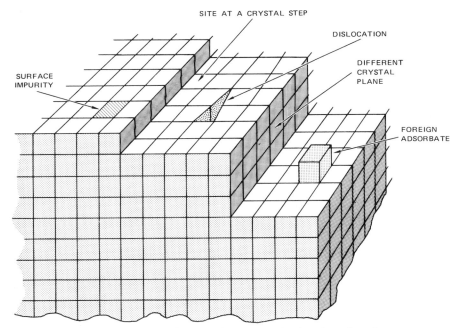

Figure 1.4. Some sources of heterogeneity in adsorption sites.

This case is illustrated in Figure 1.4; a crystal step is one or more extra half-planes on the surface. The step in Figure 1.4 is shown straight. Further active sites will arise because the step is crooked, which leads in particular to highly active "kink" sites, an inside corner site arising because the surface step changes direction, a site where an adsorbate can form bonds to many host atoms simultaneously. Another inevitable defect on the surface is due to the emergence of dislocations at the surface. The edge dislocation is easiest to visualize: it is an extra half-plane of atoms through the bulk, a linear bulk defect. Where the dislocation "pipe" emerges at the surface (as in Figure 1.4) the crystal structure is badly perturbed, and very active surface atoms can result.

Many other sources for heterogeneity can be visualized, depending upon the pretreatment of the material. We have already discussed the emergence of acid or basic centers on a partially dehydrated surface. Similar effects can occur with other adsorbed materials. At times adsorption occurs in islands, and the edges of the islands have characteristics different from the center.[36] Impurities from the bulk can precipitate at the surface during a heat treatment when the temperature is high enough for diffusion but low enough that the impurity is supersaturated. Etching of a crystal in a preferential manner or cleavage which is imperfect can expose crystal planes other than the one intended.

When evaluating a piece of surface research in the current literature, one does not ask whether or not the sample surface showed heterogeneity. Rather, one asks about the degree of heterogeneity and what influence this heterogeneity has on the interpretation of the experimental results. Unfortunately it may turn out in the future that many chemical and electronic phenomena that have been attributed to uniform surfaces (albeit with various crystal planes exposed) will be found to be dependent on imperfections such as steps or dislocations.[37]

Crystal growth depends on the presence of steps, and indeed in general it depends on steps due to dislocations. Atoms from the gas phase deposit at steps during the growth process because there is a much stronger bond at such sites. Etching of crystals often originates at dislocations. The highly active atoms at the point where a dislocation pipe emerges are more easily attacked than surface atoms on a "uniform" region. After the active atom is removed, its neighbor becomes unstable and is attacked, followed by the next neighbor and so on. In adsorption and catalysis, evidence of heterogeneity effects is becoming more obvious. Recently in studies of adsorption processes substantial evidence of "spillover" has been demonstrated, in which a molecule is preferentially adsorbed and dissociated at a surface additive or impurity, and only when so activated can it migrate onto the "uniform" surface of the host solid. Again there is evidence emerging from Somorjai's group[38] using a technique (Section 3.2.7) to provide a crystalline face of Pt with a well-defined distribution of steps, that such steps are required for adsorption processes heretofore attributed to the uniform surface. Here also the adsorption needs the extra bonding power of the step to initiate the interaction.

In the following chapters many experiments will be reviewed and interpreted in terms of a uniform surface. In many cases future work will prove such interpretation wrong. However, it is not planned to point out in each case the possible role of surface heterogeneity. In some cases the possible contribution of surface flaws will be so obvious that it will be discussed. In other cases it will be left to the reader to maintain some reservation regarding the complete accuracy of the interpretation, for it would be tedious to qualify every discussion with the remark that the possible contribution of special sites has been neglected.

2 | Space Charge Effects

2.1. General

Separation of charge into double layers is a common occurrence at a solid surface. At times the phenomenon can be interpreted simply in terms of two plane sheets of charge. On a metal with an adsorbed layer of oxygen, where the oxygen has a large negative partial charge and the surface layer of metal an equal and opposite positive charge, the system can be viewed to a good first approximation as such a double layer composed of two such charged planes, as sketched on the left in Figure 2.1. If an insulating oxide layer grows between the positively charged metal and the now ionosorbed oxygen ions, the separation of the two sheets of charge becomes greater but to a reasonable approximation, with a coherent oxide and a thickness of less than a few hundred angstroms, the system can still be viewed as two planar sheets of charge.

More often, however, at a semiconductor or insulator surface or even across a metal oxide layer on a metal, the double layer is better represented by a sheet of charge on the one side and a distributed charge, a space charge, on the other. In the example above of a metal/adsorbate system with a growing oxide, there is normally distributed charge in the oxide contributed by the interstitial ions or vacancies migrating through the oxide that cause the oxide growth. When the charge (per unit area) from these ions becomes of the order of the charge on the adsorbed oxygen, then the effects due to this space charge begin to dominate.

At a semiconductor or insulator surface, charge transfer to form a double layer (for example ionosorption) generally must be interpreted[1-3] in terms of a space charge due to distributed ions within the solid as on the right in Figure

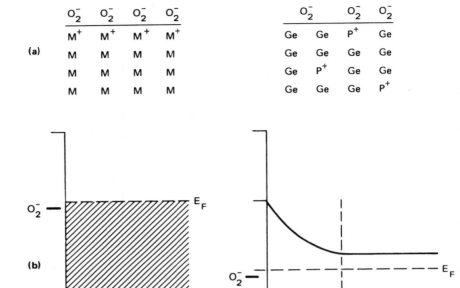

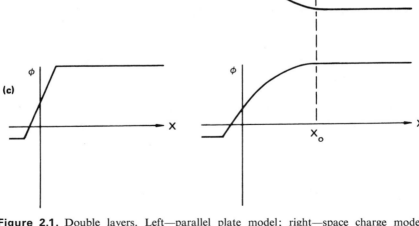

Figure 2.1. Double layers. Left—parallel plate model; right—space charge model (Ge = germanium; P = phosphorous). (a) Atomic model, (b) band model, and (c) potential.

2.1. In some cases, the charge in the semiconductor is associated with mobile electrons, but in such cases the density of energy levels available at the band edge is so low that the electrons are spread a substantial distance into the crystal and so must still be considered a space charge. Analysis of semi-conductor double layers in terms of two charged planes is permissible only in the case of (a) a strictly local heteropolar chemical bond between the sorbed species and the surface and (b) the Helmholtz double layer at the semiconductor/electrolyte interface (to be discussed in Chapter 8).

The behavior of the solid, both its electrical and chemical charac-teristics, are strongly affected by the development of double layers at the surface. The electrical properties of a semiconductor or insulator can be dominated by such double-layer effects in several ways. For example, the formation of the double layers represents injection or withdrawal of charge to or from the bands of the semiconductor, and the density of current carriers (particularly near the surface where electrical contacts are strongly affected) can be changed by a large fraction. Other electrical properties such as work function are directly affected on all solids. A second example is the photoresponse of semiconductors. Photoproduced electrons, excited from the valence band to the conduction band, are often returned to the ground state via centers (recombination centers) at the surface, and the double layer near the surface has a dominating effect on the rate.

The chemical properties of the solid surface are also dominated in many cases by surface double layers. For example, the availability of electrons for a reaction is controlled by such double layers. In adsorption, when electronegative species are adsorbed, the double layer forming will, by electrostatic repulsion, decrease the density of electrons near the surface. In turn this decrease will lower the rate and energy of further adsorption.

The mathematical description of the double layer is substantially simplified because only one dimension need be considered, the direction perpendicular to the surface. There are cases where the fact may be con-sidered that the charged layers are really made up of point charges, and so a one-dimensional analysis is inadequate. However, normally either the distance between the sheets of charge are greater than the separation of point charges in one of the sheets, or orbital overlap within the sheet of charge permits a reasonable description in terms of a uniform sheet. We will consider that only one dimension is important.

The potential ϕ must of course obey the basic Poisson equation (in one dimension):

$$\frac{d^2\phi}{dx^2} = -\frac{\varrho}{\varkappa\varepsilon_0} \tag{2.1}$$

where ϱ is the charge density (C/m³), $\varkappa$ is the dielectric constant, and ε_0 is the permittivity of free space.

Another fundamental equation is the Gauss relation between the charge and the associated electric field. Under all circumstances of interest in this book, the charge at the surface of the solid will be considered a plane sheet of charge, most often due to charges in surface states at the surface. Then, independent of whether the countercharge is distributed (semiconductor or insulator) or planar (metal), the surface electric field can be given by the Gauss relation:

$$\frac{d\phi}{dx}\bigg|_{x=0} = -\frac{Q}{\varkappa\varepsilon_0} \tag{2.2}$$

with Q the net positive surface charge density (in surface states, for example). Equation (2.2) permits determination of the surface charge when the parameters of the double layer are experimentally determined.

In discussions of semiconductors, it is useful to define a "band bending" function V such that eV is related to the potential energy of an electron:

$$V = \phi_b - \phi \tag{2.3}$$

where ϕ_b is the potential in the bulk semiconductor. In terms of the parameter V, the Gauss relation becomes

$$\frac{dV}{dx}\bigg|_{x=0} = \frac{dV}{dx}\bigg|_{V=V_s} = \frac{Q}{\varkappa\varepsilon_0} \tag{2.4}$$

where V_s is the surface band bending, often termed the surface barrier.

2.1.1. The Double Layer Involving Two Planar Sheets of Charge

If we are dealing with a double layer that can be described as two planar sheets of charge, Poisson's equation (2.1) can be solved for the region between $x = 0$, the position of the surface charge, and $x = x_0$, the position of the countercharge. The charge ϱ is zero in the region between the two charged layers, and a double integration of Equation (2.1) with $\varrho = 0$ yields

$$\phi = c_1 x + c_2 \tag{2.5}$$

where the two constants of integration are determined by use of the Gauss relation, Equation (2.2), and by defining $\phi = \phi_s$ at the surface, and $\phi = 0$

at the position of the second plane, $x = x_0$. Then

$$\phi_s = Qx_0/\varkappa\varepsilon_0 \tag{2.6}$$

which of course is the simple law for a parallel plate capacitor.

A typical numerical example is informative and helps indicate the orders of magnitude of the parameters for such a double layer. There are of the order of 2×10^{19} surface atoms/m² on a typical crystallographic plane. If we assume one adsorbate atom can bond to each surface atom, as in Figure 2.1, so that one monolayer is adsorbed and assume the charge on each adsorbate species is one electronic charge, then with a dielectric constant of unity, the equation reduces to

$$\phi_s = 3.6 \times 10^{10}x_0 \tag{2.7}$$

and with a minimum value of x_0, $x_0 = 1$ Å (10^{-10} m), we have a 3.6-V double-layer potential at the surface. As will be recognized more fully in later sections, this is extremely high; it would require a highly reactive species to induce such a huge double-layer potential. For monolayer covalent bonding to a clean metal, the observed changes in surface potential are of the order of tenths of a volt. The parameters chosen above are not entirely realistic for the following reasons. First, in a covalent bond we are seldom dealing with partial charges on the sorbate approaching unity. Second, in general there will be some relocation, movement of the surface atoms to decrease the apparent separation of the layers (the initiation of another phase). Third, if there is no relocation, the buildup of the surface potential will prevent adsorption to the extent of a full monolayer.

2.1.2. The Space Charge due to Immobile Ions: The Depletion Layer

If we are dealing with a distributed space charge, as in the case, for example, with ionosorption on a semiconductor or insulator, the charge ϱ in Equation (2.1) is no longer zero. Analysis of the distributed space charge is more complex than that of the two plane sheets of charge, and most of the remainder of the chapter will be devoted to space charge analysis. A particularly simple, but most commonly assumed case for semiconductors (and a satisfactory approximation for our purposes for insulators with trapped charge) is the Schottky model. This simply assumes that the space charge in the solid near the surface is immobile and is independent of distance in the entire space charge region.

Consider a crystal which has N_D bulk donor atoms and N_A acceptor atoms per unit volume, both of which are completely ionized (the donor ions have a charge $+1$, the acceptors -1). With n_b bulk electrons and p_b bulk holes in the conduction and valence bands (all quantities per unit volume), then in the bulk material by charge neutrality

$$N_D + p_b = N_A + n_b \tag{2.8}$$

The simplifying assumption leading to the Schottky relation is that in the space charge region the density of minority carriers is negligible and that the surface states capture the majority carrier. The surface region is depleted of both types of carriers to a distance x_0 (Figure 2.1b). For example, with an n-type semiconductor ($p_b \sim 0$), N_A electrons from the donors are captured by the acceptors, leaving $n_b = N_D - N_A$ electrons in the bulk conduction band. Near the surface these electrons are captured at the surface sites, leaving a charge per unit volume in the space charge region of $e(N_D - N_A)$. In general (p-type or n-type bulk) the charge in the space charge region by this model is $\varrho = e(N_D - N_A)$ for $x < x_0$, and $\varrho = 0$, for $x > x_0$. Substituting into the Poisson relation, Equation (2.1), and integrating yields successively

$$d\phi/dx = [e(N_D - N_A)/\varkappa\varepsilon_0](x_0 - x) \tag{2.9}$$

and

$$\phi_b - \phi = [e(N_D - N_A)/2\varkappa\varepsilon_0](x - x_0)^2 \tag{2.10}$$

where the boundary conditions have been applied that $\phi = \phi_b$ at $x = x_0$ by definition and $d\phi/dx = 0$ at $x = x_0$. The latter boundary condition arises because there is no electric field beyond the space charge region in the bulk of the crystal.

On the ordinate in a band diagram it is customary to plot the energy rather than the potential of an electron, so a switch in parameters to the V parameters [Equation (2.3)] is convenient.

The value of V at the surface is V_s, and from (2.10) we immediately obtain the Schottky relation

$$V_s = e(N_D - N_A)x_0^2/2\varkappa\varepsilon_0 \tag{2.11}$$

Another form of this important relation is obtained by recognizing that the number of charges per unit area on the surface, N_s, arising from electrons or holes exhausted from a distance x_0 from the surface, is given by

$$N_s = -(N_D - N_A)x_0 \qquad (2.12)$$

and the Schottky relation [Equation (2.11)] can be expressed in terms of N_s:

$$V_s = eN_s^2/2\varkappa\varepsilon_0(N_D - N_A) \qquad (2.13)$$

Again it is valuable to examine typical numerical values. In particular it is of interest to determine the value of N_s possible with a heavily doped semiconductor. As will be discussed below, a value of V_s the order of one volt is reasonable (if a little high) for a surface barrier arising from adsorption. With a dielectric constant of 8 (reasonable for a semiconductor) and a value for $N_D - N_A$ either 10^{20} m^{-3} (very pure material) or 10^{25} m^{-3} (fairly impure), then Equation (2.13) gives a value of N_s either 3×10^{14} m^{-2} or 10^{17} m^{-2}, respectively. This corresponds to about 1.5×10^{-5} or 5×10^{-3} monolayers, respectively. Clearly only a small surface charge can be accommodated without leading to a very high double-layer potential.

Weisz[4] first pointed out this limitation on surface coverage. Because N_s varies as the square root of the impurity concentration, no matter how impure the sample, one is limited to about 10^{-3}–10^{-2} monolayers of equilibrium ionosorption when a depletion layer is present.

2.1.3. The Double Layer in the Band Diagram, Fermi Energy Pinning

The Schottky space charge region on an energy level diagram is illustrated in Figure 2.2. An n-type semiconductor is shown with surface states at an energy level E_t that can acquire a negative charge. The energy of an electron at the bottom of the conduction band is higher by an amount eV [from Equations (2.3) and (2.10)] than the energy of a bulk electron, and this is indicated by the curvature of the conduction (and valence) bands. From Equation (2.10) the shape of the bands in the space charge region is parabolic. In this subsection the various qualitative features of such a band representation of the space charge region will be introduced because the representation is referred to so often in discussing models of solid surfaces. We will discuss first the position of various energy levels as the double layer develops, then the occupancy of the various levels by electrons.

Two equivalent diagrams, differing only in the zero of potential that is adopted, are shown in Figures 2.2a and 2.2b. In Figure 2.2a is shown the space

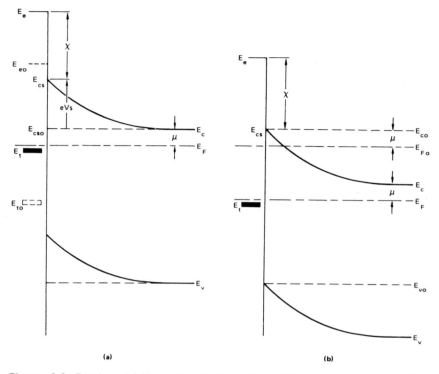

Figure 2.2. Band model illustrating the formation of the space charge region in a semiconductor. In (a) the potential in the solid is used as a constant reference. In (b) the potential of the electron in vacuum (E_e) is used as a constant reference. The dashed lines and subscript zero show the model before charge is transferred to the surface. The solid lines show the equilibrium double layer with a negative surface charge.

charge region as it is usually drawn, with the potential deep in the semiconductor (at $x = x_0$) designated as the constant reference potential. The dashed lines indicate the bands before a charge is moved to the surface (the "flat band" case). The solid lines indicate the bands as bent due to electrons being captured by the surface state E_t, leaving a positive space charge due to immobile donor impurities. As the surface charge develops, E_e, E_t, E_{cs}, and E_{vs} vary. In Figure 2.2b is shown the same picture, with now the constant reference potential designated as ϕ_s, the potential at $x = 0$. Again the dotted lines indicate the bands before a negative charge is transferred to the surface state. In this representation the surface energy levels are fixed, the bulk band edges and E_F vary. With no surface charge the conduction and valence bands are at E_{c0} and E_{v0}. The solid lines marked

E_c and E_v are the bands as bent due to electron capture on the surface state. The Fermi energy being a thermodynamic function (the electrochemical potential for electrons), it is invariant through the system. Since it is defined by the availability of electrons in the bulk material, it moves with the bulk bands.

Figure 2.2a is the model most useful if one is interested in the solid and prefers the bulk properties of the solid to appear unchanged by the development of the space charge region. The representation of Figure 2.2b is sometimes helpful if one is interested in bonding of species to the surface. This latter representation indicates clearly that the energy levels at the surface (E_e, E_t, E_c, and E_v) are unaffected by the fact that an electric field exists inside the solid. In other words, the scientist interested in bulk properties recognizes from Figure 2.2a that an electron deep in the bulk is not affected by this thin surface layer (as long as the electron does not try to cross it), but the scientist interested in the energy of surface bonding orbitals notes from Figure 2.2b that relative to the adsorbing species all the surface parameters except the Fermi energy are unchanged by the development of the double layer.

The more accurate diagram is Figure 2.2b, which shows the expected constant value for E_e, the energy of an electron at infinity. As electrons move to the surface, the potential in the bulk of the solid varies, so the bulk values for the conduction band edge and the valence band edge shift. The Fermi energy also shifts, as discussed above. To the scientist interested in the availability of electrons for reaction with a surface species, the shift of the Fermi energy at the surface is of dominating importance.

The occupancy of various levels is strongly affected as the double layer develops. Consider first the occupancy of the surface state energy level E_t. In the flat band case (dashed lines) the surface state energy as shown is far below the Fermi energy E_F, and from the Fermi distribution function

$$f = 1/[1 + \exp(E - E_F)/kT] \tag{2.14}$$

which describes the fractional occupancy f of any nondegenerate level at the energy E, the surface state at energy E_t should be almost completely occupied with $E_F - E_t$ large as indicated in Figures 2.2. If the surface state is an acceptor and thus unoccupied before charge transfer, then the system is not at equilibrium with the bands flat, and electrons will move from the solid to the surface state. The double layer, as represented by the surface barrier V_s, will develop as indicated. By either Figure 2.2a or 2.2b, as the surface state becomes charged, it moves toward the Fermi energy (or the Fermi energy moves toward the surface state). As they move together, the fractional occupancy required for equilibrium by Equation (2.14) becomes lower, and of course the actual occupancy becomes greater as charge is transferred. As the level nears and crosses the Fermi energy, there is, by Equation (2.14), a very rapid drop in the required fractional occupancy. At some point, often when E_t is within a few kT of E_F, the Fermi condition becomes satisfied, and the system is at equilibrium.

This movement of E_t relative to E_F leads first to a limitation on V_s and second to an important phenomenon, termed "pinning," of the Fermi energy. From the above discussion it is clear that the surface barrier is limited at equilibrium roughly to a value such that the surface state energy level has moved near the Fermi energy. Thus V_s is limited by the initial difference between E_F and E_t, and this difference is associated with the relative chemical donor/acceptor characteristics of the semiconductor and the adsorbates. Values of V_s up to ± 1 V will be common; greater than 1 V would imply highly reactive species such as potassium or fluorine as the adsorbate. Second if there is a high density of surface states, it is seen by these qualitative arguments [made quantitative in Equation (6.1)] the Fermi energy becomes "pinned" at the energy of the surface states. Its value becomes essentially independent of bulk impurity concentration. We have already used the first qualitative feature (the expected value of V_s) in the numerical evaluations of Equations (2.6) and (2.13). Both features become of dominant significance in ionosorption, as discussed further in Section 7.2.

The occupancy of the conduction and valence band by electrons is also described by the Fermi energy. As indicated in Figure 2.2, the bands in the surface region move relative to the Fermi energy. The conduction band in this example becomes far from the Fermi energy. From Equation (2.14), the density of electrons near the surface must be low. This is all self-consistent; the double layer is a "depletion" layer. The valence band, on the other hand, has moved closer to the Fermi energy. From Equation (2.14) the fractional occupancy of the band by valence electrons will decrease (the density of holes will become significant). If the band is still far from the Fermi energy at equilibrium, the concentration of holes (unoccupied levels) will still be negligible. If the valence band approaches the Fermi energy, as is common in semiconductors with a narrow band gap, the density of holes near the surface can become substantial, and when the hole density is greater than the electron density, the surface is termed "inverted," and the double layer termed an "inversion layer" as discussed below.

2.2. Space Charge Effects with Reactive Surface Species

In this section a more complete model is outlined to describe the space charge region when both immobile ions and free holes and electrons contribute to the space charge. The present analysis is of primary interest for semiconductors which have highly reactive surface states, for example very electropositive adsorbates that will inject electrons into the conduction band or very electronegative adsorbates that will inject holes into the valence band (extracting valence electrons). Then the analysis of the space charge region must be extended to cases with an excess of electrons in the conduction band or an excess of holes in the valence band.

In this more general analysis[1,2] the restrictive assumption will still be made that the Fermi energy is always far from the band edges, so that in the Fermi function, Equation (2.14), $E_c - E_F \gg kT$ for the conduction band, and $E_F - E_v \gg kT$ for the valence band. Then the Fermi function applied to the bulk conduction band is approximated by a Boltzmann distribution and yields

$$n_b = N_c \exp[-(E_c - E_F)/kT] \tag{2.15}$$

and, in the space charge region where there is an added potential ϕ, from [Equation (2.3)]:

$$n = N_c \exp[-(E_c + eV - E_F)/kT] \tag{2.16}$$

where N_c is the effective density of states in the conduction band.

Analogous expressions obtain for p and p_b. From such expressions the hole and electron density as a function of V become

$$p = p_b \exp(eV/kT) \tag{2.17}$$

$$n = n_b \exp(-eV/kT) \tag{2.18}$$

Equations (2.17) and (2.18) state that in the space charge region where the potential is no longer zero, the hole and the electron density differ from the bulk values by the appropriate Boltzmann factors. With space charge provided by both immobile ions and by free electrons and holes by Equation (2.17) and (2.18) Poisson's equation, Equation (2.1), now becomes

$$d^2V/dx^2 = e[n_b - p_b + p_b \exp(eV/kT) - n_b \exp(-eV/kT)] \tag{2.19}$$

where Equation (2.8) has been used to replace $N_D - N_A$, the space charge due to ionized impurities, by $n_b - p_b$.

We distinguish two cases. First is an accumulation layer that arises when the majority carrier is injected by a reactive surface state. In this case either the third or last term in Equation (2.19) becomes dominant depending on whether the majority carriers are holes or electrons, respectively. The band model for an accumulation layer on an n-type semiconductor is sketched in Figure 2.3a. The second case is the inversion layer, when the majority carrier is captured at the surface, but the surface states are so reactive that they also inject minority carriers. On the band picture the bands are bent substantially in this case, so far that for n-type material the third term of Equation (2.19), for p-type material, the fourth term becomes important. The band picture for an inversion layer on an n-type semiconductor is sketched in Figure 2.3b.

The general equation (2.19) can be integrated once analytically by multiplying each side by $2(dV/dx)$ and applying the boundary condition that $dV/dx = 0$ when $V = 0$ (where $V = 0$ in the bulk for $x \geq x_0$). The integration yields

$$\left(\frac{dV}{dx}\right)^2 = 2\left(\frac{e}{\varkappa\varepsilon_0}\right)\left\{(n_b - p_b)V + \left(\frac{kT}{e}\right)n_b\left[\exp\left(-\frac{eV}{kT}\right) - 1\right]\right.$$
$$\left. + \left(\frac{kT}{e}\right)p_b\left[\exp\left(\frac{eV}{kT}\right) - 1\right]\right\} \tag{2.20}$$

This equation thus defines the electric field in the material.

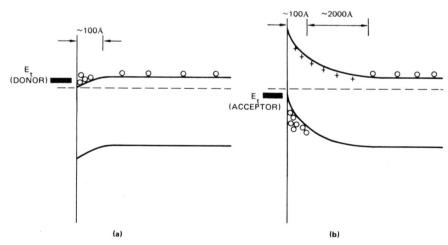

(a)   (b)

Figure 2.3. Band bending on an n-type semiconductor with reactive surface states: (a) accumulation layer and (b) inversion layer. Typical dimensions are shown, but (b) is obviously not to scale.

Without approximation, Equation (2.20) is not analytically soluble. It can be converted to a more convenient form for numerical solution by introduction of the definitions

$$v = eV/kT$$
$$u = (q/kT)(E_F - E_i)$$
$$n = n_b \exp v \qquad (2.21)$$
$$p = p_b \exp(-v)$$
$$L = [\varkappa\varepsilon_0 kT/e^2(n_b + p_b)]^{1/2} \quad \text{(the effective Debye length)}$$

where E_i is the value of the Fermi energy for the particular case of $n_b = p_b$, in which case the equation can be written

$$dv/dx = \pm(2/L^2)[\cosh(u_b + v)/(\cosh u_b) - v \tanh u_b - 1]^{1/2} \qquad (2.22)$$

Numerical solutions for various values of the parameters of interest are given in Ref. 1. The reader who is involved in numerical interpretation of data is referred to that source or the references therein.

2.2.1. The Accumulation Layer

An accumulation layer (Figure 2.3a) arises when a strong reducing agent injects electrons into an n-type semiconductor or a strong oxidizing agent injects holes (extracts valence electrons) into a p-type semiconductor. For illustration, if we consider an n-type semiconductor with an accumulation layer, the last term of Equation (2.19) dominates and Equation (2.20) becomes

$$dV/dx = (2kTn_b/\varkappa\varepsilon_0)^{1/2} \exp(-eV/2kT) \qquad (2.23)$$

and the solution (with the boundary condition that $V = V_s$ at $x = 0$)

$$x = (2kT\varkappa\varepsilon_0/e^2 n_b)^{1/2}[\exp(eV/2kT) - \exp(eV_s/2kT)] \qquad (2.24)$$

Again it helps to introduce typical numerical values. With n_b about $10^{24}/\text{m}^3$, and the dielectric constant about 5, the coefficient of Equation (2.24) is 1.2×10^{-8} m. At room temperature the distance from the surface at which the surface barrier has decreased to $-kT/e$, with $|V_s| \gg kT/e$, is only 73 Å. Such a dimension is therefore indicated in Figure 2.3a.

Such characteristic distances are reasonable, from an intuitive point of view. The charges of interest here that make up the space charge are mobile, so will tend to move to the surface. However, the density of available states in the conduction band (or valence band) is low near the edge of the band, so without surface states the charge cannot reside right at the surface. The case differs from the double layer at a metal surface primarily in this available density of states. On a metal surface the double-layer charge is induced at the Fermi energy, but the Fermi energy is somewhere in the midband region, where the density of states is high. Thus the charges can reside within an angstrom of the surface.[5,6]

2.2.2. The Inversion Layer

The other extreme, the inversion layer, appears when a very strong oxidizing agent (acceptor surface state) appears on an n-type semiconductor or a very strong reducing agent (donor) is provided on a p-type semiconductor. Such a case is shown in Figure 2.3b. For an inversion layer to be created at all, the semiconductor is usually of reasonably narrow band gap, say less than 1.5 eV. This is not a quantitative rule, it is simply that the wider the band gap, the stronger the reactant must be to bend the bands so far, and in the case of wide band gap semiconductors a chemical reaction is likely to happen before the bands get so perturbed.

Near the surface, in the region where the minority carrier density is higher than any other form of space charge, Equation (2.20) can be simplified. For example with an n-type semiconductor the equation reduces to

$$(dV/dx)^2 = (2e/\varkappa\varepsilon)(kTp_b) \exp(eV/kT) \qquad (2.25)$$

where it must be recognized that with an n-type semiconductor p_b is very small, so V must be relatively large to make this term [from Equation (2.20)] dominate. From solid state theory p_b is given by

$$n_b p_b = N_c N_v \exp(-E_g/kT) \qquad (2.26)$$

where N_c and N_v are the effective densities of states in the conduction and valence bands, respectively, and where E_g is the energy gap in the semiconductor. The larger is E_g, the larger V in Equation (2.25) must be to permit the approximation. Assuming, however, that the energy gap is small enough that the approximation is chemically possible, we can integrate Equation (2.25) and with the usual boundary condition, $V = V_s$ at $x = 0$, the solution is

$$x = (2\varkappa\varepsilon_0 kT/ep_b)^{1/2}[\exp(-eV/2kT) - \exp(-eV_s/2kT)] \qquad (2.27)$$

where the minus root of Equation (2.25) was used so that when V is less than V_s, x is positive (Figure 2.3b).

Equation (2.27), based on the approximation that the hole density is large, is only valid for a short distance from the surface, not even to the point where the Fermi energy crosses the midgap energy, for by then the hole density is small. We will define the limit of the inverted region of the inversion layer as the value of x where the minority carrier density in the surface layer is decreased to a magnitude equal to the density of ions n_b in the bulk. The thickness of the inverted region is thus x', where x' is the distance where

$$p_b \exp(eV/kT) = n_b \qquad (2.28)$$

Substituting this value for V, and using the same values for the parameters as used in the calculation following Equation (2.24), we find $x' \sim 73$ Å for the thickness of the inverted region of the inversion layer.

Thus both the accumulation region and the inverted region of an inversion layer are the order of 100 Å in thickness.

2.3. Electron and Hole Transfer between the Solid and Its Surface

From both a chemical and physical point of view, the process of electron transfer to or from the surface of a solid is of dominating technical importance. Most of this book is dedicated to analyzing such processes. At the solid/liquid interface the importance is of course clear. Electrochemical corrosion, electrochemical reduction, or oxidation of species in solution, and even catalysis, depend fundamentally on steps involving electron transfer. At the solid/gas interface, again such corrosion processes as oxidation, and many catalytic processes, as well as processes of more direct physical interest such as trapping by surface species, dissociation of compound semiconductors, and recombination of excess hole/electron pairs, all depend upon the kinetics of hole and electron capture or injection at surface states (sites).

A discussion of electron transfer to or from a surface energy level must include two factors, one that can be classified as electronic, the other more chemical. The first is the availability of electrons and of occupied levels, both in the solid and on the surface group. This is the aspect we consider in this chapter, where the surface state is considered a simple, uncomplicated electronic energy level. The second factor, which will be considered more thoroughly in later chapters, particularly Chapters 5, 8, and 9, is the chemical transformation of the surface species after (or before) electron or hole transfer occurs. This can include such a simple process as movement of the resulting surface species to a more favorable site (which can be less than an angstrom away) or can include a very complex chemical reaction or can include dissociation or desorption of the resulting species. Both these factors, the electronic and the chemical, are of great importance in describing the process, whether it be a simple electron–hole recombination or a complex catalytic process. A successful synthesis of the two factors is required to describe surface behavior.

In this section we address ourselves to the question of the role of the space charge layer in controlling electron transfer. Three models are useful under various approximations to describe electron and hole transfer between the solid and the surface. Only the first is needed if the worker deals only with a surface free of volatile species and with a low density of surface states (as for example at a Si/SiO_2 interface). This first model is the classical physical model, where the density of surface states is assumed low and constant. In the second model the surface state density is assumed high but constant so that the transfer of electrons to the surface does not change appreciably the density either of occupied or unoccupied surface states. The third model is one where the energy level of the surface states fluctuates appreciably with time, as when a polar material is present. Such fluctuations in energy level change the mathematics completely. In the first two models we will assume a single surface state energy E_t, with the surface state partially occupied. In the third case, different energy levels must be assumed depending on whether an electron occupies the level. The need to include this possibility is related to Franck–Condon effects; the details will be developed in Section 5.5. Other cases, such as the case when the density of surface states is not constant (the surface state is due to a volatile adsorbate) will be considered as variations of the above cases, and will not be dealt with until Chapter 7.

2.3.1. Basic Physical Model of Electron and Hole Capture or Injection

We will consider the rate of exchange of electrons between a surface state and a single band, say the conduction band, of a semiconductor. Concurrent hole and electron capture will be discussed in Section 9.2. We will assume throughout this book that the rate of electron transfer is first order, that is that the rate of electron capture is proportional to (a) the density of carriers in the conduction band at the surface, n_s, and (b) the density of available final states. Similarly, the rate of electron injection to the conduction band is proportional to the density of occupied surface states, n_t, and to the density of available conduction band states, N_c.

The rate of change of electron density on the surface state, n_t, due to exchange with the conduction band is then

$$dn_t/dt = K_n[n_s(N_t - n_t) - n_1 n_t] \qquad (2.29)$$

where the first term represents the rate of electron transfer to the surface state, the second term the rate of electron injection with the parameter N_c included in the "emission constant" n_1. Here N_t is the total density of surface states (assumed constant), K_n is the rate constant for the process of electron capture by the surface state, and is given[1] by $K_n = \bar{c}\sigma$, where $\bar{c}$ is the thermal velocity of electrons and σ the capture cross section of the unoccupied surface state level. The product $K_n n_1$ is the rate constant for the process of electron injection.

An expression for n_1 is obtained in terms of equilibrium parameters. At thermodynamic equilibrium

$$dn_t/dt = 0 \qquad (2.30)$$

and using the subscript zero to indicate equilibrium values, we have

$$n_1 = n_{s0}(N_t - n_{t0})/n_{t0} \qquad (2.31)$$

Now the Fermi distribution [Equation (2.14)] can be expressed in a more convenient form to represent the equilibrium ratio of occupied (n_t) to unoccupied (p_t) surface states at a certain energy E_t. The convenient expression for the Fermi distribution function is

$$n_t/p_t = \exp[-(E_t - E_F)/kT] \qquad (2.32)$$

where

$$p_t = N_t - n_t \qquad (2.33)$$

is the number of unoccupied levels (trapped holes). From Equation (2.16), using $n = n_s$ and $V = V_s$, we have

$$n_s = N_c \exp[-(E_c + eV_s - E_F)/kT] \qquad (2.34)$$

when the Fermi energy is well below the conduction band. The conduction band edge at the surface is

$$E_{cs} = E_c + eV_s \qquad (2.35)$$

so the parameter n_s becomes

$$n_s = N_c \exp[-(E_{cs} - E_F)/kT] \tag{2.36}$$

Substituting from Equation (2.32) and Equation (2.36) into Equation (2.31), we finally arrive at a useful expression for n_1:

$$n_1 = N_c \exp[-(E_{cs} - E_t)/kT] \tag{2.37}$$

Equation (2.29) with n_s given by Equation (2.18) and n_1 by Equation (2.37) represents the electron transfer process. That is, from Equation (2.18) we have

$$n_s = n_b \exp(-eV_s/kT) \tag{2.38}$$

and Equation (2.29) becomes

$$\frac{dn_t}{dt} = K_n\left[(N_t - n_t)n_b \exp\left(-\frac{eV_s}{kT}\right) - n_t N_c \exp\left(-\frac{E_{cs} - E_t}{kT}\right)\right] \tag{2.39}$$

Figure 2.4 illustrates the important parameters. Electron transfer to the surface is an activated process because the factor n_s has an activation energy $E_{cs} - E_F$ [Equation (2.36)] that varies directly with V_s. Electron injection from the surface state to the conduction band is also an activated process with the activation energy arising through the factor n_1, as given by Equation (2.37). Note that n_1 is independent of V_s, so the activation energy for electron injection $E_{cs} - E_t$, is independent of V_s, as is clear, for example from the sketch in Figure 2.2b.

The analogous expressions can be developed for hole capture and injection. The rate of hole exchange is given, analogously to Equation (2.29) by

$$dp_t/dt = K_p[p_s n_t - p_1 p_t] \tag{2.40}$$

with the first term representing the rate of hole transfer to the surface state, the second representing the rate of hole injection by the surface state into the valence band. Again the value of the hole density at the surface p_s can be immediately stated in terms of V_s, and the equilibrium conditions can be used to obtain an expression for p_1.

$$p_s = N_v \exp[(E_{vs} - E_F)/kT] \tag{2.41}$$

where N_v is the effective density of states in the valence band, and E_{vs}

is the position of the valence band edge at the surface. Again

$$p_1 = N_v \exp[-(E_t - E_{vs})/kT] \qquad (2.42)$$

with an overall picture for hole transfer completely analogous to that described for electron transfer above.

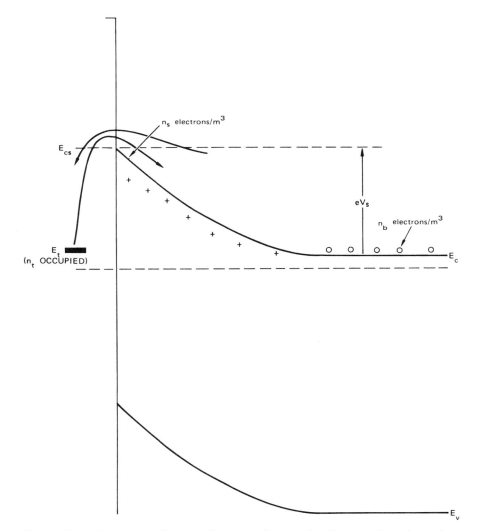

Figure 2.4. Electron transfer to or from a surface species. The rate depends on the parameters shown. The rate of electron capture depends on n_s, which decreases exponentially with V_s. Electron injection is activated with activation energy $E_{cs} - E_t$.

As discussed above, K_n and K_p are given by $\bar{c}\sigma$, where $\bar{c}$ is the mean thermal velocity of the electron or hole and σ is the appropriate capture cross section. An order of magnitude value for $\bar{c}$ is 10^5 m/sec, and a typical value for capture cross section is the geometrical area of an atom, 10^{-19} m^2. However the capture cross section can vary from about 10^{-26} to 10^{-16} m^2; it depends on factors such as the charge in the surface state, for the charge determines whether the position is attractive or repulsive to electrons or holes. Thus a working value for K_n or K_p is 10^{-14} m^3/sec, but it must be recognized that this value can vary by many orders of magnitude.

In most surface physics calculations one is interested in a very low density of surface states, such as surface states at a semiconductor/oxide interface. In this case a shift of n_t from its equilibrium value will not lead to an appreciable shift of n_s or p_s, and so n_t and p_t $(= N_t - n_t)$ will be the dominant variables in Equations (2.29) and (2.40). We can estimate the magnitude of N_t below which n_t and p_t must be the dominant variables. Consider again Equation (2.13). If, for example, $N_D - N_A = 10^{21}$/m^3 and the dielectric constant is 8, a change from $n_t = 0$ to 4×10^{14}/m^2 $(2 \times 10^{-5}$ monolayers) changes V_s by less than 0.01 V, so that n_s and p_s [Equation (2.38)] will change a negligible amount at room temperature (where kT is 0.025 eV). Thus if N_t is the order 10^{15}/m^2 or less, so charge transfer must be low, then at room temperature or above, charge transfer will usually change either n_t or $N_t - n_t(= p_t)$ much more than $\exp(-eV_s/kT)$.

If n_t and p_t are the dominant variables, the integration of Equation (2.29) or (2.40) leads clearly to exponential transients of n_t or p_t with time. This model has been used extensively in surface state analysis of measurements of the rate of recombination of excess holes and electrons. Such recombination expressions will be analyzed further in Section 9.2.

2.3.2. Electron and Hole Transfer with Large Changes in the Surface Barrier

It there is a very large density of surface states, and the states are partially occupied, then it may well be that a change Δn_t may represent a small perturbation on n_t or $N_t - n_t$ while affecting $\exp(-eV_s/kT)$ substantially. As $\exp(-eV_s/kT)$ is so temperature sensitive, this is particularly probably at low temperature.[7–12] Thus as suggested at the end of the last subsection, if the density of active surface states N_t is significantly higher than 10^{15} states/m^2, it is to be expected that $\exp(-eV_s/kT)$ will become the dominant variable even at room temperature.

A high density of surface states can occur for example when the surface states are intentionally added in substantial quantities. For example suppose that the compound $FeCl_3$ is added to provide Fe^{3+} ions as surface states that can accept electrons, and simultaneously $FeCl_2$ is added to provide Fe^{2+} ions as surface states that can donate electrons [Equation (1.1)]. Such additives can be deposited with no net charge on the surface, obviously, because the counterions, the chloride ions, compensate. Thus quantities of each the order of tenths of a monolayer or more can be added. But as electron transfer can occur to a small fraction of a monolayer only, it is seen that the densities of Fe^{2+} and of Fe^{3+} are essentially unchanged by such a small electron transfer. However, V_s is affected drastically; in the example described in the preceding section 3×10^{-4} monolayers converted from Fe^{+3} to Fe^{+2} suffices to change the surface barrier V_s from zero to 0.42 eV.

If in Equation (2.39) $N_t - n_t$ and n_t are less sensitive to n_t than is $\exp(-eV_s/kT)$, then the equation reduces to

$$dn_t/dt = B'[\exp(-eV_s/kT) - \exp(-eV_{s0}/kT)] \qquad (2.43)$$

where V_{s0} is the surface barrier at equilibrium and B' is a collection of constants. In particular, for a change ΔV_s from the equilibrium value, the equation becomes

$$dn_t/dt = B[\exp(-e\Delta V_s/kT) - 1] \qquad (2.44)$$

where $B = B' \exp(-eV_{s0}/kT)$.

With the exponential factors in Equations (2.29) and (2.38) dominating, leading to equations of the form (2.44), the integral of the equation with time is no longer a simple exponential. If we assume that for a transient shift from equilibrium ΔV_s is small compared to V_{s0} and is proportional to Δn_t, Equation (2.44) becomes[10,12]

$$d\,\Delta n_t/dt = B[\exp(-\beta\,\Delta n_t) - 1] \qquad (2.45)$$

with the integrated form

$$1 - \exp(\beta\,\Delta n_t) = A_0 \exp(-\beta Bt) \qquad (2.46)$$

where B and β are constants and A_0 is the constant of integration. The rate of return to equilibrium from Equation (2.45) is highly assymmetric, depending dramatically on the sign of Δn_t. If Δn_t is positive and greater than β^{-1}, then the last term of Equation (2.45) dominates and the rate of electron injection is independent of time. If Δn_t is negative and somewhat greater than β^{-1}, the equation describing the initial electron transfer rate to the surface becomes the highly nonlinear Elovich equation

$$d\,\Delta n_t/dt = B\exp(-\beta\,\Delta n_t) \tag{2.47}$$

to which we will often refer when discussing experimental and theoretical models.

Again it is of value to introduce numerical values for expected rates of electron transfer to surface states. A case where the above rate equations can be expected to apply is that of ionosorption, for example the transfer of electrons to adsorbed oxygen molecules at the surface. We will be discussing in Chapter 7 the possibility of the rate of such ionosorption decreasing to a negligible value, not because the increase in surface barrier brings about equilibrium, but because the increase in surface barrier slows the electron transfer to a negligible rate. This effect provides an interesting application of the rate equations. The case of interest is that where the first term of Equation (2.39) is measureable when $V_s = 0$, but where the temperature is so low that the last term is negligible. Thus adsorption will occur on a clean surface ($V_s = 0$), but as adsorption occurs the barrier V_s grows, and the rate of adsorption will decrease exponentially.[7] As an example of numerical values appropriate for Equation (2.29) and the subsequent equations, an estimate will be made to show that such a "pinching off" of adsorption may easily occur, using the example of the adsorption of oxygen molecules to form O_2^- on an n-type semiconductor.

To calculate the band bending required to reduce the rate of ionosorption to a negligible value, it is necessary to define a negligible value. Now a measureable surface coverage of adsorbed species may be the order of $10^{14}/m^2$, or about 10^{-5} monolayers of adsorbed species. We will calculate the band bending necessary to decrease the rate of further adsorption to 10^{-5} monolayers/hr. We will assume the number of neutral oxygen molecules adsorbed is $10^{17}/m^2$, or about 10^{-2} monolayers, and assume this represents $N_t - n_t$, the density of unoccupied surface states. This number is substantially higher than is normally expected at room temperature for a species adsorbed with primarily van der Waals forces (see Chapter 7), but with stronger bonding it may be realistic. Substituting these values into Equation (2.29) the room temperature rate will be 10^{-5} monolayers/hr when n_s is decreased to $3 \times 10^7/m^3$. From Equation (2.36), if we assume $N_c \sim 2 \times 10^{25}/m^3$, this corresponds to a band bending such that $E_{cs} - E_F = 1.0\,eV$. If for example the bulk Fermi energy is 0.2 eV below the conduction band, $V_s = 0.8\,eV$ causes pinch-off (stops further adsorption) at room temperature. At lower temperature pinch-off will occur at much lower values of V_s (much lower coverage by the adsorbed species). Such "pinch-off" is one possible mechanism leading to irreversible chemisorption as discussed in detail in Chapter 7.

2.3.3. Charge Transfer to a Surface Species in a Polar Medium: The Fluctuating Energy Level Mechanism

In certain cases a third model can dominate electron and hole transfer to and from the surface even for steady state conditions. This is when the energy level of the surface state fluctuates with time over a substantial energy span. For some cases, and particularly if a polar medium is in contact with the surface states, very large temporal fluctuations in the energy levels can occur and the kinetically most probable mechanism for electron (or hole) transfer can change completely from that of the preceding models. Specifically the kinetically most active mechanism for electron exchange with a conduction band can be one in which the surface state energy level fluctuates to an energy equal to the conduction band energy, and the electron transfer occurs with no change in the energy of the electron. We will call this the fluctuating energy level mechanism. In this section we show qualitatively that such fluctuations will occur, analyze the probability that a surface state will fluctuate to the conduction band edge, and then calculate the rate of electron transfer across the double layer based on such a model.

The fluctuation of the surface state energy is caused by fluctuation in the polarization field of the medium. Consider the surface site surrounded by fluctuating dipoles. As the dipoles fluctuate, the electrical potential at the surface site, and therefore the electronic energy level of the surface state, fluctuates. In our discussions this dipole fluctuation will be considered the only variable, the only "coordinate" causing energy level fluctuations.

Such fluctuations, however, can only affect electron transfer from (or to) the solid if the polarization is "slow," dependent on ionic motion rather than electronic motion. If the fluctuations stem from polarization of electron shells, the fluctuations will be so rapid they will average out during an electron transfer process. Only when the polar effects stem from slow ionic motion is the surface state energy "frozen" during the electron transfer. Thus in the mathematics below, the factor $(\varkappa_{op}^{-1} - \varkappa_s^{-1})$ appears, where $\varkappa_{op}$ is the optical dielectric constant, $\varkappa_s$ the static.

We will argue in later sections that a kinetic model depending on surface state fluctuation can dominate in cases where the polar medium affecting the surface state is either a polar liquid, the semiconductor itself, or an adsorbed polar gas. However, because the dielectric constant changes abruptly at the surface, analysis is difficult for the latter two cases. The case which has been thoroughly analyzed is that where the polar medium is a liquid (water is the best example, with its polar molecules and its high dielectric constant of about 80) and where the electron transfer is to species near the surface (so tunnelling can occur) but not specifically adsorbed. Thus the ions can in a sense be considered surface states and yet be analyzed as if they were in a homogeneous dielectric medium.

To help clarify the quantitative calculation below, we will first introduce a sketch of the band picture to be derived for ions in a polar medium and describe it qualitatively.

An energy level on an ion in solution will have a characteristic most probable value, its energy when the free energy of the system as a whole is minimized. As described above, the energy will fluctuate about this most probable value. When we plot the probability $W(E)$ that the energy level will have a given energy, we expect therefore a distribution as suggested in Figure 2.5b. E_{ox} is the most probable energy of the electron acceptor (oxidizing agent), E_{red} for the electron donor (reducing agent). Figures 2.5a and 2.5c compare the fluctuating energy levels to the band picture for a metal and a semiconductor, respectively. Electron transfer from the solid to the oxidizing agent can only occur when the "surface state" energy level of the oxidizing agent fluctuates to the energy of an occupied level in the solid, and conversely electron transfer to the solid can only occur if the energy level of the reducing agent fluctuates to the energy of an unoccupied level in the solid. In the figure we indicate the E_{red}, the most probable energy for the occupied level (reducing agent), is at a lower energy than E_{ox}. This is due to Franck–Condon splitting of the surface state energies; the theory for this splitting will be developed in Section 5.5.2.

The equilibrium location of the energy levels relative to the Fermi energy, as shown in Figure 2.5, comes about because the rate of charge transfer at equilibrium must be equal in both directions. Qualitatively, it can be seen in the case of a metal that if the levels E_{red} and E_{ox} were much lower, the electron transfer rate from the metal to the surface would become very large, and the reverse transfer rate small. Similarly unequal transfer would result for the two levels much above the Fermi energy. In the positions indicated for E_{ox} and E_{red}, electron transfer in both directions is low and about equal. Gerischer[13] has shown the same situation occurs at a semiconductor electrode. Thus in general for equilibrium the Fermi energy of the solid should be between the two levels in solution.

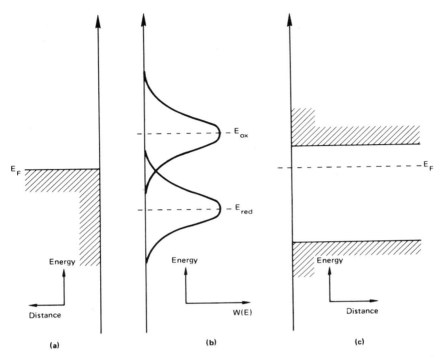

Figure 2.5. Fluctuating energy levels of states in a polar solution: (a) metal; (b) solution; (c) semiconductor. Here $W(E)$ is the probability that the state E_{ox} or the state E_{red} has fluctuated to the energy E. The ordinate shows the approximate position of E_{ox} and E_{red} at equilibrium for equal concentrations of oxidizing and reducing agents, relative to the Fermi energy in the solids.

2.3.3.1. Analysis of the Fluctuating Energy Level Mechanism.

There have been two derivations of the probability of electron transfer between a solid and an ion in aqueous solution, both of which have yielded essentially the same result. In both it is assumed (a) electron transfer occurs between two states of equal energy, (b) the medium can be treated as a uniform dielectric, (c) there are no changes of bonding of the species (in its inner coordination sphere) associated with the electron transfer, and (d) the temperature is reasonably high.[14] The first derivation, by Marcus,[15] involved a complicated method of evaluating the energy of nonequilibrium polarization. He used a variational technique which permitted him to prove that the activation energy for electron transfer that he calculated was the minimum possible with this one "coordinate," the polarization. Another analysis was used by Dogonadze *et al.*[14] and Christov,[16] who treat the fluctuations in terms of phonons, rather than dielectric polarization. Gerischer[13] has applied the energy level fluctuation model to describe rate processes at a semiconductor surface.

Consider the energy ΔE_p necessary to cause fluctuations from the equilibrium polarization of the dielectric around an ion without changing the charge on the ion. In its lowest-energy configuration, the dielectric medium is polarized an amount corresponding to the charge Ze on the ion. We want an expression for the energy increase in the polarized dielectric medium when the polarization fluctuates from its equilibrium configuration to a new configuration which would normally correspond to a charge $(Z + \gamma)e$ on the ion. Thus for example if $\gamma = 1$, the polarization would be exactly the equilibrium configuration with an extra electron on the ion. The energy ΔE_p is given to a good approximation by

$$\Delta E_p = \gamma^2 \lambda \tag{2.48}$$

where the "reorganization energy" λ is given by

$$\lambda = (e^2/8\pi\varepsilon_0 a)(\varkappa_{\mathrm{op}}^{-1} - \varkappa_s^{-1}) \tag{2.49}$$

Here a is the ionic radius, $\varkappa_{\mathrm{op}}$ is the optical, $\varkappa_s$ the static dielectric constant of the medium, and ε_0 the permittivity of free space. This expression was derived by Marcus[15] and by Dogonadze et al.[14] As mentioned above, Marcus showed that such a collective mode of polarization leads to the lowest activation energy for the process. Dogonadze et al. developed a slightly refined expression for λ. We will omit the details of the derivation of Equation (2.48). It derives from simple electrostatics, a calculation of the energy in a polarized dielectric with the polarization induced by a charged sphere, but the derivation is cumbersome.

With the help of Equation (2.48) we can obtain an expression for the energy needed to reduce an ion in solution, and thence determine the net energy required to shift the surface state energy by polarization. We calculate the energy necessary to reduce the ion in three steps. First we let the medium prepolarize, expending an energy as given by Equation (2.48). This prepolarization shifts the energy level from the energy E_{ox} to an intermediate energy E as yet undefined. Then we let the electron transfer from the solid, releasing an energy $E_{\mathrm{cs}} - E$. Now the ion has a new charge, and we let the polarization of the dielectric relax to its new equilibrium value, the release of energy again being determined by Equation (2.48). We use the subscript i to indicate the configuration of the system when the surface state is at the intermediate energy E. The three steps in the reduction described above become

Reaction Step	Thermal Energy Required	
$S_0 A^Z \rightarrow S_i \cdot A^Z$	$\gamma_i^2 \lambda$	(2.50)
$S_i A^Z + e \rightarrow S_i A^{Z-1}$	$(E - E_{cs})$	(2.51)
$S_i A^{Z-1} \rightarrow S_f A^{Z-1}$	$-(1 - \gamma_i)^2 \lambda$	(2.52)

Reaction (2.50) represents the first step, the change in the polarization or change in dipole configuration around the ion A, shifting the energy level from its equilibrium value E_t ($= E_{ox}$ if the ion is the oxidizing agent) to a value E. This shift is accomplished by a change in polarization γ_i, requiring [Equation (2.48)] energy $\gamma_i^2 \lambda$. Reaction (2.51) is the electron transfer, during which the polarization is frozen at the configuration denoted as S_i. Equation (2.52) is the relaxation of the polarization of the medium to its new equilibrium value. As the charge on the surface state (on the ion) is now $(Z - 1)e$, and the polarization of the surrounding medium corresponds to an ionic charge $(Z - \gamma_i)e$, the increase in polarization required to bring the system to its new equilibrium configuration is $1 - \gamma_i$. When the polarization field shifts to the new equilibrium value, [from Equation (2.48)] the energy released is $(1 - \gamma_i)^2 \lambda$.

The total energy change of Equations (2.50)–(2.52), which represents the energy required to transfer an electron to a surface state with an appropriate change of equilibrium polarization of the medium, is

$$\Delta = \gamma_i^2 \lambda - (1 - \gamma_i)^2 \lambda + E - E_{cs} \qquad (2.53)$$

The energy Δ must be independent of the energy E of the intermediate energy level, and we can therefore use the relation to calculate the relation between E and ΔE_p. First we determine Δ by calculating the energy change for the case where $\gamma_i = 0$ and $E = E_t$, where E_t is the mean energy of the surface state (either E_{ox} or E_{red}) and find Δ is $-\lambda + E_t - E_c$. Inserting the value for Δ into Equation (2.53) and solving for γ_i, we find

$$\gamma_i = (E_t - E)/2\lambda$$

whence the energy to complete step (2.50) is

$$\Delta E_p = (E_t - E)^2/4\lambda \qquad (2.54)$$

Equation (2.54) gives the thermal energy necessary to shift the energy level from E_t to an arbitrary energy E.

If we consider thermal fluctuations where the energy $\Delta E_p \sim kT$, we have from Equation (2.54) for a mean square fluctuation

$$(E - E_t)^2 = 4\lambda kT \tag{2.55}$$

Because λ is so large [from Equation (2.49) it is the order of 1 eV for an ion of about 3 Å radius dissolved in water], the energy levels should fluctuate by energies large compared to kT. By substituting $\lambda = 1$ eV into Equation (2.55) it is seen that mean square surface state fluctuations of the order of 0.3 eV or more may be anticipated.

The probability that the energy level of the species will have a value E is less than the maximum probability by an amount given by the Boltzmann factor $\exp[-\Delta E_p/kT]$ or $\exp[-(E_t - E)^2/4\lambda kT]$, and the probability function as illustrated in Figure 2.5 becomes

$$W(E) = (4\pi\lambda kT)^{-1/2} \exp[-(E_t - E)^2/4\lambda kT] \tag{2.56}$$

where the pre-exponential is the normalizing constant making the integrated probability unity.

An equation of the form of Equation (2.56) was developed by Gerischer[13,17] on the basis of a harmonic oscillator approximation and used in his theory of electrode reactions. In this theory he considers the time fluctuating energy levels (Figure 2.5) as bands of energy levels. Such a model, equating the temporal fluctuation to a band of energy levels, is very convenient but must be used at times with caution. Such "bands" arising from polarization fluctuations have different properties from the fixed bands usually encountered in solid state discussions. There is an essential difference in concept between on the one hand electron/phonon interactions causing a fluctuation of electron energies in a static distribution of levels and on the other hand ion/phonon interactions causing a fluctuation of the energy levels themselves. For example, one cannot have optically induced transitions between the occupied and unoccupied levels shown in Figure 2.5b because the levels do not overlap in real space.

The above formulations are restricted to the case where the chemical species is imbedded in the dielectric medium. For surface states very close to the surface, such as adsorbed species, modifications must of course be made to account for the discontinuity in the dielectric constant. The qualitative concepts (in particular the possibility of shifting the surface state to a substantial degree with little thermal energy) should still apply, and it seems that an expression similar to Equation (2.56) with an effective λ lower than given by Equation (2.49) should be useful.

2.3.3.2. The Rate of Electron Transfer to a Fluctuating Energy Level.

The actual rate of electron or hole transfer is the integral of the density of electrons (or holes) times the probability that the surface state will have an energy equal to that of the electron or hole. This integral is, in principle, taken over all energies, but, as was the case for the earlier models, with a semiconductor we can assume that all the electrons are within $2kT$ of the energy of the conduction band edge and all the holes are within $2kT$ of the energy of the valence band edge. From Equation (2.29) the rate of electron transfer from the solid to the surface is given by

$$dn_t/dt = \bar{c}\sigma[n_s(N_t - n_t) - n_1 n_t] \tag{2.57}$$

where the factor $\bar{c}\sigma$ is the mean thermal velocity of the electrons (about 10^5 m/sec) times the cross section for electron capture (about 10^{-19} m^2). The factor n_s is again the density of electrons at the surface, as in Equation (2.38). The factor $N_t - n_t$ is the density of unoccupied states. The density *per unit volume* $\bar{W}$ of unoccupied states within $2kT$ of the band edge [Figure 2.5 and Equation (2.56)] is given by the concentration of ions times the probability the energy level has fluctuated to the energy of the band edge:

$$\bar{W}(E_{cs}) = c_{\text{ox}} \int_{E_c}^{E_c+2kT} \frac{1}{(4\pi\lambda kT)^{+1/2}} \exp\left[-\frac{(E - E_{\text{ox}})^2}{4\lambda kT} \right] dE$$

$$= c_{\text{ox}}\left(\frac{kT}{\pi\lambda} \right)^{1/2} \exp\left[-\frac{(E_{cs} - E_{\text{ox}})^2}{4\lambda kT} \right] \tag{2.58}$$

The parameter c_{ox} is the oxidizing agent concentration in the solution. Thus the volume density of unoccupied states is as given in Equation (2.58). To convert $\bar{W}$ into $N_t - n_t$, the density of states per unit area accessible to the electrons from the solid, we assume that all ions within a distance d from the surface can be reduced by tunneling electrons from the solid. (We will use $d \sim 5$ Å in the numerical illustration below.) Then the density of available surface energy levels per unit area is $\bar{W}d$. It should be noted that the factor σ in Equation (2.57) above includes a tunneling probability factor. In principle one should integrate over σd as a better representation of the coefficient. However, we will omit that refinement.

With these considerations the rate becomes

$$\frac{dn_t}{dt} = \bar{c}\sigma\left\{ n_s\, d\left(\frac{\pi\lambda}{kT} \right)^{-1/2} c_{\text{ox}} \exp\left[-\frac{(E_{cs} - E_{\text{ox}})^2}{4\lambda kT} \right] - n_1 \right\} \tag{2.59}$$

The value of n_1 can be calculated as was done with Equation (2.29) by recognizing that at equilibrium $dn_t/dt = 0$. Then with n_{s0} and n_{t0} as the equilibrium values of n_s and n_t, the rate becomes

$$\frac{dn_t}{dt} = \bar{c}\sigma\, d\left(\frac{\pi\lambda}{kT}\right)^{-1/2} c_{\text{ox}}\left\{\exp\left[-\frac{(E_{cs} - E_{\text{ox}})^2}{4\lambda kT}\right]\right\}\left(n_s - \frac{n_{s0}n_t}{n_{t0}}\right) \quad (2.60)$$

Typical numerical values are $\bar{c} = 10^5$ m/sec, $\sigma = 10^{-19}$ m^2, $d = 5\times10^{-10}$ m, $\lambda = 0.7$ eV, $kT = 0.025$ eV (room temperature), $E_{\text{ox}} - E_c = 0.5$ eV, and $c_{\text{ox}} = 10^{25}$/m^3. With these values, the first term in dn_s/dt is about $(3\times10^{-2}n_s)$ m^{-2} sec^{-1}.

We can compare this value to the rate of electron capture by an acceptor level using the earlier model of Section 2.3.1 with Equation (2.29). However, if the level E_{ox} is 0.5 eV above the conduction band edge E_c (Figure 2.5), then the density of electrons with sufficient energy to transfer to the surface state is $n_s \exp[-(E_{\text{ox}} - E_{cs})/kT]$. Then Equation (2.29) becomes

$$dn_t/dt = \bar{c}(N_t - n_t)n_s \exp[-(E_{\text{ox}} - E_{cs})/kT] \quad (2.61)$$

Again we can use $\bar{c} = 10^5$ m/sec, $\sigma = 10^{-19}$ m^2, and $N_t - n_t = 5\times10^{15}$/m^2. Then $dn_t/dt = (10^{-7}n_s)$ m^{-2} sec^{-1}, obviously much lower than the rate based on the fluctuating surface state model with λ large. However, it must be recognized that normally (without the polar medium), capture by a surface state far above the Fermi energy is not even considered.

In our discussions we have emphasized calculation of the rate of electron capture at the surface state, to provide better comparison with the preceding models based on nonfluctuating levels. In the literature, however,[17-19] the more usual approach is to estimate the rate of electron injection into the solid and assume zero current at equilibrium to give the expression for the rate of capture. It is instructive to mention this approach briefly, since it emphasizes further that the fluctuating model can give much faster electron transfer for deep levels and large λ.

The rate of electron transfer into the solid has been estimated by Hale[18] using the theory of Marcus[15,20] and assuming that if the energy level of a donor (reducing agent) at the surface fluctuates to an energy where there is a band of unoccupied levels in the solid, the electron will transfer with a probability of unity. The rate at which ions inject electrons turns out to be proportional to their thermal velocity Z (about 10^2 m/sec) times their concentration c_{red}. If we assume no work is required to move ions to the surface, and that all other chemical transformation energies are much less than λ, the rate of electron injection is calculated to be

$$dn_t/dt = Zc_{\text{red}} \exp[-(E_{cs} - E_{\text{red}})^2/4\lambda kT] \qquad (2.62)$$

A more detailed discussion of this analysis is presented in Section 8.2.2. Now let us assume that the energy level of the reducing agent E_{red} is 1 eV below the band edge (Figure 2.5). With $\lambda = 0.7$ eV, and the concentration ions in solution $c_{\text{red}} = 10^{25}/\text{m}^3$, we find $dn_t/dt = -6 \times 10^{20}$ m^{-2} sec^{-1}. This can be compared to the rate of electron injection from a nonfluctuating surface state of concentration $10^{16}/\text{m}^2$, with $K_n = 10^{-14}$ m^3/sec and $N_c = 2 \times 10^{25}/\text{m}^3$. With the last term of Equation (2.39), the rate of electron injection is $dn_t/dt = -10^{10}$ m^{-2} sec^{-1}, a much lower rate than that associated with the fluctuating energy levels in a polar medium. Thus if the parameter λ is large, the dominant electron transfer mechanism will in general change to the fluctuating energy level mechanism.

3 | Experimental Methods

There are three classes of measurements which have been used to gather a large fraction of our knowledge about surface chemical physics, and in particular knowledge about bonding orbitals on the surface that are of interest as surface states or surface sites. The first class, electrical measurements, usually on a semiconductor, in general offer detailed information about surface states with energy close to the Fermi energy of the solid. The information is most valuable when one is considering application to semiconducting materials; trying to control electrical properties and trying to control chemical properties involving electron transfer. However, it must be noted that ionic semiconductors are similar to insulators, so that many of the results can be extrapolated to the description of insulators, and covalent semiconductors are similar to metals, so that many of the results can be extrapolated to the description of metal surfaces. The second class, a group of measurement techniques we will call the surface spectroscopies, are measurements where particles are beamed at and/or emerge from a surface. Measurements of the particles and/or related photons provide information about the surface states, but over a wide energy range. The resolution is not generally adequate to use the information for the control of the electrical properties and electron transfer (chemical) properties of semiconductors, but the information is extremely valuable for investigating chemical bonding of adsorbates to solid surfaces, and developing theories of chemisorption. These measurements have been applied primarily to metal, often to semiconductors, and a few measurements have been made[1] on insulator surfaces. The third class, a group of chemical measurement techniques, is one step more qualitative than the spectroscopies, but much more sensitive to low densities of states (sites). These latter measurements are ideal for examining states on the surface involved in adsorption/desorption, in acid/base chemical properties, and in bonding of diatomic or polyatomic molecules to the surface. They are also more sensitive to sites of low density but high activity. Finally, they are our major source of information about states on insulators, aside from special measurements and from extrapolation from semiconductor behavior.

Between the three types of measurements, together with some others that fall into none of these classes, we can build up a reasonably good picture of the behavior of solid surfaces in general.

3.1. Surface Measurements Based on Electrical and Optical Techniques

The discussion will initially center on methods of measuring surface state energy and density, and of measuring electron transfer processes between the solid and its surface based at least in part on the measurement of the double layer V_s on a semiconductor or on a metal. Because the double layers are caused by charge capture at surface groups (surface states), a measurement of the magnitude or rate of change of V_s provides valuable information about electron transfer processes occurring at the surface. When V_s is known, some of these electrical techniques can also provide quantitative information about surface state density and energy. To illustrate the use of electrical measurements to measure surface states, consider the field effect measurement. As will be discussed below, in this technique extra electrons are externally induced at the surface and a known number are captured by surface states, changing their occupancy by a known amount. When the surface barrier V_s is known, the position of the Fermi energy at the surface is known (Figure 2.2). Thus the occupancy of the surface states can be plotted as a function of Fermi energy position. Comparison of the experimental observation with the Fermi distribution function may enable analysis of the energy and density of the surface states in simple cases.

Since most of the methods to be discussed have been in use for some time, the books by Many, Goldstein, and Grover,[2] and by Frankl,[3] together with the recent reviews by Frankl[4] and Reed and Scott[5] are valuable supplementary reading.

3.1.1. Work Function

One of the properties that is simple to understand but difficult to measure reliably is the work function. The work function ϕ is the difference between the free electron energy E_e and the Fermi energy E_F, which, from Figure 2.2 is given by

$$\phi = \mu + eV_s + \chi \tag{3.1}$$

where μ is the energy difference between the bulk conduction band and the Fermi energy, and χ is the electron affinity of the conduction band relative to the free electron energy. Thus the work function increases linearly with V_s, the surface barrier. At a metal surface also, a double-layer potential simply adds to the work function of the clean surface, if no surface phases (oxides, for example) are present.

The work function of a solid can be measured by many of the modern "spectroscopies" (see Section 3.2), but a more accurate method is the classical Kelvin method.[6,7] A typical apparatus schematic is shown in Figure 3.1, with ZnO shown as the material to be examined. A vibrating electrode of known work function is located near the surface to be studied, forming a parallel plate capacitor. The work function difference between the reference and the sample appears as a difference in surface potential between the two surfaces. By the simple capacity law,

$$Q = CV_a \tag{3.2}$$

where V_a is the difference in work function, the "contact potential difference" (cpd), between the sample and the reference electrode, Q is the charge on the surfaces, and C is the capacity. As the reference electrode is vibrating the capacity varies periodically with time, so

$$i = dQ/dt = V\, dC/dt \tag{3.3}$$

and the derivative of the charge becomes a measurable ac current i. Figure

Figure 3.1. The Kelvin method of surface potential (work function) measurement. The vibrating electrode causes an ac voltage which can be adjusted to null by the variable dc voltage. When the ac voltage is zero, the meter reads the surface potential.

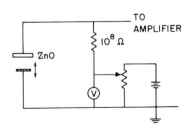

3.1 suggests a circuit, where the amplifier is locked into the signal driving the vibrating electrode. Normally it is operated as a null device: a dc voltage, as indicated, is varied until the measured current i goes to zero. At null the dc voltage is equal and opposite to the cpd. An excellent review of work function measurements is given by Riviere.[6]

The work function measurement [with Equation (3.1)] in principle measures double-layer potentials such as V_s fairly directly, or at least changes in V_s, and so is a particularly valuable measurement. Unfortunately, the measurement has the disadvantage of extreme sensitivity to contaminants so that measurements on anything but noble metals or clean surfaces (Chapter 4) are fraught with difficulties.

3.1.2. Surface Conductivity

The conductance G of a semiconductor crystal along a direction parallel to the surface of interest will vary due to changes of the carrier concentration in the space charge region. The measured conductance is given by

$$G = G_0 + \Delta\sigma W/L \qquad (3.4)$$

where $\Delta\sigma$ is the surface conductivity, and G_0 is the conductance expected if the bands are flat so the bulk properties extend to the surface. L and W are the length and width of the sample. The surface conductivity is given by

$$\Delta\sigma = (\mu_n \Delta N + \mu_p \Delta P)e \qquad (3.5)$$

where the μ's are the mobilities of the electrons and holes, respectively, and the values ΔN and ΔP are the changes from the values with flat bands in the electron and hole density (per unit area) contributed or removed by the surface. The units of surface conductivity (mhos per square) differ from those of bulk conductivity (mhos per meter). Because the units of surface conductivity turn out to be mhos [Equations (3.4) or (3.5)], the same as for the conductance itself, the term "per square" is used to avoid confusion when discussing surface conductivity.

The values of ΔN and ΔP as a functions of V_s are calculated by integrating $n - n_b$ and $p - p_b$ through the space charge region:

$$\Delta N = \int_0^\infty (n - n_b)\, dx$$

$$\Delta P = \int_0^\infty (p - p_b)\, dx$$

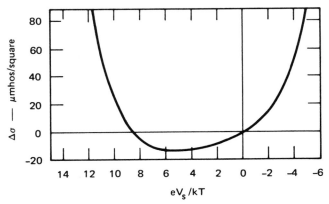

Figure 3.2. The room temperature surface conductivity of n-type germanium with 2×10^{20} donors/m³ as a function of the surface barrier V_s. Bulk electron and hole mobilities have been used in the calculation.

where n and p (as functions of x) are determined with Equations (2.17) and (2.18), respectively, with V as a function of x given by the computer solution of Equation (2.19).

The qualitative features of the $\Delta\sigma$ versus V_s relation as illustrated in Figure 3.2 for an n-type semiconductor are easily understood. With flat bands, by Equation (3.4) $\Delta\sigma = 0$. With an accumulation layer (V_s negative) there are extra electrons to carry current, so $\Delta\sigma$ is positive. With a depletion layer ($V_s > 0$ for Figure 3.2) there are fewer conduction electrons, so G is lower, $\Delta\sigma$ is negative. As V_s increases further, however, and the surface is inverted, conduction by holes begins and $\Delta\sigma$ increases again.

A p-type material is analogous, also showing a conductance minimum as the depletion layer moves to an inversion layer.

For a quantitative analysis of a conductivity measurement another problem arises. In a potential well such as an inversion or accumulation layer, the mobility of carriers is lower than in the bulk material. The carriers are restricted in their motion and are scattered by the surface, which decreases their mobility (their velocity under unit electric field). Thus in Equation (3.5) the surface conductivity depends on V_s through not only the carrier density but also the mobility. Several analyses have been made, and the interested reader is referred to a book on surface physics[2,3] for details.

When a measurement G in Equation (3.4) is made, the G_0 contribution is large and cannot be neglected. This "calibration," (the determination of G_0 for a sample) is done by determining G under one of several conditions where V_s is known, and hence $\Delta\sigma$ is known from theory. Many of the following measurement techniques can be used. The two common "calibration" points are the values where $\Delta\sigma$ is zero (the bands are flat) or where $\Delta\sigma$ is a minimum.

The conductance measurement itself can be used for calibration if the value of V_s can be changed enough to reach the minimum in conductance. In the case of narrow band gap conductors such as germanium the conductance minimum is relatively easily reached by varying V_s by gas adsorption or by normal electric fields.

Even without calibration of the absolute value, changes in conductance with time often provide valuable information about transients in surface reactions, where electrons or holes are injected into or extracted from the semiconductor (or metal) by an adsorbing species for example.

3.1.3. Electroreflectance

Historically the development of the electroreflectance was made after the space charge models were satisfactorily worked out, so little work has been done with this tool. However, the approach shows promise.[8] Its great interest arises because it shows a null measurement (so a "calibration point" for the conductivity) when the bands are flat ($V_s = 0$), and such a band condition is usually attainable even with very wide band gap materials.

In the electroreflectance measurement band gap radiation (photon energy greater than E_g) is reflected from the sample. The reflectance varies with the electric field due to the Franz–Keldysh effect. A greater reflectance obtains at a higher electric field, a lower reflectance at a lower electric field. Experimentally a small ac electric field is applied normal to the surface. The ac field is superimposed on the field of the space charge region, and the reflectance varies with this applied potential. On the negative swing of the potential, the electric field will be increased if the bands are bending up, but will be decreased if the bands are bending down. Thus the phase of the ac component of the reflected light will depend on the sign of the surface barrier. The ac component of the reflected light shows a null at $V_s = 0$, so it is easily determined when the bands become flat.

Electroreflectance measurements can be made at the gas/solid interface using a counterelectrode to produce the electric field. It is most convenient at the liquid/solid interface[9–11] where, as discussed in Section 3.1.6, the surface barrier can be shifted substantially with very low applied voltages.

3.1.4. Field Effect

The field effect measurement provides information both about the space charge region and about surface states. In this measurement, an electric field is applied normal to the surface of the sample, inducing a known [from Equation (3.2)] charge at the surface, δQ, as indicated in Figure 3.3, and the resistance change due to these induced charges is measured.

The quantity $\delta(\Delta\sigma)/\delta Qe\mu_n$ is determined, where $\delta(\Delta\sigma)$ is the measured change in surface conductivity, and $\delta Qe\mu_n$ is the conductance change that would be observed if all the induced charge entered as conduction electrons of bulk mobility.

Now for a perfect surface, where all the induced charge goes to the valence or conduction band, an expression for $\delta(\Delta\sigma)/\delta Qe\mu_n$ as a function of V_s can be calculated from the theory of Chapter 2. Such calculations have been made.[2,3] However in practice all of the induced charge does not remain in the bands, but some moves to surface states. Then the observed conductance change is much less than the theoretically calculated value. If we know V_s from another measurement (say the conductance measurement), the difference between the actual and the calculated field effect signal can give information about surface states. Assume that δQ_{ss} is the amount of charge trapped by surface states. The change in the electron or hole density in the space charge region, δQ_{sc}, is the difference between that induced and that trapped by surface states:

$$\delta Q_{sc} = \delta Q - \delta Q_{ss} \tag{3.6}$$

Then if V_s is known from other measurements, so that δQ_{sc} can be calculated from theory, then δQ_{ss} can be calculated, providing valuable information about surface states.

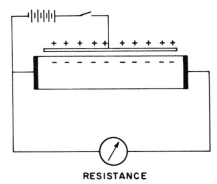

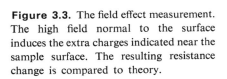

Figure 3.3. The field effect measurement. The high field normal to the surface induces the extra charges indicated near the sample surface. The resulting resistance change is compared to theory.

RESISTANCE

In addition the field effect measurement provides a "calibration" point for the surface conductance measurement. If the surface is strongly n type, δQ_{sc} will appear as electrons. If a negative charge is induced by the field, the conductance will increase. If on the other hand the surface is strongly p-type, δQ_{sc} will appear as holes, and when a negative charge is induced the conductance will decrease. Thus by the sign of the conductance change it can be immediately determined whether the surface is n or p type (whether or not there is an inversion layer). When the field effect signal goes to zero, the surface has a value of V_s corresponding to the minimum in Figure 3.2, and this can be used to "calibrate" the surface conductivity. Thus often the field effect measurement has been combined with surface conductivity measurement, with the surface conductivity providing V_s and the field effect providing δQ_{ss}.

3.1.5. Surface Photovoltage

Another measurement that gives information about the sign of V_s is the surface photovoltage measurement. When the surface of a semiconductor is illuminated with band gap radiation (light of sufficient energy to excite an electron from the valence band to the conduction band), then an electron/hole pair is produced. The electron and hole each move under the influence of the electric field in the space charge region. The electron, being a negative charge, moves downward in the band diagram. The hole, being positive charge, tends to move toward higher V in the band diagram. Thus if the bands bend down (Figure 2.3a), photoproduced electrons move to the surface, photoproduced holes move away from the surface. If the bands bend up, (Figure 2.3b or Figure 2.2), electrons move away from the surface, holes move toward the surface. The double-layer potential always tends toward zero. A measurement of the change in surface potential with illumination (as for example by the Kelvin method described above) gives a different sign for the two cases. There is a slight complication termed the Dember effect,[2] which we will not go into. However, in principle, the surface photovoltage can be used to determine approximately the flat band condition and thus provide a "calibration point" for the surface conductivity measurement. The absolute value of the photovoltage depends upon the surface recombination velocity (Chapter 9), so depends upon parameters other than the surface barrier. In principle, however, with enough illumination intensity, V_s can approach zero, and if the assumption is made that this is so, then the surface photovoltage gives a measure[12,13] of V_s.

The surface photovoltage has been used extensively[14] in recent years as a technique to detect and describe surface states on semiconductors. In this case the sample is illuminated with light of wavelength less than the band gap energy. The energy of the surface state (relative to the conduction band, say) is determined by the wavelength response. At the threshold wavelength ($h\nu = E_c - E_t$) carriers are excited from surface states into the band. Movement of these carriers in response to the space charge electric field reduces the double-layer potential, and, as above, a photovoltage is detected. Heiland and Mönch and Lüth[15,16] have introduced a slight variation in this technique by measuring the conductance rather than the voltage change induced by the photo-injected carriers. Maltby, Reed, and Scott[17] analyze the effect assuming bulk trapping occurs.

As with the photovoltage associated with band gap radiation described above, the absolute photovoltage values depend on recombination rates so are very sensitive to contamination.[18,19] But even when a photovoltage threshold is observed, its interpretation in terms of surface states must be made with caution. First it should be recognized that the chance of a surface state excitation by visible light is low. For such excitation the extinction coefficient of the surface state for the photons must be anomalously high, as the light must be absorbed by less than a monolayer. Thus one must be extra cautious to eliminate the possibility that the photovoltage observed may be associated with excitation of traps within the space charge region, rather than excitation of surface states. Second, if the excitation is due to surface state excitation, there is the possibility that the excitation may be from the ground state to the excited state of the surface group, rather than from the ground state of the surface group to the conduction or valence band. For example, organic dyes adsorbed on semiconductors have the necessary extinction coefficient, but the electron transfer is not to the semiconductor band, but always to the excited state of the dye. This is determined by the observation that the absorption spectrum is unchanged by the adsorption of the dye (see Section 9.5). The electron reaches the bands of the semiconductor, but only via the excited state of the dye, so the wavelength response does not give the surface state energy but the absorption spectrum of the dye.

If the above possible problems can be shown to provide no difficulties, then the photovoltage provides a very simple and direct way of measuring very accurately the energy levels of surface states in the band gap region of a semiconductor.

3.1.6. Capacity of the Double Layer

The measurement of the capacity of the double-layer region has been particularly valuable in studies at the semiconductor/liquid interface.

A capacity measurement at the semiconductor surface is a measure of the thickness of the semi-insulating depletion layer plus the thickness of any other external insulators pressed to the surface. One "plate" of the capacitor is the conducting bulk of the semiconductor, the other is a counterelectrode provided outside the semiconductor surface. The measurement monitors the differential parallel plate capacity

$$C = dQ/dV_a \qquad\qquad (3.7)$$

where dQ is the charge flowing into the counterelectrode (and of course into the sample), and V_a is the applied potential.

Two complications arise in the capacity measurement at the gas/solid interface. One complication arises because usually (but not always[20]) a thick insulating film is needed. As the depletion layer is only the order of 1000 Å thick, to obtain a good sensitivity the insulating layer must be very thin indeed (or be a material of a high dielectric constant). Thus the thickness of the insulating layer introduces a dominating unknown into the measurement. Another complication is that all the charge dQ does not go to the bands, but some (dQ_{ss}) goes to surface states. This introduces another unknown, the "surface state capacity" C_{ss}. The charge introduced is $dQ = dQ_{ss} + dQ_{sc}$, where dQ_{sc} is the charge remaining in the bands. Then

$$dQ/dV_a = dQ_{ss}/dV_a + dQ_{sc}/dV_a \qquad\qquad (3.8a)$$

or

$$C_m = C_{ss} + C_{sc} \qquad\qquad (3.8b)$$

and the measured capacity C_m only partially reflects the space charge capacity C_{sc} that is interpretable (from theory) to give V_s. Information about surface states can be extracted from capacity measurements, as was the case in the field effect measurement.

Oddly enough, it turns out that in many cases these problems do not arise at the semiconductor/electrolyte interface. The "counterelectrode," the electrolyte, is effectively separated only by the thickness of the Helmholtz double layer (see Chapter 8), which is an angstrom or so thick, and usually negligible. And for reasons to be discussed in Section 8.3.4.3, trapping of charge by surface states often does not occur at the semiconductor/electrolyte interface. Thus the surface state capacitance does not appear. So the capacity measurements have been a very effective method of analyzing the space charge configuration at the semiconductor/liquid interface.

Because the measurement is so powerful a tool for studies of the semiconductor/electrolyte interface we will develop the mathematics, but for simplicity we will consider here only the case of a depletion layer, assuming for specificity an n-type semiconductor. Then all terms in Equation (2.20) are assumed negligible that include p_b or $\exp(-eV_s/kT)$. Equation (2.20) evaluated at $V = V_s$ (the surface) becomes

$$\left(\frac{dV}{dx}\right)^2\bigg|_{V_s} = 2\left(\frac{e}{\varkappa\varepsilon_0}\right)(V_s - kT)n_b \qquad (3.9)$$

The factor n_b arose in Equation (3.9) because we assumed $n_b = N_{sc}$, the charge in the space charge region. This restriction is not necessary, and we will generalize Equation (3.9), replacing n_b with N_{sc}. However, the analysis is only valid if N_{sc} is independent of x. Recalling from Gauss' theorem that the surface charge is given by

$$Q = -A\varkappa\varepsilon_0 \frac{d\phi}{dx}\bigg|_{x=0} = A\varkappa\varepsilon_0 \frac{dV}{dx}\bigg|_{V=V_s} \qquad (3.10)$$

where A is the area, we can replace the derivative in Equation (3.9) by a factor proportional to Q^2, and then are in a position to determine the capacity from Equation (3.7). The result is

$$1/C^2 = (\tfrac{1}{2}eN_{sc}kT\varkappa\varepsilon_0 A^2)^{-1}(V_s - kT/e) \qquad (3.11)$$

In a typical measurement at the semiconductor/electrolyte interface, the capacity is used to measure V_s and the applied voltage (in the electrochemical cell) is used to vary V_s at will. Theoretically, as discussed further in Section 8.2.1, any change in the applied voltage V_a should appear entirely across the space charge region for a moderately doped semiconductor. The capacity extrapolates to infinity [by Equation (3.11)] when $V_s = kT/e$, and if we call V_{fb} (the flat band potential) the value of V_a extrapolated to $1/C^2 = 0$, then

$$V_s = V_a - V_{fb} - kT/e$$

determines V_s for each value of V_a. As an additional benefit the slope of $1/C^2$ versus V_a plotted according to Equation (3.11) yields N_{sc} directly.

The above simple theory is found applicable to a wide range of semiconductors. For example, Dewald[21] was able to show that the space charge theory as applied to the ZnO/aqueous solution interface accurately applies.

He was able further to show for this system that the more detailed space charge theory with an accumulation layer present is also valid. Gomes and his group[22,23] review the application of the technique to solid/liquid systems.

Brown and his co-workers[24,25] have applied capacity measurements to the free surface with success. However, the difficulties of the measurement make it less popular in free surface studies than in studies at the semiconductor/liquid interface.

3.1.7. Channel Measurements

A particularly convenient method of studying surfaces with inversion layers, only applicable to materials which can be made both n- and p-type, was developed by Statz and his co-workers.[26] This is the channel resistance measurement. A pnp configuration is prepared as illustrated in Figure 3.4. (Channels on p-type materials are examined with the corresponding npn configuration.) An inversion layer as indicated is induced in some way, perhaps by adsorption of some active gas. This p-type surface layer, the inverted region at the surface of the n-type material, is the "channel" connecting the two p-type regions. The two junctions are biased in the reverse direction using the dc voltage indicated, so that current cannot flow from one p region to the other through the n region. However, the two p-type regions are connected electrically by the p-type channel, and a conductance measurement monitors the conductance of the channel. The conductance measurement can be made, for example, using ac methods.

The important novelty here is that the conductance measured is only due to carriers in the channel, so that there is no bulk contribution such as G_0 in Equation (3.4) to decrease the sensitivity to surface effects. The possibility of varying the bias across the junctions permits an extra degree of freedom, because the bias voltage appears across the n region and its surface channel and causes a variation in the charge in the channel (induces a charge just as the field effect measurement does). With these two features

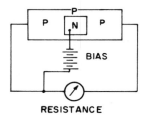

RESISTANCE

Figure 3.4. The channel measurement. When an inversion layer is present (the p layer on the n-type region) this "channel" establishes a low-resistance path between the two outer regions of the pnp structure. With a voltage to reverse bias the two junctions, the resistance of the channel becomes a very sensitive probe of surface changes.

—the high sensitivity of the conductance measurement and the ability to induce large changes in charge near the surface—a complete analysis of the charge in the surface states, δQ_{ss} [obtained from the difference between the actual conductance change and the expected conductance change, with Equation (3.6)] can be made as a function of V_s over a wide range of V_s.[26] The variation in Q_{ss} with V_s can be compared with theory [Equation (2.32), (2.33), and the solution of (2.22)] to yield information on the surface state energy and density.

A recent and extremely important technical development based on such channeling phenomena is the MOS transistor in silicon technology. The term MOS stands for metal/oxide/semiconductor (e.g., $Al/SiO_2/Si$). In this device two n regions are prepared by diffusion at nearby spots on a p-type silicon substrate, giving the npn configuration, and a channel is induced between the n regions in the following way. A thin oxide (the SiO_2) is grown over the surface between the n regions and the metal overlayer film is evaporated on the oxide. Application of a strong voltage V_a to the metal film leads to a high electric field at the surface of the p region. V_s can easily be controlled to form an n-type inversion layer (a channel) at the surface that provides a conducting path between the two n-type spots. In the usual terminology, the metal film is called the gate, the two n-type spots are called the source and the drain. A small change in gate voltage changes V_s and affects the conductance between the source and the drain substantially. Since the current through the gate is negligible, there is a power gain between the input and the output. The device is of particular value because it is an extremely high-input impedance amplifier, suitable for electrometer use. The above device depends on effects associated with the solid/solid interface, which we are not emphasizing in this book. Nonetheless the technique does provide a good example of the use of the concepts of surface space charge layers.

3.1.8. Powder Conductance

In some cases measurement of the conductance of a pressed pellet of a semiconducting powder can provide information about the energy level of surface states, in particular when the surface states are associated with additive species deposited on the powder surface.[27] Such a conductance measurement can be of particular value to provide supplementary information in chemical studies where the use of powders rather than crystals is preferred—powders have the large surface areas needed to provide a high rate of chemical reaction.

The conductance of a pressed powder pellet is often dominated by intergranular contacts: the transfer of current carriers across the surface barrier at the intergranular contacts is the current limiting process. For such cases a measure of conductance versus temperature yields interpretable information about the surface barrier and the surface states that induce it. The rate at which electrons in an n-type semiconductor can cross the surface barrier at an intergranular contact (and thus the conductance) is proportional to the density n_s of electrons in the conduction band at the surface (Figure 3.5). The value of n_s is given by Equation (2.36), which describes n_s as a function of the surface Fermi energy. However, the Fermi energy at the surface is controlled by surface states and is given by Equation (2.32). On combining these, the conductance is given either by

$$G = G'f_1(A)f_2(d)N_c \exp[-(E_{cs} - E_F)/kT] \tag{3.12}$$

$$= G_0 \exp[-(E_{cs} - E_F)/kT] \tag{3.13}$$

or, with Equation (2.32), by

$$G = G_0(n_t/p_t) \exp[-(E_{cs} - E_{ss})/kT] \tag{3.14}$$

where Equation (3.12) expresses the conductance as a product of a constant factor G', a function f_1 of the intergranular contact area A, a function f_2 of the particle diameter d, and the density of electrons at the surface n_s. If the pre-exponential factors are considered temperature insensitive compared to the exponential factor, then all these pre-exponential factors can be collected into a factor G_0 and the conductance be expressed in terms of the position of the Fermi energy relative to the conduction band energy of the powder semiconductor, as in Equation (3.14). If the Fermi energy is determined by a surface state at the energy E_{ss}, then Equation (3.14) applies and the slope of an Arrhenius plot of conductance versus temperature gives the surface state energy directly. In this measurement the absolute conductance is not of interest, but only the slope of the Arrhenius plot. Thus the variable of contact area, usually a problem with pressed powder measurements, is bypassed. The approach requires (a) the bulk conductance of the powders must be high enough that it does not contribute to the overall resistance measured, (b) the surface state density and occupancy must be relatively temperature insensitive [clearly from Equation (3.14)], thus volatile gases cannot be studied in general, and (c) the electron transfer to and from the surface species must be rapid, so that the thermodynamic equilibrium is reached at all temperatures of the experiment.

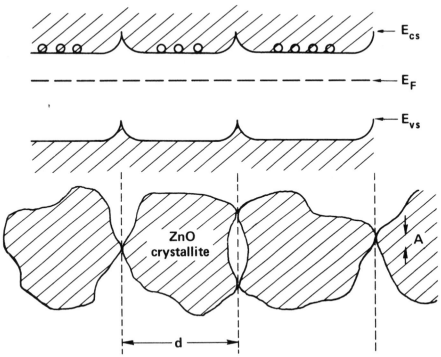

Figure 3.5. A schematic showing how the intergranular contact resistance of a pressed pellet is extremely sensitive to the surface barrier (and hence to surface states) at the surfaces of the grains. Bulk electrons must cross this surface barrier (top diagram), and the number of electrons with sufficient energy varies rapidly with temperature.

The approach has been used primarily[27] for the measurement of the Fermi energy or the surface state energy of nonvolatile surface additives deposited in significant quantities on a powder.

3.1.9. Ellipsometry

Ellipsometry is a very sensitive measure of surface properties as observed by optical reflectance. In an ellipsometry measurement plane polarized light, with the plane of polarization rotated 45° from the plane of incidence of the light, is reflected from a surface. The two components of the polarization, a component polarized perpendicular and a component polarized parallel to the surface are then equal in amplitude in the incident beam. Now if there is a film on the surface, the two components of polarization in the reflected light will not be equal either in phase or amplitude.

The difference in phase between the two reflected components is termed Δ and the ratio of the two amplitudes is termed ψ. By a suitable arrangement of polarizers, compensators, and analyzers, the values of Δ and ψ can be accurately determined. The thickness and the index of refraction of the film can be determined with the Drude theory.

In much recent work, particularly by Bootsma, Meyer, and their colleagues,[28–31] this approach is applied to surfaces (even metal surfaces[32]) on which there is not a film, but just a small fraction of a monolayer of adsorbed material. Interesting values of Δ and ψ are observed which apparently are providing information about the adsorbed species. Unfortunately the exact interpretation is not clear, although possible models have been suggested to relate the signals to adsorption with submonolayer coverage.[29] Certainly the technique has proven valuable in monitoring adsorption of gases on solid surfaces.

In has been suggested[30,31] in some cases there are effects arising from intrinsic surface states (surface states on the clean surface). Large values of Δ and ψ occurring in a certain wavelength region are interpreted in terms of optical transitions between these surface states. The fact that the observed Δ and ψ are sensitive to adsorbed gases is considered supporting evidence for the model, in that adsorbed gases should remove such surface states (Section 4.4.2). However caution must be exercised in using this interpretation: it has been pointed out that if the absorption is in the energy region associated with bulk band-to-band transitions the effect may be due to shifts in such absorption with changes in the double-layer electric field.[33,34]

3.1.10. Other Electrical and Optical Measurements

Other measurements of the space charge layer can be made, for example photoconductivity,[2,15,16] luminescence,[35] or transport properties,[36–39] including either the Hall effect or the magnetoresistance measurement. Such magnetic measurements are of particular value in the case of an accumulation layer to study the quantization of the electron or hole wave functions normal to the surface. In general, however, the magnetic measurements are complex and will not be dealt with in the present discussion.

Other optical measurements which can be applied more directly to the analysis of surface states and surface bonding will be reviewed below. We will deal with photoelectron emission in Section 3.2 as one of the "surface spectroscopies," and with infrared (ir) absorption in Section 3.3 as a "chemical measurement."

3.2. The Surface Spectroscopies

A form of measurement that is not limited to semiconductors (as most electrical measurements are) is a measurement where some particle or photon is beamed at the surface, and a particle or photon emerges with characteristics that provide information about the surface. We will use the term "surface spectroscopies" to refer to those measurements where a particle is either incident or emergent or both and discuss such measurements in this section. The information obtained from the surface spectroscopies includes the chemical composition, the geometrical structure of the surface, the distribution of occupied surface state energy levels, the distribution of unoccupied energy levels, the directional characteristics of the orbitals, the oxidation state of the surface species, the reactivity of the surface or species present on the surface, characteristics of solid/adsorbate bonds, and even such parameters as the vibrational modes (phonon levels) in the solid.

In Table 3.1 is a listing of the most important of these spectroscopies. This will provide a preliminary comparison of the techniques available for measurement of the various parameters. The column marked in/out refers to the composition of the incident and emergent beams, respectively. The depth sampled is usually determined in an obvious way by the particles involved. A low-energy electron has a mean free path of 5 to 100 Å before it is scattered (depending on its energy) and ions cannot penetrate the solid appreciably, so reactions involving ions must be with the surface layer. If the emergent beam is a photon, it provides no limitation, and the depth limitation is associated with the incident beam.

The table orders the surface spectroscopies in four groups. The first group, (a)–(f), includes those techniques which appear to have the capability to measure surface states and bonding orbitals characteristic of the clean surface and characteristic of the interaction of foreign species with a surface. As mentioned above, the resolution of the techniques is inadequate to predict electrical properties in detail, but the measurements are of great value in comparing surface state energies with theoretical models and thus in understanding how adsorbates interact with surfaces.

The second group, (g), consists of only one technique, low-energy electron diffraction, a highly valuable addition to the tools of surface science which has provided us with information about the structure of surface layers and in particular how the surface atoms are displaced from their expected lattice sites.

The third group, (h)–(l), includes a series of measurements that are used primarily to investigate surface composition and detect impurities.

Table 3.1. The Surface Spectroscopies

Measurement	$\dfrac{\text{in}}{\text{out}}$	Approximate depth sampled, Å	Parameters measured	References
(a) UPS (uv photoelectron spectroscopy)	$\dfrac{\text{uv}}{\text{electrons}}$	50	Surface states (occupied) Direction of bonds Valence state of ion	40–43
(b) ELS (electron loss spectroscopy)	$\dfrac{\text{electrons}}{\text{same electrons}}$	10	Transitions between surface states	44, 45
(c) SXAPS (soft x-ray appearance potential spectroscopy)	$\dfrac{\text{electrons}}{\text{x-rays}}$	30	Surface states (unoccupied) Composition Valence state of ion	46–48
(d) FEM, FEED (field emission microscopy, field emission energy distribution)	electron emission	5	Surface state (occupied) Adsorbate levels (unoccupied) Crystal structure Direction of bonds	49–51, 146
(e) FIM (field ion microscopy)	$\dfrac{\text{atoms}}{\text{ions}}$	3	Surface states (unoccupied) Crystal structure	49, 52–56, 147, 148
(f) INS (ion neutralization spectroscopy)	$\dfrac{\text{ions}}{\text{Auger electrons}}$	5	Surface states (occupied)	56–58
(g) LEED (low-energy electron diffraction)	$\dfrac{\text{electrons}}{\text{diffracted electrons}}$	10	Crystal structure Steps on surface	52, 59–64

(h) AES (Auger electron spectroscopy)	$\dfrac{\text{electrons}}{\text{Auger electrons}}$	10	Composition Valence state of ion	65–67
(i) XPS, ESCA (x-ray photoelectron spectroscopy, electron spectroscopy for chemical analysis)	$\dfrac{\text{x-rays}}{\text{core electrons}}$	100	Composition Valence state of ion	41, 65, 67–69
(j) SIMS, IMMA (secondary ion mass spectroscopy, ion microprobe mass analysis)	$\dfrac{\text{ions}}{\text{ions}}$	50	Composition (as a function of depth)	70–72
(k) ISS (ion scattering mass spectroscopy)	$\dfrac{\text{ions}}{\text{ions}}$ (usually He/He)	3	Composition Relocation of atoms	73–76
(l) EMP (electron microprobe)	$\dfrac{\text{electrons}}{\text{x-rays}}$	10^4	Composition	77
(m) MBRS (modulated beam relaxation spectroscopy)	$\dfrac{\text{chopped ion beam}}{\text{product of reaction}}$	3	Reaction kinetics	78–80, 149
(n) ESD (electron- or uv-induced desorption)	$\dfrac{\text{electrons or uv}}{\text{ions}}$	3	Composition Bond direction	81–83
(o) Ultrasonic-induced desorption	$\dfrac{\text{sound energy}}{\text{ions}}$	3	Composition	83

This is a valuable group from a practical point of view, and the development of these techniques has permitted substantial improvements in the past few years in the reliability of surface experiments.

The fourth group, (m)–(o), is a group of three spectroscopies which provide information more of a chemical kinetic nature. The first is the use of molecular beams to study sorbate/sorbate or sorbate/surface reactions, and the other two are physical methods of induced desorption, which are new. Their advantage over thermal desorption as described in Section 3.3 has not been clarified completely.

3.2.1. Ultraviolet Photoelectron Spectroscopy (UPS)

In UPS, ultraviolet radiation is beamed at the surface, and the emitted photoelectrons are observed. The wavelength of the uv radiation is the primary variable, with the angle of incidence and the polarization of the light as secondary variables. The energy spectrum of the emitted photoelectrons is measured, and in some experiments the angular dependence is also determined, perhaps as a function of a parameter such as the polarization of the incident beam.

In order for photoelectrons to be produced, the electron must be excited from an occupied energy level to an energy beyond the free electron energy. If it is assumed that the fraction of electrons emitted is constant for all electrons excited to an energy higher than the free electron level, E_e (provided they are created near the surface, so that a collision cannot de-excite them), then the measured photoelectron energy distribution reflects the density of occupied energy levels, both bulk and surface levels. Consider Figure 3.6, which is intended to represent a density of states function for the bulk energy bands in a solid. If the energy of the incoming photon is $h\nu_i$, then photoelectrons will be observed emanating from all the occupied states above E_i, where $E_i = E_e - h\nu_i$. If the density of states as a function of energy is $n(E)$, then to a first approximation the energy distribution of the photoemitted electrons will have the shape of $n(E)$ between E_i and E_F. Freeouf and Eastman[84] suggest that to avoid structure due to the conduction band, $h\nu_i > 25$ eV is preferred.

If there is a high density of occupied surface states, the electrons from these too can be excited to the vacuum level or above and the spectrum of surface states may appear riding on the spectrum of bulk states. Observation of both intrinsic (clean surface) surface states and peaks associated with adsorption have been claimed.

It is difficult to distinguish whether structure in the photoelectron

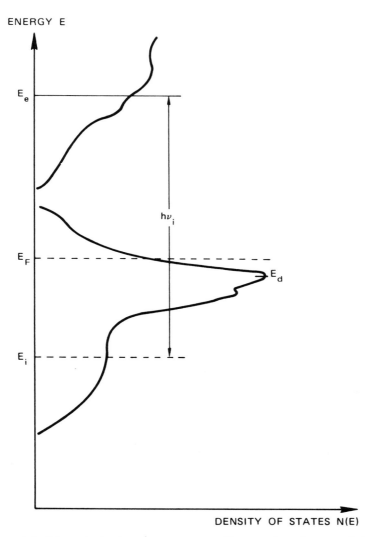

Figure 3.6. Schematic density of states curve to illustrate photoelectron emission. Here E_i is the lowest energy state from which photoelectrons can be emitted with photon energy $h\nu_i$. E_F is the highest occupied state. The energy distribution of photoelectrons will mirror $N(E)$ between E_i and E_F. The peak at E_d is intended to represent a high-density d band.

spectrum arises from optical absorption by the bulk material, by surface states on the clean surface, or by surface states due to foreign species. As will be discussed in the next chapters, intrinsic surface states on the clean material disappear in general when an adsorbate is introduced, and the disappearance of a UPS peak from the clean surface upon adsorption can be interpreted to mean that the signal observed was a surface rather than a bulk state. However, as Lewis and Fischer[85] comment, such "fading" of peaks during adsorption also occurs for peaks associated with the d bands of Ni upon adsorption of oxygen, so there is still some question with respect to the interpretation of intrinsic surface state measurements on clean surfaces. Another method of distinguishing bulk transitions[86] is applicable if the bulk transition is "direct" (conservation of wave vector **k** during the transition). It turns out for reasons beyond the scope of this book that for such transition the energy is a function of $h\nu_i$. Thus photoelectron peaks that shift in a certain way with $h\nu_i$ can be assigned to bulk transitions.

When the spectrum changes due to adsorption the peaks added (or removed) can be associated with the sorbate/solid complex. These changes are complicated and the interpretation of each measurement must be examined with caution to be confident that the interpretation of the new photoelectron peaks in terms of a particular model has adequate supporting evidence.

The sources of uv illumination are either synchrotron radiation,[87] or hydrogen, helium, or neon resonance lamps. The resolution of the emission peak from these lamps is the order of 0.3 eV, which provides a lower limit on the resolution of the distribution of photoelectron energy, a lower limit that may lead to apparent broadening of bands of energy levels and uncertainty about the position of the edge of the bands. There are other effects contributing to uncertainties in quantitative interpretation. As with other techniques where electrons are excited from deep levels and thus produce high-energy holes, there may be peak shifts associated with screening effects, where reorganization of the valence electrons around the hole releases extra energy to the emitted electron.[88] Such screening may be different for bulk levels than for surface levels. Evaluation of the possible effect of variations in this "relaxation energy" on the resolution and even on the accuracy must await further analysis. Other effects such as heterogeneity of the surface or resolution limitations in the analyzer may decrease the resolution further.

On the whole, however, UPS studies of surface states associated with adsorbate/solid bonds are providing valuable information about chemisorption. In the development of such information the resolution limitation is not so critical, since here chemical bonding concepts rather than semiconductor electrical effects are being explored. The UPS technique provides exactly the information needed, a reasonably quantitative picture of the bond energies to match with the emerging theoretical models of gas/solid bonding. There is no question that UPS is the most important tool of the surface spectroscopies in providing us with information on surface states and adsorbate bonding.

3.2.2. Energy Loss Spectroscopy (ELS)

As suggested by Rowe et al.[44] energy loss spectroscopy is a natural complement to UPS, in that where the latter shows the density of occupied states, the former (ELS) shows the energy of transitions from these occupied states to unoccupied states.

In this technique low-energy electrons (E_0 about 100 eV) are beamed at a surface and the energy of the scattered electrons is analyzed. Many of these electrons will suffer only one inelastic collision, in which they lose a definite amount of energy ΔE exciting a dominant transition in the solid. Then there will be a peak in the energy distribution of the emerging electrons at energy $E_0 - \Delta E$. A spectrum of peaks in energy is observed which is interpreted as favored electronic transitions in the solid. In order to use the ELS technique to study surface transitions, it must be determined which of the observed peaks is due to bulk transitions. Two types of bulk transitions are observed, a "plasmon" transition and bulk electron excitation transitions. To identify bulk transitions, various surface treatments are applied. The peaks that are independent of surface treatment are identified with bulk effects and the rest are assumed due to surface states. The use of the technique depends upon the skill of the investigator in correctly distinguishing between surface states and bulk effects near the surface which are affected by the treatments used.

As pointed out by Ludeke and Esaki,[89] the interpretation of the energy loss peaks (to give a density versus energy of surface states) is simple only if it is assumed that the initial state is a sharp narrow band such as a d band. With such an assumption, the intensity/energy plot should mirror the density of states versus energy spectrum of the final state. If both the initial and final energy level bands are broad, the interpretation becomes very difficult.

The energy half-width of the incident electron beam is the order of 0.3 eV or larger, so even if the initial state is a very sharp band, the resolution of the energy distribution in the final state will be limited to about 0.3 eV, and usually the resolution is significantly poorer due to analyzer limitations.

Other problems in interpretation of the ELS spectrum are associated with the directionality of the path of the electron, which relates to the particular excitation stimulated. Because electron diffraction occurs, the direction of the electron in the solid, when it is undergoing its inelastic scattering event, is not always obvious. Sickafus and Steinrisser[90] have shown that for a clean surface or a surface with an ordered adsorbate structure, the ELS spectrum changes with primary beam energy (diffraction angles change), and difficulties in resolution and/or interpretation follow.

3.2.3. Soft X-Ray Appearance Potential Spectroscopy (SXAPS)

In these measurements electrons are beamed at a solid, the electron energy needed to stimulate particular x-rays is determined, and the x-ray intensity versus the energy of the impinging electron is followed. The technique is old, but modern electronic methods increase its sensitivity manyfold.

The excitation energy necessary to produce x-rays is that required to excite an electron from an inner level to an unoccupied level (the conduction band in the solid, for example). An inner level on an atom (a core level) is a relatively narrow band because there is little overlap with the equivalent orbitals of neighboring atoms in the solid to cause band formation. To excite an electron from this core level, the core electron must go to an unoccupied level (above E_F) so the impinging electron must have energy $E_F - E_{core}$ plus enough remaining energy (E_F) to itself go to an unoccupied level. When this energy requirement is met, then the characteristic x-rays are observed associated with the atom whose core electron is being ejected.

As the energy of the incident electron increases above the appearance energy, the intensity of emitted x-rays should be related to the density of unoccupied states to which the core electron can be excited. Thus for the energy level distribution shown in Figure 3.6, the intensity should be very high for impinging electron energy corresponding to $E_F - E_{core}$, but should decrease for the higher energy because the density of available states decreases with increasing energy. In addition, because the unoccupied levels depend in some cases on the oxidation state of the atom, changes in spectrum can be associated with oxidation state. Such behavior is observed qualitatively. Unfortunately there are complications related to the lifetime of the hole in the core level, to screening of the positively charged atom, and to the requirement that unoccupied levels for two electrons must be available that reduce the resolution.[47,88]

With adsorbed atoms bonded to the surface, similar arguments obtain. For example appearance potential measurements of oxygen on nickel show very broad structure, with peak widths the order of 10 eV.

Thus at the present time, SXAPS is inadequate in resolution to provide the information regarding the structure of bands of unoccupied levels, pending a better understanding of the theory. A more detailed description of the problems is presented by Park.[47,91] A discussion of apparatus is given by Musket and Taatjes.[92]

The use of SXAPS to determine surface composition is obvious, as the x-ray lines emitted will be characteristic in general of the atom whose core electron is excited, particularly for deep core levels (as in the electron microprobe technique). However, Musket's study[93] comparing SXAPS and Auger electron spectroscopy suggests that in composition analysis the sensitivity of low-energy SXAPS is unreliable, since it depends too much on the density of unoccupied states, so the sensitivity, say, to carbon on germanium, would be entirely different from the sensitivity to carbon on iron. Thus AES is a far better tool for measurement of surface composition.

3.2.4. Field Emission Microscopy (FEM)

Field emission is one of the older of the spectroscopies in its use as a surface tool, being introduced by Müller[94] in 1937, but only recently has it been used successfully to study surface state bands.[95-97] The sample to be studied is prepared as a fine point (tip) such that a strong electric field can be induced at the surface with a high negative voltage applied to the tip. Under these conditions tunneling of the electron from an occupied state below the Fermi energy into the vacuum can occur. In Figure 3.7a an illustration of the effect is shown where the band picture of a metal is shown to the left, and a plot of the energy of an electron in vacuum is shown to the right. Due to the high electric field, the energy of a free electron decreases rapidly as it moves away from the surface (except for an initial rise associated in part with image forces at the metal surface). If the field is high enough so that the electron only has to tunnel a few angstroms to emerge, tunneling will occur as indicated by the arrow. The term "microscopy" is used because different faces emit in different amounts and in different directions so that beams from the different faces can be identified on a fluorescent screen.

By a careful study of the energy distribution of the emitted electrons as a function of electric field, the density of states function of the band below the Fermi energy can be determined.[95,98] In particular, there is evidence that emission from occupied surface states, if in high density, can be observed.[85,99] However surface state emission can be complex[100] for the surface state charge can be strongly affected by the field.

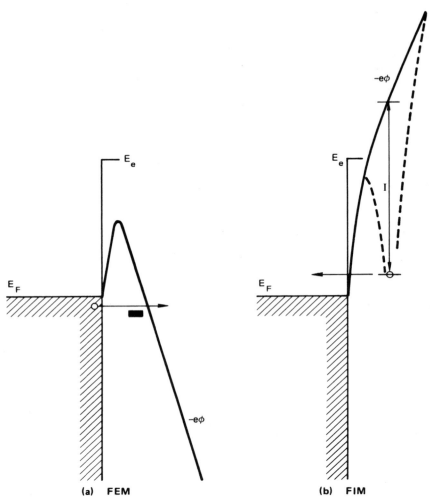

Figure 3.7. Emission of electrons (a) or of ions (b) with a strong field $-d\phi/dx$ at the surface. Here $-e\phi$ is the energy of an electron. In (a), FEM, if $d\phi/dx$ is large enough to narrow the barrier to a few angstroms, electrons can tunnel out. An unoccupied surface state such as shown can enhance the tunneling (see text). In (b), FIM, if the field is strong enough, the ground state of an atom can be raised as shown above the Fermi energy so electrons can tunnel in. The resulting positive ion is then expelled.

Of particular interest for the study of surface states and bonding is the observation that an unoccupied energy level associated with an adsorbed species, which is located in the middle of the potential barrier (Figure 3.7a) can offer the electron an easy path to the vacuum. The unoccupied surface state can be shifted to a position below E_F by the electric field. An effect termed "resonance" occurs which substantially increases the electron emission for electrons from the solid which have the same energy as the unoccupied surface state. This enhancement of electron emission can be used to detect unoccupied surface states associated with adsorbate/solid bonds.

3.2.5. Field Ion Microscopy (FIM)

In field ion microscopy as in field emission microscopy the sample is in the form of a sharp tip to permit high electric fields. In FIM electrons from gas-phase atoms tunnel into the solid. Thus where FEM usually gives information on occupied levels in the solid, FIM provides information about unoccupied energy levels in the solid. In the FIM technique[101] gas-phase atoms at a low pressure are made available. They are ionized at the positively charged tip when their higher-energy electrons tunnel through a field barrier into the solid. The ion is accelerated away from the tip, and the resulting beam of ions is observed on a fluorescent screen. The more probable the electron tunneling (the more states available to which the electron from the atom can tunnel), the higher the intensity of the resulting ion beam.

A band picture illustrating the energy of a negative charge as a function of distance is shown in Figure 3.7b, where an energy level associated with one of the gas-phase atoms near the surface is indicated, I being the ionization potential of the atom. A shift of the energy levels of the atom (relative to the levels in the solid) occurs due to the strong electric field. Because of this shift the occupied level on the atom is raised sufficiently for electron tunneling to occur, as indicated by the arrow. The tip is made highly positive, to attract the electrons, and repel the resulting ions. Each crystal plane or imperfection will emit ions at a different rate and in a different direction, so observation of the ion beams striking a fluorescent screen provides a picture of the surface, the FIM image. A feature on the surface as small as 3 Å can be resolved.[102]

If desired, the field can be increased to the point of "field evaporation," where the atoms of the host crystal are ionized and peeled off layer by layer. The resulting motion of steps can be observed on the screen. It should be noted that true FIM is limited to a few materials (such as W) where field evaporation occurs at a higher field than atom ionization. Atom ionization is required to identify planes and develop a picture of the surface before field evaporation causes the tip to become eroded.

In principle the measurement of intensity of the resulting ion beam as a function of electric field should provide information about the density of the unoccupied states to which the electron can tunnel. In practice the interpretation of the results is still somewhat complex.

There is evidence that unoccupied surface states as well as bulk states can be detected by FIM. For example, the work of Ernst and Block on germanium,[103] which will be discussed in Section 4.4.3, suggests there is information in the spectrum about the acceptor states at the germanium clean surface.

An interesting variation on the FIM technique is the incorporation of an "atom-probe hole" in the observation screen[104-107] The hole in the screen is the entrance port to a mass analyzer. As each crystal face provides a spot on the screen due to the beam of ions emanating from that face, it becomes possible, with manipulation of the direction of the tip, to direct a spot from any particular crystal face to the hole and mass analyze the particles emanating from that crystal face. Such a modification has enabled workers to study, for example, the tendency of gas ions to be adsorbed under the very high electric fields.[106,108] In addition, after orientation to observe a particular crystal face, the field can be increased to the point of field evaporation of the metal from the tip, and the composition of the host material can be studied.

3.2.6. Ion Neutralization Spectroscopy (INS)

Ion neutralization spectroscopy is based on a very complex process that has been chiefly studied by Hagstrum et al.[58] In these measurements, ions (usually helium ions) are beamed at a solid surface. Two electronic processes occur. First, an electron in an occupied level at the surface neutralizes the impinging He ions, releasing energy. Then second, the excess energy is given to another electron at the surface which becomes an Auger electron (an Auger electron is an electron that absorbs the total energy released when another electron drops into an unoccupied low energy level). The Auger electron may have enough energy to escape from the solid, and its energy can be measured.

Thus two electrons are involved in the INS event and analysis must consider the density of states at the initial energy of the neutralizing electron and also the density of states at the initial energy of the Auger electron. A single distribution parameter is measured, the energy distribution of the Auger electrons. Analysis is made possible by the fact that the energy of the two electrons are related; the energy lost by one is assumed to be acquired entirely by the other. With this relation introduced into the mathematics, one of the energy parameters can be eliminated and the concentration of Auger electrons at each energy can be expressed as an integral over a single energy variable. Computer analysis of the energy distribution of Auger electrons can then be "deconvoluted" to provide an expression for the density of states at the surface.

Many approximations must be made in the analysis of this rather complex process. For example, the ionization potential of the incident ion must be assumed independent of position, and the two electrons involved in the interaction must be assumed to arise from the surface layer (to have the same local density of states). It must be also assumed there is no energy loss by inelastic processes, so all emitted electrons are true Auger electrons. Such assumptions make it very difficult to estimate the resolution of this spectroscopy with respect to energy of the surface state peaks. Hagstrum and Becker[57] concede that the resolution is worse than UPS, but it is very difficult to estimate to what extent. The results reported in general show peak half-widths of 0.5 eV or larger.

The major advantage of INS, from a surface point of view, is that the process by definition has to take place very near the surface layer of the solid. Electrons from the solid must first be transferred to the helium ion, and indeed those electrons will be most active whose orbitals are directed normal to the surface. The second step, the production of the Auger electron, can occur further in, but the transfer of energy almost requires that this step also occur very close to the surface.

3.2.7. Low-Energy Electron Diffraction (LEED)

Low-energy electron diffraction is in a class by itself in terms of the information it provides with respect to the properties of a surface. As normal with diffraction measurements, LEED provides data on lattice spacing, but because very low-energy electrons are used, which do not penetrate, the information provided concerns primarily the top layer or two of atoms at the surface. Thus (a) it is used to determine qualitatively the removal of layers of foreign species in a cleaning procedure, by observing when the lattice structure of the host material becomes clearly developed, (b) it is used to provide information about reconstruction (where the surface plane of the lattice is periodic, but does not have the same structure as the substrate lattice), (c) it is used to provide information about ordered spacing of adsorbed species, and (d) it is used to provide information on regular surface steps (that can be induced by cutting a sample a few degrees off a plane of low Miller indices).

Recently LEED has also been found useful for detecting the amplitude of vibrations in the surface atoms in the host crystal[62,109] by monitoring the width of the diffraction spot. Surface plasmons in the conduction electrons of the solid have also been investigated (by a technique closely related to ELS). We will not be describing these results as the effects are of secondary interest in our understanding of the chemical/physical properties.

The technique was developed primarily by Farnsworth et al.,[110] who also were the first to observe the surface reconstruction of the topmost layer of some crystals. A bibliography of pre-Farnsworth work and a complete bibliography of work to early 1970 has been compiled by Haas et al.[111]

In electron diffraction measurements the incident electron beam is usually in the range of a few tens to a few hundred electron volts. In Farnsworth's original measurements the diffracted beam was detected by a Faraday cup, but with present day commercial LEED apparatus a fluorescent screen is used whereon all the diffracted spots are simultaneously observed and photographed.

The interpretation of the LEED pattern during adsorption is straightforward if one is not interested in details. Suppose for a clean surface a spot pattern of the symmetry expected and diffracted angles expected (Bragg) for substrate are observed. Addition of an adsorbed species may (a) mask the pattern, (b) leave it unchanged, or (c) introduce new spots. In the case (a) it can be concluded that an amorphous layer is formed covering the periodic host lattice; in the case (b) it can be concluded that either no adsorption occurs, or the electron beam causes desorption,[112] or the periodicity of the adsorbed species is the same as that of the substrate; and in the case (c) it can be concluded that the adsorbate forms a periodic structure which does not have the same lattice or lattice spacing as the substrate. For example, suppose the clean substrate shows a pattern corresponding to a simple cubic lattice, but when adsorption is permitted new spots are introduced (half-order spots) between the original spots. In this case it is concluded that the spacing of the adsorbate follows that of the substrate, but the atoms are at double the spacing of the host lattice. This is termed a $c(2 \times 2)$ or $p(2 \times 2)$ overlayer, the letter referring to the location of the overlayer mesh, either "centered" or "primitive" with respect to the substrate lattice.[113] May[64] and Strozier et al.[114] give a more complete review of the various common symbols in usage to describe reconstruction and superstructure formation.

Changes in spacing as low as 5% can be resolved.[115] The complete interpretation of LEED data,[116–118] fitting the observed spot spacing, intensity, and half-width to a model of the surface, becomes more complex particularly for complicated patterns. Not only single events but multiple scattering must be included in such a model. For adsorbate-free surfaces reasonable representative models have been developed. With adsorbed overlayers where the detailed location of the adsorbed species would provide very valuable information about bonding characteristics, the problems unfortunately are even harder to overcome.

The observation of stepped surfaces, showing regular steps and terraces, is possible because the steps lead to a particular variation of the spot intensity with beam voltage. The steps and terraces can be analyzed in detail. We will not develop the model, but refer the reader to the original work by Ellis and Schwoebel[119] and to the reviews by Henzler and Klabes[120] or Perderau and Rhead.[121]

3.2.8. Methods of Chemical Composition Determination for the Surface

The group of measurements in Table 3.1 labeled (h), (i), (j), (k), and (l) are all primarily of interest for their ability to analyze surface composition. The most popular of these is Auger electron spectroscopy (AES). Although AES provides information about composition a few layers into the sample, whereas ISS (ion scattering mass spectrometry) examines only the topmost plane of atoms, AES is far more useful than ISS because the latter does not have the versatility or the sensitivity of AES.

In Auger analysis[122–124] electrons are beamed at the sample with an energy the order of 1000–5000 eV, high enough to dislodge core electrons from atoms near the surface. There are many ways in which electrons can be restored to the core level, including x-ray emission and including complicated processes whereby some of the energy is given to an emitted electron, some to phonons, and some to radiation. But there is one process which leads to an emitted electron which has a characteristic energy, the Auger process. In this process an electron from an outer shell drops down to the vacant core level, and all energy thus released is given to an electron from another outer shell orbital. The latter electron, the Auger electron, is emitted from the sample with a very characteristic energy. Note that the energy of the emitted Auger electron does not depend in any way upon the energy of the incident electron—the energy of the Auger electron is determined entirely by the energy level spectrum in the solid.

For example, the electron in the incident beam could eject an electron from the K shell of a certain atom. An electron from the L_1 shell drops into the vacant K level and the energy released in this transition is given to an electron from an L_2 shell which is emitted with a very characteristic energy. The operator observes a peak in the spectrum of emitted electrons, and that peak is denoted as the KL_1L_2 transition for that element. Each element will have a characteristic Auger spectrum, corresponding to various preferred combinations of the three contributing levels, and composition analysis is made by identifying energy peaks in the Auger electron spectrum.

As there is a flood of secondary electrons emitted in addition to the Auger electrons, the Auger peaks must be separated from the background. Fortunately the Auger peaks are very sharp, and differentiation of the intensity/energy curve for the emitted electrons eliminates the broad spectrum of background electrons and leaves the Auger peaks clearly visible.

Auger spectroscopy determines composition to a depth corresponding to the mean free path (escape depth) of the Auger electron. The escape depth[61] increases from about 5 Å for electrons of 50 eV to about 20 Å for Auger electrons of 500 eV. Thus the part of the Auger spectrum in the low-energy region (if a characteristic line exists there) is more likely to indicate a surface species.

The minimum area that can be examined in an AES analysis is not as small as could be desired. In principle, of course, it would be desirable to explore the composition of areas a few angstroms in diameter to observe, for example, whether impurities segregate at dislocations or steps or to examine dispersed catalyst particles. Unfortunately, in order to produce a reasonable current of Auger electrons for detection, a substantial incident electron beam must be used. Thus the minimum spot size is limited by heating effects which could change the sample during the measurement if not controlled. In practice it is difficult to study areas much less than a tenth to a hundredth of a millimeter in diameter.

The sensitivity of AES is very dependent on the element to be detected. A rule of thumb is that the characteristic peaks may appear if the fractional concentration on the surface is 10^{-3}–10^{-2}.

The use of AES to study adsorbed species may not be reliable in some cases, according to Joyce and Neave[112] because of electron stimulated desorption (ESD) as discussed in Section 3.2.9.

Shifts in the Auger spectrum may occur associated with changes in bonding,[125] and the use of AES to study bonding is of substantial interest. Unfortunately there are other causes for shifts in the energy of an Auger line either up or down[67] to give "satellite" peaks. In particular, a loss of energy due to plasmons (as in ELS) is commonly observed with measurements on a metal surface.[126]

Secondary ion mass spectrometry (SIMS, IMMA) is another means of composition determination. It has the disadvantage that it removes the material it is detecting. The concept is straightforward; the surface is sputtered using a beam of ions, and the sputtered material is observed by mass spectrometry. The sensitivity varies widely with the atom sputtered, so care must be taken before quantitatively interpreting a mass distribution in terms of relative density of the various species.

Ion scattering spectrometry (ISS) is another interesting technique, introduced as a technique for surface analysis by Smith.[127] A beam of noble gas ions of mass near the mass of the species to be detected is scattered from the surface. The peak due to elastic scattering is observed. At a given angle of scattering, the energy with which the gas ion is scattered depends upon the mass of the scattering ion.

If the ions are scattered at 90°, the expression for the energy E_1 of a particle scattered from a surface atom of mass M_s is particularly simple:

$$E_1/E_0 = (M_s - M_0)/(M_s + M_0) \tag{3.15}$$

where M_0 is the mass of the noble gas ion in the beam and E_0 is the energy of the incident beam. Thus a scan of the energy of the beam emerging at 90° will show a series of peaks corresponding to the masses of the surface atoms, interpretable by Equation (3.15). The peaks are fairly broad, however, so only a few different species should be present. The sensitivity to atoms of mass near that of the impinging beam may be high enough to detect a few thousandths of a monolayer.

Niehus and Bauer[75] in their evaluation of ISS studied the scattering of He+ from W with varying amounts of oxygen on the surface. They found the relation between oxygen coverage and the W and O signals was not linear. They concluded that AES is more sensitive and more linear in its monitoring of surface composition.

The use of EMP (electron microprobe) is conceptually straightforward. A beam of high-energy electrons produces an x-ray spectrum showing lines characteristic of the elements in the target. Unfortunately the measurement is not restricted to composition very near the surface. The high-energy electrons can penetrate deep into the crystal, and of course there is no distance limitation on the emerging x-rays, so the technique is almost a bulk composition measurement. However, the beam diameter can be made small (say 5 μ), which permits good lateral resolution.

ESCA or XPS (x-rays in, photoelectrons out) is the reverse of the electron microprobe and much more a surface tool. The emitted photoelectrons must be produced within a mean free path of the surface. A monochromatic x-ray beam is used such that the electrons emitted from a given core level of a given atomic species in a given environment will all have the same energy. The presence of this peak in the energy distribution of photoelectrons then indicates the presence of the particular chemical species. The ESCA technique is particularly valuable in the study of organic molecules because the incident beam is x-rays. This avoids to a great extent the disruption of the molecules that can be induced by beams of particles.

Small shifts of peak energy are often observed as the valence of the atom in the solid changes. The binding energy of the core electron becomes greater as the charge on the ion becomes more positive. Such shifts enable changes in the chemical surroundings of the atom to be determined by ESCA. As the resolution of the ESCA peaks is the order of 1 eV, peak shifts must be greater than this to enable the change in chemical environment to be detected. Peak shifts with changes in the chemical environment can occur also because of relaxation (screening) effects. The positive charge associated with the newly created hole in the core level becomes "screened" by electrons in surrounding polarizable orbitals, and the energy released by this polarization is added to the energy of the photoelectron. Changes in the properties of the solid will lead to changes in relaxation energy. With this complication, quantitative analysis of ESCA peak shifts is not yet possible.

The observation of surface atoms can apparently be emphasized greatly by measuring only those photoelectrons that are emitted almost parallel to the crystal surface. Photoelectrons traveling in this direction from several layers deep must traverse more than a mean free path and thus would not contribute to the observed peak. Shirley[69] suggests that with a low-energy x-ray source and this technique, observation can be restricted to the surface layer of atoms.

3.2.9. Studies of Chemical Reactions due to the Impinging Beam

In this section we mention briefly studies where particles or photons are beamed at a surface and the products of a surface reaction are monitored. We will discuss first the use of photon or electron beams to cause electron stimulated or photon stimulated desorption. Second, we will discuss the use of an impinging beam of particles which can react with the surface or with a surface adsorbate.

At the present time these techniques really should be classified as individual research studies, which do not yet belong in the category of tools for surface studies. Thus in future chapters such experiments will be described individually in terms of the contribution of each to our picture of the solid surface. However, as the techniques and the models develop, it is felt that they will become useful as general tools for studying reactions, comparable to the tools described above. Anticipating such standardization, we will include them in this chapter on general techniques.

Electron stimulated desorption (ESD) has been reviewed by Menzel[81] and by Leck and Stimpson.[82] As they point out, the mass difference between the electron and an adsorbate precludes the possibility of direct sputtering of the adsorbate. The accepted theory for ESD is therefore based on an effect where a bonding electron is excited to an antibonding orbital, and a Franck–Condon shift in atom position occurs of such a magnitude it leads to desorption. While the adsorbate is within 10 or 20 Å from the surface, electron exchange by tunneling can occur. It is not clear whether the initial excitation produces an ion or an excited atom, but the latter may be more probable because for emission an ion would have to work against a strong image force in the initial stages of desorption. However, the excited electron can (and, from the observed isotope effect below, must) tunnel back into the empty levels of the conduction band, forming an ion. The observation of ions is of particular interest as these are much easier to detect and are the species normally monitored. However, only about 10% of the desorbing particles are ions, suggesting that a final electron tunneling process from the filled levels of the metal to the ground state of the ion is highly probable. Since the velocity of the atom from the surface depends sensitively on its mass, and the tunneling probability from the metal to the ground state of the ion depends upon the time the atom remains within a few angstroms of the surface, it is clear there should be a strong isotope effect. Such an effect is observed for the desorption of hydrogen and deuterium, with up to a factor of 150 higher desorption cross section for H^+ than for D^+ [using 100-eV electrons beamed at hydrogen and deuterium adsorbed on the tungsten (100) surface].

A particularly valuable use of the ESD technique is the indication of the position of the adsorbate atoms relative to the bonding substrate atoms on the surface plane of a solid. Yates and his co-workers[128] have shown that during ESD, the desorbed species emerges at well-defined angles from a surface. From a simple momentum conservation argument it can be seen that if the adsorbate atom has a single bond normal to a substrate atom, the adsorbate will emerge normal to the surface. More rigorous arguments are presented by the above authors to infer the position of the adsorbate by such measurements.

Photodesorption from metals by uv may be, as suggested by Lichtman,[83] entirely associated with semiconducting overlayers, rather than be effects at the metal surface itself. Such photodesorption has been examined for many years and is a powerful tool in characterizing adsorption on semiconductors. The subject is discussed in detail in Section 9.3.

Modulated beam relaxation spectroscopy (MBRS) is a fascinating tool which will provide much information[129,149] about reaction kinetics on surfaces. A beam of reactive species is directed at a surface which may or may not be covered by an adsorbate, and the products of the reaction with the surface or with the adsorbate are monitored. In some cases, with a very fast reaction, the reaction can be characterized by the angle of the emerging beam.[130] If the reaction occurs as the beam is scattered off the surface, the emerging beam of product molecules will show some specularity of reflected intensity. If the species in the beam must be adsorbed before a reaction can occur, the products will have a cosine distribution with respect to angle of emergence. Another technique, from which the title MBRS comes, is to chop or modulate the incident beam. Then the phase lag between the time when the beam strikes the surface and the time when the products desorb provides information about the reaction kinetics.[131]

These chopped beam techniques are limited to cases where the reaction probability is reasonably high. For example, as reported by Somorjai and Brumbach[80] studies were made by Smith[132] of the hydrogenation of ethylene on platinum. No product emerged, and the negative result was ascribed to a low estimated reaction probability, 10^{-4}, in the temperature range of interest.

3.3. Chemical Measurements

In this section we examine several general chemical methods that provide information about sites or states on a clean surface or about the bonding between a surface and an adsorbate. The division of methods into "chemical" and "physical" measurements is to a great extent arbitrary. For example, the techniques described in the first two subsections, infrared methods and thermal desorption, could equally well be included in physical methods. They are included here because a substantial amount of work has been done on systems of primary interest in adsorption and catalysis. The techniques to be described in the third subsection, methods of detecting and classifying acid and basic centers on a solid have historically received more emphasis in catalytic studies. However, the observations and interpretations discussed in Chapter 4 show, as expected, a close parallel to studies on ionic solids with electrical techniques. The two approaches to the study of the surface of ionic solids provide complementary information, with the electrical observations of particular value on semiconducting solids, and the "chemical" measurements showing overlap but having the greatest success in describing the surface of insulating ionic solids.

The techniques of the first two subsections below describe methods of primary interest in examining adsorbate/adsorbent bonds. The techniques of the third subsection describe methods of interest in understanding the clean surface of ionic solids and in understanding these surfaces with protons present (say, from adsorbed water).

3.3.1. Infrared Absorption

The detection of surface species by its characteristic infrared (ir) absorption requires that enough light be absorbed to change appreciably the intensity. Absorption by a single monolayer is usually inadequate, so measurements with the beam transmitted through a single crystal with the species adsorbed on its surface are usually not rewarding. Three techniques are used to increase the sensitivity. One is the powder technique, absorption of light transmitted through a pressed powder pellet, where the species of interest is adsorbed on the surface of each powder grain in the pellet, thus providing many opportunities for absorption of the ir as the beam passes through. A second is the technique of multiple reflection, where two parallel surfaces are placed a millimeter or so apart, and a beam of ir is reflected back and forth between the two surfaces. A third is the technique of total internal reflection, where a crystal which transmits in the ir region of interest is cut with two parallel faces, and a beam of ir is reflected back and forth within the crystal, emerging at the other end after many reflections from the two surfaces. In the latter case the gas of interest is adsorbed on the surface, and at each reflection a small fraction of the characteristic ir is absorbed.

The absorption of light transmitted through a pressed pellet can be analyzed by the Beer–Lambert law:

$$I = I_0 \exp(-kcd) \tag{3.16}$$

where c is the concentration in moles per liter of absorption centers, k the molar extinction coefficient (with dimensions liter per centimeter per mole) and d is the path length in centimeters. At the absorption maximum of CO, for example, the molar extinction coefficient is 3.3×10^5 liter cm^{-2} mol^{-1}. Consider a powder of surface area 100 m^2/g, pressed to a thickness d of 0.01 cm. We can estimate the effective concentration of adsorbed CO in the pellet and calculate whether the peak should be resolved. If the density of the pellet is 2 g/cm^3, the total surface area available per unit volume of the pellet is 200 m^2/cm^3. If the CO adsorption were a complete monolayer of 2×10^{19} absorbate particles/m^2 ($= 3.2 \times 10^{-5}$ mol/m^2), there is an effective molarity of 6.4×10^{-3} mol/liter. If only 0.1 monolayer of CO is adsorbed, which may be more realistic, we have from Equation (3.16) $I/I_0 = 0.12$, which is a substantial reduction in intensity. Most adsorbates of interest have a lower molar extinction coefficient than CO, so the decrease in the intensity may not be so high. But the pressed pellet technique is capable of detecting under favorable circumstance substantially less than 0.1 monolayer of adsorbed species. The pressed pellet technique is limited to particles less than a few microns in diameter[133] because scattering of the ir becomes a problem with larger particles.

Naturally the adsorbent must be transparent in the ir range of interest. However, it should be pointed out that substantial work has been done using the pressed pellet transmission method where metal particles of the order of 100 Å diameter have been supported on an inert, ir transmitting substrate such as silica, and the adsorption of gases on these metal particles has been monitored by the pressed pellet technique.

The reflection/absorption technique, where the adsorbate is present on the surface of two parallel plates, and the ir is reflected back and forth between the two plates, is of value particularly in studies of metal surfaces. The reflection/absorption technique is only a modest improvement over the single reflection technique, as only three or four reflections can be used.[134] If there are too many reflections, even the signal from the adsorbate-free surface is not transmitted. If there are too few reflections, the advantage of the technique is lost. Kottke et al.[134] define an optimum number of reflections by maximizing

$$R_F = (R^0)^N - R^N \tag{3.17}$$

where R^0 is the reflectivity for an adsorbate-free surface, R is the reflectivity in the center of the absorption band of the adsorbate, and N is the number of reflections. As mentioned, for reasonable parameters, R_F is maximum with three or four reflections.

The total internal reflection technique, an older technique developed by Harrick,[135] is suitable for studies of adsorption on materials transparent in the ir. This technique has the great advantage over the above reflection-absorption technique that in principle the reflection is total (if the angle of incidence is greater than the critical angle, Brewster's angle) unless absorption occurs by the adsorbate of interest. Thus in practice the order of 100 internal reflections can be used with the intensity still about 50% of the original.[135] The point of the method is that although the reflection is total, an exponential tail of the light beam does extend into the vacuum (or gas) the order of 10^{-5} cm, and an adsorbed species will have the opportunity to absorb the light.

The sensitivity of this method is not very high, requiring the order of a monolayer of adsorbed species. Thus it has been most useful in studies of second phases on the surface of semiconductors, such as oxide phases, and for the study of coatings and their interaction with materials.[133]

Infrared methods can provide a substantial amount of information about the sorbate and about the sorbate/solid bond. In particular, shifts in the ir spectrum can be used to interpret the bonding character. The different vibrational frequencies depending upon which atom in the adsorbing molecule bonds to the surface, and the interpretation of these vibrational frequencies, has provided a substantial store of information regarding how various molecules are adsorbed. An extra degree of freedom which often helps distinguishes the origin of absorption bands is the use of isotopes to replace atoms in different position on the gaseous molecule. An analysis of isotope effects in the ir vibrational spectrum can often confirm or dismiss a particular model of adsorption proposed.

3.3.2. Temperature-Programmed Desorption

A fairly straightforward way of determining bond energies of adsorbed species on a surface is to increase the temperature in vacuum until they desorb. If the temperature is increased at a constant rate, the species with lowest activation energy of desorption will desorb first, followed successively by species with higher and higher activation energies of desorption.

It can be shown,[136,137] that with a single activation energy of desorption and with a desorption rate first order in the concentration of adsorbed species, a peak in the desorption rate will occur at a temperature T_p given by

$$E/RT_p^2 = (\nu/\beta) \exp(-E/RT_p) \qquad (3.18)$$

where E is the activation energy (here in cal/mol), R is the gas constant, ν is a pre-exponential frequency factor, β is the heating rate in deg/sec.

The frequency factor is often assumed to be 10^{13}/sec, and the activation energy of desorption estimated from this.

Falconer and Madix[138] and King[139] review techniques by which the frequency factor need not be estimated. Models are developed relating the peak temperature to the heating rate by the following expression:

$$\ln(\beta/T_p^2) = A - (E/RT_p) \tag{3.19}$$

where A is a constant. By varying the heating rate to obtain different peak temperatures T_p, the value of E can be determined. The heating rate should be varied over two orders of magnitude to obtain reasonable accuracy, which presents experimental difficulties in the use of this analysis.

Clearly the technique is quantitatively applicable to cases where only one or two monoenergetic adsorption sites are presented to the adsorbate. The results can be very complex to analyze[150] if a spectrum of desorption activation energies is present.

3.3.3. Adsorption of Gaseous Acids and Bases or of Indicators

The chemical activity of clean surfaces of highly ionic solids often arises because sites which are present are able to accept or to donate electron pairs or protons to an adsorbing species as discussed in Section 1.2.3. In this section some of the methods of measuring such acid and basic sites on solids are described. Most techniques for measuring the acid strength and the number of acid centers cannot distinguish between Lewis and Brønsted acids. Some can, and these will be indicated.

There is a broad gray region of definition which further complicates the picture. Sites which have a high electron affinity and can share an electron pair can often also accept a single electron. Thus the site can act either as a Lewis acid or as a surface state. For example, results purporting to measure acid strength by monitoring the appearance of electron spin resonance lines, which of course monitors the appearance of unpaired electrons, are really measuring surface state behavior, although such behavior may well be directly related to acidity in the cases studied. The connection between surface states and acid/base centers will be discussed further in Section 4.2.4, but it is clear that the two types of surface sites are not distinctly separable.

3.3.3.1. Hammett Indicators. The earliest and experimentally the simplest way of measuring the distribution of acid or base strengths on a solid surface is by colored indicators. The method of estimating the

Brønsted activity of a surface is straightforward, based on Equation (1.5) or (1.7). For example, to measure acid strength, indicators in their basic form that have various pK values are suspended in an inert solvent and the solid is added to each in turn to see whether the indicator is adsorbed in the acid form. If, for example, phenylazonaphthylamine ($pK = 4.0$) is not converted to its acid color upon adsorption by the solid of interest, but methyl red ($pK = 4.8$) is converted to its acid color, the acid sites on the surface have an acid strength (electron affinity) corresponding to H_0 between 4.8 and 4.0. Note that a greater acid strength corresponds to a lower value of the acidity H_0. A completely analogous technique obtains for basic surface sites (electron pair donors or proton acceptors).

Suggested lists of indicators are provided by Tanabe.[140] Indicators for acid sites are preferred which are almost colorless when in the basic form but show a strong color when adsorbed and converted to the acid form. Thus when the color of the indicator is changed, the adsorbed colored indicator is clearly visible through the carrier liquid. Indicators for basic sites are preferred which are almost colorless in the acid form but show a strong color when adsorbed and converted to the basic form. Titration techniques can be used to determine the acid or base site density.[140,141]

3.3.3.2. Adsorption of Gaseous Acids or Bases.

A somewhat more quantitative way to determine the distribution of electron affinities (or proton affinities) associated with acid or basic sites on ionic solids is the adsorption of gases. Acid sites adsorb basic gases, basic sites adsorb acid gases, and the strength of the bond between the adsorbate and the site provides exactly the information desired.

Most commonly, the gaseous bases ammonia or pyridine are used to probe the acidity of surface acid sites, and phenol to probe basic sites. Temperature programmed desorption (Section 3.3.2) provides a curve of amount adsorbed as the temperature increases, and thereby a measure of the distribution of the surface sites in terms of their acidity (or basicity). A differential scanning calorimeter measurement together with a measurement of amount desorbed can provide more quantitative information about the bond strength versus concentration of sites.

3.3.3.3. Infrared Measurement of Lewis or Brønsted Sites.

Infrared measurement of the vibration frequencies of ammonia or pyridine when adsorbed on the surface sites can be used to distinguish Brønsted and Lewis acid sites[142] and to measure the acid strength.[143] If the site is a Brønsted site, the adsorbed ammonia is an NH_4^+ species, if on a Lewis site, an NH_3 species. The ir bands are easily distinguished. Similarly pyridine is bonded differently on Lewis sites than on Brønsted sites. Other adsorbates have also been used for ir studies of the sites.[144] Optical absorption by pyridine in the uv region has also been used to explore acidity.[145]

3.3.3.4. Electron Spin Resonance (ESR).

Studies have been made (Ref. 140, pp. 24–25) showing the formation of free radicals on strongly acid solids such as the mixed oxide silica/alumina. Both optical absorption and ESR measurements have shown the presence of these radicals. There are indications that the same sites are responsible for the removal of a single electron from these radical forming adsorbates (p-phenylenediamine, anthracene, perylene) as share an electron pair with gaseous bases such as ammonia. Thus in these ESR measurements we are observing the "gray area" between true surface states (one-electron transfer) and Lewis acids (electron pair sharing). Despite the fact that the same sites react both ways, there is some value in Tanabe's[140] suggestion that "we should maintain the distinction between those sites which accept one electron (Lewis acids in the widest sense), and those sites which accept an electron pair (Lewis acids properly so called)."

4 | The Adsorbate-Free Surface

4.1. Introduction

We are very fortunate in recent years that excellent theoretical models and excellent experimental techniques have been developed that permit a reasonable description of the solid surface with no foreign species adsorbed or reacted. Although from most practical points of view the more interesting system is that with foreign species present, a knowledge of the behavior of the clean surface is necessary as a starting point in understanding surface phenomena and as background to discuss the bonding of the foreign species. From the chemical point of view (of surface reactivity) one is interested in the available "intrinsic" surface orbitals present on the clean surface. We want to know their chemical binding characteristics, whether they can provide or accept single electrons in a level of energy suitable for covalent bonding or can provide or accept electron pairs in levels of energy suitable for acid/base bonding, and the directional characteristics of such bonding orbitals. From the physical point of view (the electronic properties of the solid), the energy of the orbitals (surface state energy) and whether they donate or accept electrons is of first importance. But it is also important to know what surface states will result when these energy levels are altered by reconstruction or by foreign species adsorbing and interacting with the surface sites.

4.1.1. The Classification of Solids

We will classify solids in two groups according to their ionicity (electronegativity difference between the cation and anion). Ionic solids will be considered one group, ranging from highly ionic materials such as alkali halides through decreasing ionicity to where the covalent character may be dominating. In this group the electrical properties will vary from insulators to moderate band gap semiconductors. Highly ionic materials will be expected to show dominantly Lewis activity (electron pair bonding) or electrostatic (polar) effects in bonding of adsorbates. "Ionic surface states,"[1] able to capture or donate single electrons, may appear if the cation has a high electron affinity or the anion has a low electron affinity. A second group will be solids which are best treated as covalent or metallic. This group will include some semiconductors including both elemental and a few III–V semiconductors, and will include metals. Bonding to the surface of such materials can occur through dangling bonds, unpaired electrons in orbitals directed toward the surface that would be bonding orbitals of the crystal were the crystal not terminated. Localized covalent bonds will be shown to be present on covalent solids, and can easily be induced on metallic solids.

Such an ionic/covalent classification has a great advantage over the more normal metal/semiconductor/insulator classification for the discussion of surface effects. Surface effects on ionic semiconductors, where a wide variety of electrical tools is available for experimentation, become naturally related to surface effects on ionic insulators, where the investigation is more difficult. Similarly, surface effects on covalent semiconductors, where again a wide variety of electrical tools is available, become naturally related to effects on metal surfaces where the investigation is more difficult.

In addition to the advantages of this classification as described above in helping to visualize the expected chemical bonding properties of the surface, there is substantial evidence that such a classification into ionic versus covalent solids is also best if one wants to consider bulk or surface electrical properties of the material. The interpretation of bulk band gaps and other properties in terms of the ionicity of binary compounds is reasonably well established.[2–5] The interpretation of surface states in terms of ionicity of the bulk material is being developed.[6,7] For example, Mark[8] points out that the activity of the surface in adsorbing oxygen with a covalent bond increases by orders of magnitude between highly ionic semiconductors such as ZnO and highly covalent semiconductors such as Si.

There is usually little question about whether a given solid is a covalent or ionic crystal. However, there will be borderline cases. Kurtin et al.[9] have explored several electrical properties of semiconducting solids as a function of ionicity, and concluded there is a rather narrow transition point where these properties switch from ionic behavior to covalent. One such property is the surface barrier V_s developed at the surface when a metal film is evaporated thereon. If the semiconductor is ionic, the value of V_s obtained varies linearly with the electronegativity of the evaporated metal. A highly electronegative metal (Au) extracts electrons from the semiconductor and creates a high V_s; a highly electropositive metal (Cs) injects electrons leading to a low or negative V_s. On the other hand, if the semiconductor is covalent, the electronegativity of the deposited metal has little effect on the surface barrier. Presumably the covalent bond between the metal and the covalent semiconductor so dominates the surface and is so dependent on the semiconductor characteristics, that the influence of the metal is minimized. By this approach solids such as ZnS, AlN, SiO_2, ZnO, Al_2O_3 are clearly ionic, semiconductors such as Ge, Si, InSb, InP, diamond, and GaAs are covalent, and an intermediate group including CdS, SiC, GaP, and CdTe are in the transition region. These authors[9] point out that in addition to this surface barrier property, bulk properties related to the selection rules for optical excitation and to the optical formation of excitons also show abrupt changes in gross characteristics as the ionicity goes through this same critical region. Phillips[10] has pointed out that dielectric properties are also related to ionicity.

The relation of surface states to ionicity will be developed in the present chapter. A more detailed classification based directly on surface states stems from a suggestion of van Laar and Scheer[11] arising from their studies on GaAs. They concluded that the surface state properties of GaAs resembled an ionic semiconductor for the clean surface case, but covalent behavior was easily induced by additives on the surface. It was this induced covalent behavior that was observed by Kurtin et al. They therefore suggested a third classification to include GaAs and similar material where ionic surface states are present on the clean surface, but orbital hybridization to form covalent bonds can be easily induced by a foreign adsorbate.

In the discussion below, we will classify the solids by their ionicity, grouping ill-defined cases like GaAs with ionic or covalent (or metallic) solids in accordance with the type of surface state that seems to be dominating.

4.1.2. Preparation of a Clean Surface

There are three types of contamination that must be removed in order that the results can be described as representing a clean surface. One is a separate phase, such as an oxide on a metal. The second is adsorbed gases and impurities. The third is crystal damage, including implanted ions near the surface. The problem of crystal damage arises particularly when sputtering is used to clean the surface.[12,147] Ions strike the surface with substantial force, for the principle is to knock off surface impurities. Thus ions are striking with sufficient force to penetrate into the surface; the implanted ions themselves can alter the results in surface measurements, or, by introducing stresses, they can cause dislocations and other imperfections to form. The imperfections in turn may alter the results in surface measurements.

The formation of a surface by cleavage is the preferred method for preparing a clean single crystal surface. In a vacuum of 10^{-9} Torr or better it may take hours to adsorb a monolayer of adsorbate. Crystal damage in the form of cleavage steps and dislocations will appear during cleavage, so the method is not perfect, but it is by far the best available.

Argon sputtering surfaces is the most popular method, both because cleaving a surface in ultrahigh vacuum is awkward, and because many crystal planes are of interest besides the cleavage planes. As mentioned above, sputtering causes extensive damage to the surface of the material, and a high-temperature anneal must be used to remove implanted argon and to anneal dislocations. The anneal treatment itself can cause atom movement changing the surface properties.[13] In addition, the annealing may give rise to nonstoichiometry at the surface of a compound.[14] The usefulness of the technique depends upon the sensitivity of the measurement to be made and on the material. If the material has a high surface state density associated with the clean surface (e.g., germanium), the clean surface effects will dominate with a reasonable anneal for any measurement. If the material has few surface states near the Fermi energy, and particularly if the measurement is very sensitive to surface states, such as a surface photovoltage measurement, it may be almost impossible to anneal sufficiently. Other measurements, such as LEED, are very insensitive to damage, and one simply anneals until the spots appear—if 5% of the surface is badly damaged still, it makes little difference.

Other methods of cleaning include simple heating to desorb all foreign species, a method particularly effective on tungsten, for example. In the FIM measurements, field evaporation of the surface will strip impurities off.

Many chemical measurements will be interpreted in terms of clean surfaces because the material studied is not easily contaminated and because the interactions studied appear to be representative of expectations for a clean surface. Thus, for example, platinum or other noble metals may be cleaned by a treatment in oxygen, to remove oxidizable impurities (such as carbon) followed by a treatment in hydrogen, to remove the oxygen, and finally an outgassing treatment to remove the hydrogen. In all likelihood the properties will then reflect those of a clean surface. As another example, studies on silica or alumina, in particular when the materials are freshly prepared powders, will be considered clean surface studies. As these are usually prepared in aqueous solution, there may be substantial contamination due to impurities from the water, particularly organic impurities (carbon is a universal contaminant), but for many cases it can be concluded that the properties of the silica or alumina clean surface are dominating over the contamination, so the results can still be interpreted as representative of a clean surface.

4.2. Theoretical Models

Long before any measurements were possible on a clean surface, the presence of intrinsic surface states was anticipated. From a theoretical point of view states localized at the surface arise because surface atoms have quite different characteristics from the normal bulk atoms. In a covalent crystal or a transition metal each bulk atom has a complete set of nearest neighbors (is completely coordinated), but the surface atom has at least one neighbor missing, and so has a surface state associated with the dangling bond. Surface states on metallic crystals dominated by s or p bands are more complex as will be described later. On an ionic material the surface ion is not only unsaturated from a coordination point of view, but also from an electrostatic point of view. A cation (say Zn^{2+}) in the bulk has its charge compensated by a shell of negative anions (say O^{2-}). At the surface one or more of the neighboring O^{2-} are missing. As will be discussed below (Madelung theory) this electrostatic effect gives rise to surface states or acid/base sites.

It is the objective of this section to discuss the available theory and what predictions are made regarding the surface orbitals—their energy, type, and geometry—that should be presented and active at the clean solid surface.

4.2.1. Quantum Models

The analysis of surface states (orbitals) based on quantum mechanics can be divided loosely into two general approaches. One is the molecular orbital (MO), the other is the semi-infinite crystal approach. The MO approach depends upon detailed computer calculation to obtain quantitative data on surface states and surface orbitals. To date studies of metallic or covalent crystals have been emphasized. The semi-infinite crystal approach is understandable without detailed computer work, but unfortunately the approach suffers from a great uncertainty regarding how to terminate the crystal (mathematically speaking). It thus must remain qualitative, so detailed computer calculations with the semi-infinite crystal model are less rewarding.

4.2.1.1. Quantum Calculations Based on a Semi-infinite Crystal.

As recently as 1971 Henzler in his review of surface states[15] was able to comment that he could only describe "qualitative theoretical results, since quantitative results are in general not available." At that time the modern approach emphasizing the molecular orbital description was in its infancy, and Henzler was referring to the semi-infinite crystal calculations.

In 1970 Davison and Levine provided an excellent review[1] of surface state calculations (emphasizing the semi-infinite crystal calculations), a discussion of the semiclassical models to be discussed in the next section, and a discussion comparing theory and experiment. For a detailed analysis of the theory the reader is referred to that review. Here we will describe the highlights very briefly to extract the more interesting concepts from the view of surface chemical physics. Indeed, we will not discuss the three-dimensional calculations at all, as they add little new either qualitatively or quantitatively, but will examine only the approaches and results on "one-dimensional" semi-infinite crystals.

There are two general approaches to describing the solid with a surface. One is the "crystal orbital" method, which in its simplest form is equivalent to a Hückel approximation, common in analysis of organic molecules. The other is the "crystal potential" approach, where one calculates the wave function of an electron in a box, with a periodically varying potential corresponding to the lattice spacing introduced into the Schrödinger equation.

The Hückel calculation in one dimension begins with a linear combination of atomic orbitals (LCAO) representation of the wave function:

$$\psi = \sum_\nu C_\nu \phi_\nu \tag{4.1}$$

where ϕ_ν is the atomic wave function of the νth atom in the chain and the C's are weighting factors for each atomic wave function. Since the wave function must obey the Schrödinger equation

$$(H - E)\psi = 0 \tag{4.2}$$

multiplying the left-hand side by the complex conjugate of one of the atomic orbitals, say ϕ_μ^*, and integrating over all space, we can develop a series of simultaneous equations

$$\sum_\nu C_\nu(H_{\mu\nu} - S_{\mu\nu}E) = 0 \tag{4.3}$$

where

$$S_{\mu\nu} = \int \phi_\mu^*\phi_\nu \, dv \tag{4.4}$$

and

$$H_{\mu\nu} = \int \phi_\mu^* H\phi_\nu \, dv \tag{4.5}$$

With $\mu = \nu$, the integral $H_{\nu\nu}$ is called the Coulomb integral for the atom ν and is denoted by α:

$$H_{\nu\nu} = \alpha \tag{4.6}$$

For simplification when $\mu \neq \nu$, only nearest neighbor interaction is assumed, so $H_{\mu\nu}$ is zero unless atoms μ and ν are nearest neighbors. For nearest neighbors $H_{\mu\nu}$ is called the resonance integral, denoted by β:

$$H_{\mu\nu} = \beta \quad (\mu = \nu \pm 1) \tag{4.7}$$

Assuming that the atomic wave functions are orthogonal, so $S_{\mu\nu} = 0$, if $\mu \neq \nu$, and $S_{\mu\nu} = 1$ for $\mu = \nu$, we can solve the set of simultaneous Equations (4.3) with the H's replaced by the constants α and β.

Equations (4.1)–(4.7) describe the Hückel approximation independent of the system to be examined. For example, we shall have reference to these equations again in the next chapter when discussing the form of π orbitals on unsaturated organic molecules, preparatory to describing π bonding to a surface.

Here we will examine the Hückel model of a one-dimensional AB crystal with alternating A and B atoms. We let the Coulomb integral on atom A be different from atom B, and denote them by α_A and α_B, respectively. The parameter β then becomes related to the overlap between the atomic wave functions of adjacent A and B atoms. With this, the system of Equations (4.3) becomes

$$(X - z_1)c_\nu = c_{\nu-1} - c_{\nu+1} \quad (\nu \text{ odd}) \tag{4.8}$$

$$(X + z_1)c_\nu = c_{\nu+1} - c_{\nu-1} \quad (\nu \text{ even}) \tag{4.9}$$

with

$$X = (E - \bar{\alpha})/\beta \qquad \bar{\alpha} = \tfrac{1}{2}(\alpha_A + \alpha_B) \tag{4.10}$$

and

$$z_1 = (\alpha_A - \alpha_B)/2\beta \tag{4.11}$$

This set of equations can be solved to provide bulk electronic energy levels E of the AB crystal in terms of α_A, α_B, and β. The solution for such an infinite crystal gives two bands of allowed energies (a "conduction band" and a "valence band") with a forbidden energy gap of

$$E_g = \alpha_A - \alpha_B \tag{4.12}$$

For the present discussion the major interest is the energy of surface levels. Thus we wish to examine the semi-infinite crystal, terminating on an A or a B atom. The end ($\nu = 1$) atom, not having a neighbor on one side, is assumed to have a different electron affinity from the equivalent bulk atom. If the end atom is an A atom, the Coulomb integral is changed to $\alpha_A{}'$, or, if a B atom, its Coulomb integral is changed to $\alpha_B{}'$. The boundary condition if A is the end atom becomes, from Equations (4.8) and (4.11)

$$(X - z_A)c_1 = -c_2 \tag{4.13}$$

where

$$z_A = (\alpha_A{}' - \bar{\alpha})/\beta \tag{4.14}$$

With this boundary condition, it is found that under certain conditions complex solutions arise such that the wave functions ψ decay exponentially into the crystal (are localized). These are the surface states.

With these boundary conditions, analysis of the equations, which we will not go into, shows that surface states in the band gap region arise if

$$z_1 > z_A \tag{4.15}$$

or

$$z_1 > z_B \tag{4.16}$$

so that if the surface atom is an A atom, and its electron affinity is much greater than the electron affinity for the bulk atom A, then a state will be brought down from the "conduction band" of the crystal into the band gap region. The surface state will resemble the orbital on the atom (or ion) A. If the surface atom is a B atom, and its electron affinity is less than the electron affinity of a bulk B atom, a surface state will appear above the "valence band" of the crystal. If the valence band is occupied, the surface state associated with the surface atom B will be occupied.

The surface states arising as in the above calculation due to a difference in electron affinity between the surface atom and bulk atoms (different α's) are termed "Tamm states."

The above form of calculation provides a good fundamental understanding of why Tamm surface states will arise and how they depend on the electron affinity of the surface atom relative to that of the bulk atom. However, without detailed knowledge of the potentials at the surface, α_A' and β (or their equivalent in more complex analyses) cannot be calculated, so accurate values for the energy or even the orbital shape of surface states cannot be expected.

The other approach to calculations of the behavior of the surface of a semi-infinite crystal, the crystal potential calculation, is provided in its simplest form by the Kronig–Penney Model, or by the Mathieu approximation. Here the approach is more like the molecular orbital approach in that energy levels are first calculated and then populated by electrons. A periodically varying potential is assumed in the crystal and the Schrödinger equation is solved with this potential. The Mathieu approximation is a variation of this approach that is easily understood, and is quantitatively calculable. The periodically varying potential in the (one-dimensional) crystal is assumed to be a simple cosine function, $\cos(2\pi x/a)$, where a is the lattice constant, so that the Schrödinger equation has a particularly simple form and is soluble. The difficulty arises for the semi-infinite crystal calculation in how does one terminate the crystal? The model used is that the crystal is terminated by a potential step of energy which occurs at a distance z_0 from one of the points where the cosine function is a maximum. If the crystal is terminated at the position of a surface atom, z_0 is $\pi/2$ (the position where the cosine function is a minimum). If the crystal is terminated exactly where there is a maximum in potential, z_0 is 0. For details of the calculation the reader is referred to Davison and Levine's review. The important result deriving from the calculation is that as z_0 varies from 0 to π, there are values of z_0 where surface states occur and values of z_0 where surface states do not occur; in the cases where surface states occur, the energy level can be anywhere in the forbidden gap region depending on the exact value of z_0 chosen.

The Mathieu problem thus shows clearly how sensitive these semi-infinite crystal calculations are to the exact shape of the potential/distance curve that is chosen to represent the termination of crystal at the surface. A small error in the model can change the predicted properties of the surface state substantially.

In the more recent calculations with the more quantitative MO theory, Appelbaum and Hamann[16,17] note a similar sensitivity problem. If they artificially move the position of the surface plane (the "termination" in the Mathieu problem) by a few tenths of an angstrom, the surface band they calculate is completely wiped out.

In summary, the semi-infinite crystal approach, particularly the Hückel calculation, illustrates the appearance of Tamm surface states on heteropolar crystals to a satisfactory degree. If the Coulomb integral at the surface differs greatly from the Coulomb integral in the bulk (the electron affinity for the surface atom differs greatly from that of the bulk), surface state levels will appear out of the bands. If the atomic orbitals contributing to the conduction band are associated with the cation, the surface state levels emerging from the conduction band will have dominantly cation-like atomic orbitals and, like the conduction band, will be unoccupied. If the valence band atomic orbitals are associated with the anion, the surface state levels emerging from the valence band will reflect the anionic atomic orbital, and like the valence band, will be occupied.

These states may represent Lewis acid or basic centers, respectively; they may be acceptor or donor "ionic surface states," or they may be both, depending on details to be discussed later.

4.2.1.2. Molecular Orbital Calculations of the Surface Configuration.

The molecular orbital analysis, where a computer solution of the Schrödinger equation is obtained for a limited number of atoms, can be divided again into perhaps three distinguishable approaches. One can be termed a "pseudopotential approach," another the "tight binding method," and the third, a "Green's function model." The pseudopotential approach and the tight binding approach have been used most effectively for semiconductors and nontransition metals, while the Green's function model has been found more appropriate for the study of transition metals with dangling d orbitals.

4.2.1.2a. *The Pseudopotential Calculation for Ionic, Covalent, and Metallic Solids.* The pseudopotential approach is a calculation method which has been highly successful for the calculation of bulk energy levels. In addition to the attractive potential expected from the ion cores in the lattice, an extra artificial potential is introduced which accounts for some mathematical problems in orthogonalization of the wave functions. Then the Schrödinger equation is solved for the lattice with this "pseudopotential" replacing the potential term in the equation. The value of the pseudopotential used is chosen to fit experimental results. Now these same pseudopotentials characteristic of a given species can be used to represent the surface atoms of the crystal. For quantitative results, the Schrödinger equation is integrated numerically in a region at the surface, including a few atomic layers, and the resulting wave function is matched to the bulk Bloch wave functions in the interior.[17] By reiterative computing, the analysis is continued until the wave functions are self-consistent. That is, the wave functions for the electrons are calculated, then the resulting charge distribution is reintroduced into the Schrödinger equation, and new wave functions are calculated. When an iteration does not change the wave function, that wave function is considered a self-consistent solution.

Louis and Yndurain[18] have presented a modified version of the semi-infinite crystal approximation involving a pseudopotential approach to describe surface states on ionic solids. The difference in potential between the anion and cation is mathematically described as a phase shift. They find results qualitatively as in other schemes: an unoccupied level below the conduction band and an occupied level above the valence band, with the separation of levels increasing with increasing energy gap. Other workers[19,20,148] have also reported pseudopotential studies.

Appelbaum and Hamann[16,17,21,22] have applied the pseudopotential approach with substantial success to sodium and silicon (111) surfaces. Their analysis of the silicon surface is particularly interesting. They ignore reconstruction of the surface atoms (shift of atoms in the surface plane). They claim that very small energies and displacements are involved in the low-temperature reconstruction, so the surface states should not be altered substantially, while the calculations would be considerably more complex. However, they permit surface relaxation, a movement of atoms inward normal to the surface plane, of 0.33 Å. Actually, they find that the relaxation inward is necessary to obtain a self-consistent solution, for otherwise an excessively large surface dipole is produced, and the calculation of the surface state energy yields unacceptable (experimentally) values. The shape of the orbitals calculated for the dangling bond surface state is satisfying —a large lobe in electron density extends outward from the surface, a null occurs at the surface atom, and a secondary lobe appears behind the surface atom. This shape confirms the utility of the concept of "dangling bonds" as described in Section 1.2.

Three silicon surface states are found. The one of highest energy is the dangling bond surface state. The second and third are below the valence band edge, but are localized at the surface. They are associated with the bond between the surface atom and its neighbors, and can be localized because there is a low-energy bonding orbital between the first and second layers of atoms, arising because of the shorter bond length.

Thus the picture one obtains for the surface orbitals and surface states is much as one would expect from the most naive considerations. Quantitatively the results are in satisfactory agreement with experiment. The calculated ionization potential is 5.46 eV, and the Fermi energy when the surface state band is thus filled is 0.3 eV above the valence band edge. All these figures are in reasonable agreement with experiment, as will be discussed below.

GaAs is a semiconductor which experimentally seems to have characteristics partly ionic, partly covalent. Yndurain and Falicov[23] have analyzed the surface states associated with the polar (111) surface of GaAs, and conclude that surface states of both the ionic and the covalent type should be present. They find atomic-like surface states near both the conduction band and the valence band, associated primarily with the gallium and arsenic atoms, respectively. In addition to these surface states that have the characteristics of atomic orbitals, they find dangling bond states, similar to the intrinsic states of covalent materials. These states are centered on the surface layer of atoms independent of the face exposed. This model provides an interesting picture of a transition material between the ionic and the covalent classes of solids which, however, may be atypical (see Section 4.2.4).

4.2.1.2b. *The Tight Binding Model for Ionic and Covalent Solids.* This second approach starts with an atomic orbital model using a Schrödinger equation that describes the actual configuration of atoms (no extra "pseudopotentials" included). Such a model is somewhat more satisfying from a conceptual point of view, for the model correlates well with the concepts of chemical bonding. However, in this approach many more parameters have to be fitted by experiment.

A semiempirical tight binding calculation for semiconductors of varying ionicity has been developed by Cohen and co-workers.[24–26] In particular they studied germanium, gallium arsenide, and zinc selenide in order of increasing ionicity. They used nine parameters to fit the model to bulk behavior, analyzing a sample sixteen atomic layers in depth and of infinite (periodic) lateral dimensions. The analysis is independent of ionicity except that one parameter, which they term the potential V_0 (equivalent to the Coulomb integral of the Hückel theory), is used as the arbitrary zero of potential for germanium (homopolar), but has a value $\pm\Delta$ for the ionic crystals. The $+$ sign or $-$ sign distinguishes the potential at the anion or cation. Naturally, Δ increases with ionicity.

In Figure 4.1 is shown the results of their calculation for the local density of states (LDS) as a function of energy on the (111) surface plane. Three cases are shown: the (111) surface of germanium and the polar (111) and ($\bar{1}\bar{1}\bar{1}$) surfaces of ZnSe. The valence band edge is the zero of energy. The dotted curve is for germanium, showing LDS for the surface layer of

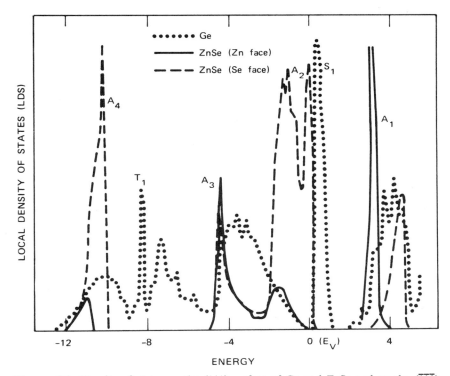

Figure 4.1. Density of states on the (111) surface of Ge and ZnSe and on the ($\overline{1}\overline{1}\overline{1}$) surface of ZnSe.

atoms on the (111) face. The band of states S_1 and the band T_1 are surface states—the rest are bulk states (they are present on layers of atoms that are deep in the crystal). The other curves are ZnSe, and, in this case ionic surface states appear. The solid curve is the LDS of the surface plane of (111) ZnSe. This plane is composed of zinc atoms.[27] An intense surface state band A_1, a weak band A_2, and low-energy bands A_3 and A_4 are found. The rest are bulk states. The dashed curve shows the LDS of the opposite surface plane, the ($\overline{1}\overline{1}\overline{1}$) plane, composed of selenium atoms. On this layer of Se atoms an intense surface state A_2, a weak band A_1, and the deep surface states A_3 and A_4 are found.

Curves such as those illustrated in Figure 4.1 showing the number of states/cm² as a function of energy are typical of the results of surface state calculations. Usually a band of dangling bond energy levels is found near the Fermi energy (for a covalent material) or near the conduction and valence bands (for an ionic material), and these surface states represent the active orbitals which are available for bonding. Deeper surface states are identified at much lower energies, corresponding to strong "back" bonds between the surface atoms and the next layer of atoms in the crystal, or between neighboring atoms in the surface layer.

The most interesting aspect of Figure 4.1 is not the LDS for germanium (the calculations of Appelbaum and Hamman using the pseudopotential approach are doubtless more accurate) but rather the changes in LDS with the ionicity of the solid. The active surface states corresponding to the Zn face are in the band A_1; these are primarily Zn states, appearing near the conduction band (the Zn-like orbitals are heavily weighted in the wave function) although some back-bonded Se states are in the same energy region. Peak A_2, which dominates the ($\bar{1}\bar{1}\bar{1}$) surface, are the active selenium surface states at the edge of the valence band. For germanium there is only one dominant dangling bond peak.

Similar results were found by this group[25] for the (110) surfaces of germanium and GaAs. In the latter case the surface plane has mixed Ga and As atoms, but still two surface states appear, one having essentially the character of an As state, the other, associated with the conduction band, is to a great extent of the Ga character. Again there are deeper-lying surface states that are of less interest here since they will normally be less active in adsorbate bonding.

Hirabayashi[28] and Pandy and Phillips[29,30] have studied silicon and germanium with the tight binding approach, terming it more "realistic" than pseudopotential in the sense that it resembles molecular orbital models. The early results[29] provide a poorer numerical fit to experiment than the pseudopotential approach, but in their more recent semiempirical fit Pandy and Phillips[30] claim agreement to experimental observations on silicon as close if not better than provided by the pseudopotential method.

4.2.1.2c. *Green's Function Model: Transition Metals.* The third approach[31,32] to a molecular orbital description of the surface is the Green's function approach. In this approach the surface atom is initially described as if it were a free atom using atomic orbitals. These atomic orbitals are then mathematically allowed to mix with the nonlocalized Bloch-type orbitals of the bands of the solid, the mathematical process having a certain similarity to the calculation of molecular orbitals in chemical bonding between two atoms. The actual calculation is done using a Green's function formalism. This approach is probably the most rewarding when the bonding electrons are d electrons, so the use of the atomic orbital model for one surface atom is more realistic than is the case with the broader s and p

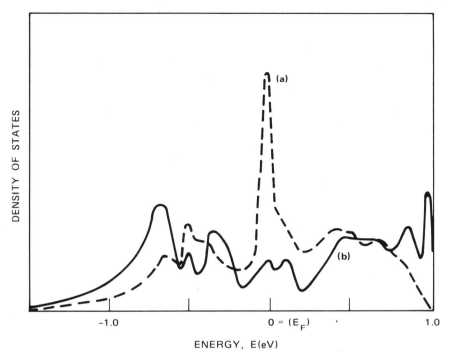

Figure 4.2. Local density of states of *d* orbitals on the surface atoms of Ni(100): (a) orbitals with two lobes protruding out of the surface, and (b) orbitals with clover-leaf lobes in the plane of the surface (source: J. R. Schreiffer and P. Soren, *Phys. Today*, April 1975).

orbitals. The success of the approach is based on the conclusion[16,17] that surface effects are localized to the first two atomic layers or so.

In Figure 4.2 is shown the density of states so calculated for *d* orbitals at a nickel (100) surface, with the solid line showing the density of states (orbitals) in the plane parallel to the surface, and the dashed line showing the density of states with orbitals consisting of two lobes directed out from the surface. The most interesting feature is the high density of states at the Fermi energy. These are the states presumably which would dominate the bonding of adsorbates to the nickel surface.

The above calculations were made with no *s* electron effects included. In an earlier calculation Forstman and co-workers[33–36] suggest that surface states should exist on *d*-band metals near the crossover points of *s* and *d* bands of the same symmetry. Waclawski and Plummer[37] and Feuerbacher and Fitton[38] come to similar conclusions. Thus, as will be discussed in

Section 4.4.3 below with respect to measured surface states on tungsten, two bands of surface states may be more realistic than the one dominant band of Figure 4.2.

To summarize, the success of quantum approximations must be considered from two points of view, first the quantitative and second the qualitative. From a quantitative point of view, it appears that great progress is being made with the new interest in the MO approach, and that even more rapid strides may be foreseen in the near future because of the stimulation of the success of this approach. From a qualitative point of view the models are even more valuable. The knowledge that is being developed about the energies of surface states, the direction and type of bonding orbitals which the surface states on a clean surface represent, and the sensitivity of the surface state energy to small changes in atom position, is of immense value in understanding the behavior of clean surfaces. In the next chapter we will be returning to quantum models, this time to use them to describe the bonding of foreign species to these surface sites, and the value of the above models will become even more apparent.

4.2.2. Semiclassical Models: The Madelung Model for Ionic Solids

The overall concept of the Madelung approach to surface states on ionic solids is very simple. By this model, the electron affinity of the cation would be that of a free ion at infinity, but is substantially modified because the electrostatic potential at its position in the crystal differs from the zero electrostatic potential of the ion at infinity. Similarly, the energy to remove the electron from a negatively charged anion is its value at infinity modified by the potential existing at its lattice position. Figure 4.3 shows the concept for the solid MX. A metal atom M has a large ionization potential I when isolated ($R = \infty$), but the ionization potential decreases to E_{bM} when the ion is in the lattice surrounded by ions X^-. On the other hand the atom X has a low electron affinity A when located at $R = \infty$, but in its normal lattice site, surrounded by the attractive positively charged cations M^+, the electron affinity E_{bX} is much higher. As shown, the energy level of the electron on X^- becomes lower than that on M. Thus in a reaction between M atoms and X atoms, the electrons leave the M atoms, are located on the X atoms, and the ionic crystal M^+X^- is formed. By the simplest model the unoccupied energy levels of the M^+ ions become the bulk conduction band ($E_{bM} = E_c$), the occupied energy levels of the X^- ions become the bulk valence band ($E_{bX} = E_v$), and the electrons on the X^- ions become the valence electrons. However, in this classical picture no band broadening is obtained.

This qualitative picture is made semiquantitative by calculating the exact potential V_{bM} at a bulk M^+ site in the lattice, summing the potential contribution from other ions in the lattice. This summed potential V_{bM} is called the Madelung potential at the site. Similarly the Madelung potential V_{bX} at a site of a bulk negative ion can be calculated. In the analysis to give the potential at the two lattice sites, a nonintegral partial charge on the ions is actually assumed. Also in the analysis to relate E_{bX} to A and E_{bM} to I, the values of I and A used are obtained by interpolating between the known values (that correspond to removal or addition of an integral number of electrons from the atom) to get values corresponding to the partial charge on the ions in the solid.

Levine and Mark[6] have extended these concepts to treat a surface ion on an ionic crystal. Again conceptually the approach is straightforward. Consider again the cation M^+. If the ion is only moved to a surface site, it is no longer surrounded by its normal complement of anions, so the potential at the site will be higher than at a bulk site and the resulting energy level will be lower as indicated by the dashed line in Figure 4.3.

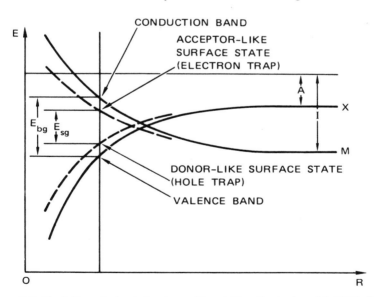

Figure 4.3. Evolution of atomic levels on M and X as the ionic solid MX is formed (the interatomic distance R is decreased to the lattice dimension). The shift in atomic levels is due to the potential arising from the other ions in the solid. The potential at the surface sites differs so surface states E_{SM} and E_{SX} are formed.

Similarly, a surface anion will require less energy to remove the electron than the bulk anions require, and thus will provide a donor state above the valence band energy, as indicated by the dashed line in Figure 4.3.

Levine and Mark[6] calculated the Madelung potential at a surface site and determined the ratio

$$E_{sg}/E_{bg} = (\gamma - \mu)/(1 - \mu) \qquad (4.17)$$

where

$$E_{sg} = E_{sX} - E_{sM} \qquad (4.18)$$

is the separation of the surface states, and

$$E_{bg} = E_c - E_v \qquad (4.19)$$

Here γ depends primarily on the crystal structure and crystallographic face and can take on values from 0 to $\frac{1}{2}$, and μ depends primarily on the partial charge z on the ions and the values of I and A interpolated for this partial charge. The parameter μ as defined can be between 0 and 1. The calculation becomes a very complex and sensitive function of the partial charge assumed for the ions. Because of this sensitivity the model is of primarily qualitative value.

From the examples calculated by Levine and Mark, some relationships may be suggested for the dependence of the energy levels on the parameters of interest. In particular it is found that the surface states become more centered in the band gap (E_{sg}/E_{bg} becomes smaller) as the partial charge on the ions, z, increases.

In the next subsection we will be examining acid and basic centers on ionic solids, so it is of interest to examine the relation of the Madelung model to Lewis acid or basic strength. Levine and Mark discuss the Madelung model only in terms of ionic surface states rather than including Lewis activity. However the concepts apply to both, because the analysis only describes the potential at the surface site, neglecting any quantum effects (including the orbitals available). A higher potential at the site will increase the electron affinity of an acceptor ionic surface state or the electron affinity of a Lewis acid site.

The Madelung analysis above is, however, incomplete if one wishes to estimate acidity, for in this case we need to know the electron affinity of the electron pair acceptor orbital toward an electron pair donated by an adsorbing molecule. When discussing the energy (electron affinity) of a surface orbital, there are two parameters: one is the energy of the surface orbital relative to the band edges in the solid; the other is the energy of the surface orbital relative to a reference energy such as E_e, the energy of a free electron at rest at infinity. Levine and Mark concerned themselves with the calculation of the former, for this determines whether trapping of electrons and holes from the bands in the solid will occur (ionic surface state activity). The acid/base properties, however, depend on the electron affinity relative to the free electron, for we are concerned with the attraction of a gaseous donor to the site. Thus in the case of the solid MX it is observed from Figure 4.3 that the energy of the surface states relative to E_e, and hence the acid or basic behavior, depends primarily upon the mean electronegativity, the value of $(I + A)/2$, and only secondarily on E_{sg}/E_{bg}. With a high value of $(I + A)/2$, the energy levels in the solid in Figure 4.3 are all low, so the solid will tend to show stronger Lewis acid character. A low value of $(I + A)/2$ conversely will lead to a high donorlike surface state (Figure 4.3) relative to E_e and will tend to promote Lewis base character.

It is unfortunate that to date the Madelung concept has not been developed for the more interesting cases of M_nX_m crystals, for the behavior of mixed crystals, or for impurities on crystals and their influence on the surface states and acid/base character. Hopefully these will be explored in the near future, for the insight to be gained in surface chemical physics about Madelung effects on the behavior of ionic solids will be invaluable.

4.2.3. Models for Electron Pair Sharing: Lewis and Brønsted Sites

In this section we will discuss the theories of the origin of acid sites that share electron pairs with gaseous donors and of basic sites that share electron pairs with adsorbing acceptors. A question of particular interest is the relation of such acid or basic centers to the surface state models as described in the preceding subsections. As before, the discussion will cover Lewis sites (sites with an affinity for or availability of electron pairs) and Brønsted sites (sites able to exchange protons). The two forms of sites, Lewis and Brønsted, are closely linked; in some cases through the simple adsorption of water [Equation (1.1)] and in other cases more remotely.

As suggested by the above Madelung analysis it is easily seen that a solid, for example an oxide, should show two types of centers even on a uniform surface. Consider the simple models of Figure 4.4. With no protons present (Figure 4.4a) the unoccupied orbital at the cation should act as an acceptor surface state or Lewis acid site and the orbital on the oxide ion should be a donor surface state or Lewis base site. If water is adsorbed

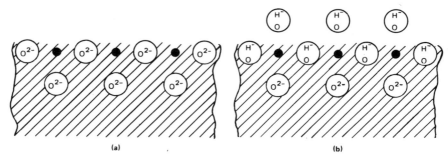

Figure 4.4. Oxide surface (a) before and (b) after hydration. The solid circles represent cations.

(Figure 4.4b), again there are two classes of acid/base centers, now Brønsted centers. If the cation is a strong Lewis acid before water adsorption, the OH$^-$ ions from the water will be held tenaciously at the cations and the remaining protons, now associated with the anion, will be easily given up. Thus the protons on the lattice oxygen ions (Figure 4.4b) become the Brønsted acid sites. If the proton from the water is attracted to the lattice oxide ion more strongly than the OH$^-$ ion is attracted to the cation, the material will be basic, tending to give up the OH$^-$ ions. Thus the OH$^-$ groups attached to the cations as shown in Figure 4.4b, become the basic sites.

The OH$^-$ groups attached to the cations can also show weak acidity; here also the proton can be given up. Thus this site too can act as a weak acid site, if a strong enough base is provided. But the acidity of the OH$^-$ group on the cation is expected always to be weaker than the acidity of the H$^+$ ion adsorbed on the lattice oxide ion.

Qualitative models have been developed relating the expected strength of the acid and basic sites on oxides to the parameters of the solid, particularly the partial charge on the oxygen. Tanaka and Tamaru[39] observing the relation[40] between the partial charge on the oxygen and the acidity for dissolved oxides, suggest the partial charge on the oxygen should be the key parameter also for oxides in the solid form. The partial charge on the oxygen is of course related directly to the electronegativity of the cation, either for the solid or the dissolved oxide cases. If the cation is highly electropositive, as for example an alkali ion, the oxygen will have a high negative partial charge. A high partial charge on the oxygen results in a strong base. If the cation is electronegative, as for example phosphorus, the bond will be almost covalent, and the oxygen will have a partial charge that is almost zero. Oxides where the oxygen has a low partial charge (for example P_2O_5) are highly acid. Wilmshurst[41] and Filimov et $al.$[42] also note the relation between the electronegativity of the metal (in the metal oxide) and the acidity, as discussed below.

The partial charge model can be compared to the Madelung concept developed in the preceding section, where the acid/base behavior was related to the quantity $(I + A)/2$, with a high $(I + A)/2$ resulting in a net acid surface and a low $(I + A)/2$ resulting in a basic surface. The Madelung formulation suggests therefore that when the anion is common (a series of oxides), a cation with a high I will most likely result in high $(I + A)/2$ and thus in a more acid solid than a cation with a low I, consistent with the present "partial charge" argument.

The acidity will be stronger with higher formal charge (oxidation number) Z on the cation at constant partial charge on the oxygen. This relation is shown by the following argument. When a water molecule adsorbs on a high Z cation the OH^- group will form a strong electrostatic bond with the cation, and the greater the value of Z (at constant oxygen partial charge) the more tenaciously is the OH^- held. As discussed with reference to Figure 4.4b, this leads to high acidity. Conversely, if the Z of the cation is low, the proton from the water, adsorbed on a lattice oxide ion, will be held more strongly than the OH^- group, and the surface will be basic.

Tanaka and Tamaru[39] suggest that Mars' expression[43]

$$a \sim (r_a/r_c)Z^2 \tag{4.20}$$

obtained by comparing the effectiveness of solids as acid catalysts, may provide a reasonable relation between an acidity function a, the cation radius r_c, the anion radius r_a, and the formal cation charge Z. Specifically, the acidity as reflected by the catalytic behavior (for example the hydration rate of propylene), of a series of metal sulfates are listed in the order

$$Fe^{3+} > Al^{3+} > Cr^{3+} > Cu^{2+} > Zn^{2+} > Co^{2+} > Cd^{2+} > Ni^{2+} > Mn^{2+} > $$
$$ > K^+ > Na^+$$

Filimov et $al.$[42] related the acidity of a series of oxides to the electronegativity of the cation. He studied the frequency change in the $950\ cm^{-1}$ line of NH_3 as affected by adsorption on acid centers, noting that Wilmshurst[41] related the frequency shift to the electronegativity of the metal. He found the acidity of oxides to be ordered as follows:

$$Al_2O_3 > Ga_2O_3 > TiO_2, Cr_2O_3, ZnO > ZrO_2 > MgO > Ni_2O_3 > NiO, CuO$$

The effect of heterogeneity on surface activity is one technologically important aspect that has been examined closely in catalytic studies, much more closely than the corresponding ionic surface state analyses. In practice the presence of foreign cations at an oxide surface, or, a step further, the use of mixed oxides, has been very effective in promoting high acidity and has been extensively studied. Mixed ionic solids have been, until now, too complex for surface state studies or for surface state theories, but as will be discussed, the very active sites observed by acid/base measurements suggest that the same configuration should lead to active ionic surface states.

Models for the high electron affinity (acidity) of sites on mixed oxides have been examined by Tanabe *et al.*[44] They suggest that an anomalously high electron affinity can arise because a foreign cation may have too high a positive charge to be compensated by the surrounding anions, assuming the surrounding anions have the charge normal for the host lattice. If the cation charge is not satisfactorily compensated and leaves an excess of positive charge in the neighborhood, one obtains a strong Lewis acid site. A measure of the relative acidities to be expected is obtained by noting the magnitude of the mismatch in coordination numbers, and assigning numbers to describe the mismatch in terms of charge. Thus if it is assumed that each cation maintains its normal coordination number (number of bonds to nearest neighbors), but the anions have the coordination number of the majority component of the compound, then an excess charge will appear at the cation for some oxide combinations. Consider a mixture of the oxide of a cation with a coordination number of 6 with the oxide of a cation with coordination number 4. Then if each has a valence of $+4$, the charge per bond of the cation of coordination number 6 is $4/6$, while the charge per bond of the cation of coordination number 4 is $4/4 = 1$. If the six-coordinated cation is in the majority, then Tanabe assumes that each oxide ion in the lattice will have $4/6$ charges per bond. However, the bonding to the four-coordinated cation only represents $4 \times 4/6 = 2.67$ negative charges from the anions, while the bonds from the cation contain $4 \times 1 = 4$ positive charges. Thus in the neighborhood of the four-coordinated cation by this simplified picture there is an excess positive charge of 1.33 electronic charges. This leads to a high electron affinity in that region—a strong Lewis acid, or a strong acceptor surface state (assuming it can accept one charge). The suggestion is made that if the charge in the neighborhood of the minority cation is positive as in the example above, the site will act as a strong Lewis acid. If on the other hand the charge is negative, the site will be neutralized by protons, and as these protons can be given up, the site will act as a Brønsted acid.

Similar but less generalized models based on coordination have been used to explain the strong acidity of the mixture, e.g., by Leonard et al.[45] on SiO_2/Al_2O_3.

Even the Tanabe model does not describe all the possible variations in acidity due to coordinatively unsaturated surface sites. Differences leading to excess charge at particular sites on the surface are associated with all the various types of heterogeneity: crystal steps, dislocations, kinks, etc. (see Figure 1.4). On the other hand, relocation of surface atoms is a phenomenon associated with surface ions which will tend to remove the excess charge at such strong sites. Near a strongly positive site, anions will shift their position toward the site, and the Madelung potential at that site will be less positive. Clearly the heterogeneous system is very complex, and sites with a whole distribution of electron or proton affinity can be expected, with coordination deficiencies and imperfections of various types causing them, and with relocation of ions tending to neutralize them.

4.2.4. Comparison of the Various Surface States and Sites

In the above discussion three types of surface states for electrons on a clean solid surface have been described. One is the acid (or basic) site, where electron pairs are accepted for sharing or offered for sharing with adsorbing species. The second is the ionic surface state, a surface state where unpaired electrons can be accepted from or offered to the bands or to an adsorbing species. If the ionic surface state is an acceptor state it is associated with the cation; if a donor state, it is associated with the anion. The term "ionic surface state" to describe such states was used by Davison and Levine,[1] although Koutecky[46] prefers the "Tamm surface state" designation to distinguish it from the "Shockley surface state" characteristic of the covalent solid. The third type is the dangling bond Shockley surface state on a covalent solid, visualized as an unpaired electron arising on the incompletely coordinated surface atom of the covalent solid. Shockley surface states can exchange electrons with the bands or form sites for adsorption with strong covalent bonding.

The orbitals associated with the first two types of surface states are described in the theoretical studies as being closely related to nonhybridized atomic orbitals, quite localized on surface ions. The third type, the dangling bond, is described by hybridized orbitals concentrated between neighboring atoms, as expected for a homopolar bond. Because the surface atoms have no neighbor on the one side, there is an unpaired electron in an orbital that is directed outward from the surface.

The purpose of the present section is to prepare for the discussion of experimental results by exploring how the various forms of surface states are related, and to describe qualitatively the transition from the one type of surface state to the next, as the character of the solid moves from strongly ionic toward that of a covalent material. Then the discussion of experimental results can be related more clearly to the type of solid and type of surface state or site.

Lewis acid or basic centers on an ionic solid are observed by adsorbing a test foreign species. The energy of the level is dependent upon the formation of a localized bonding orbital with an adsorbate. The energy level of this bonding orbital will be lower than the energy of the surface state energy level (the nonbonding orbital) before the adsorption of the test adsorbate. However, experimental measurement of surface state energy levels, by UPS for example, before the adsorption of the test acid or base adsorbate, will only reveal nonbonding orbitals. Thus measurement of the energy of orbitals associated with acid or basic sites by a technique such as UPS must be a measurement with an adsorbed species at the surface.

The second type of surface state, the ionic surface state will be represented more closely by the nonbonding atomic orbital of the surface ion. This is to a first approximation the energy at which a single electron will be captured (at the cation) or will be given up (from the anion). However, on a clean surface, these sites are not expected to have highly localized orbitals. If special sites associated with heterogeneity are excluded from the discussion, each surface ion is identical, with overlapping wave functions. Then the surface states will broaden into bands (Figure 4.5) according to quantum arguments (the Pauli principle). Thus a single electron captured at the surface will be shared between all the surface cations, and the energy level will be lowered by electron sharing. It is such bands of energy levels shown schematically in Figure 4.5a that will be measured by UPS or ELS.

The Madelung model suggests that these ionic surface states should always be present. However, as pointed out in the discussion of Madelung effects, the analysis does not take quantum effects into consideration. In particular, referring again to Figure 4.3, the model does not consider band formation in the bulk material, the formation of the conduction and valence bands. This band formation lowers the energy of the conduction band edge and raises the energy of the valence band edge substantially compared to the positions predicted in the Madelung simplification. The broadening of the ionic surface states into bands is not so great, as the overlap of these states, both with each other as above and with bulk orbitals, is less than that of the bulk cationic or anionic states. The result then is that although the Madelung

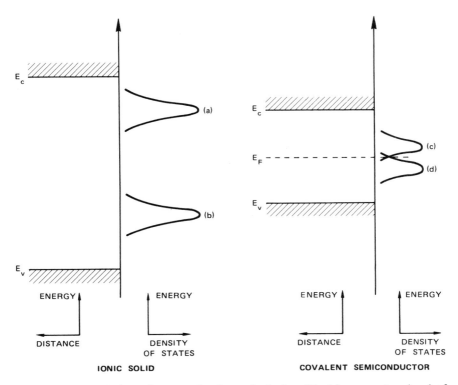

Figure 4.5. Intrinsic surface state bands on the ionic solid: (a) represents a band of surface state levels of conduction band character, (b) a band of valence band character. On the covalent solid: (c) represents dangling antibonding orbitals, and (d) bonding orbitals.

potential arguments say that with no band formation the surface state energy should be substantially lower than the conduction electron energy, band formation may negate this conclusion, and the surface state band may be found at, or above, the conduction band edge. Similarly with the valence band, the Madelung potential effect may be negated by the quantum broadening effect, although broadening of the valence band is much less than that of the conduction band (less overlap), so there may be more likelihood of the donor ionic surface states remaining above the valence band edge. The above arguments are entirely qualitative and may not be the only reason why ionic surface states are not always experimentally found in the band gap region on ionic crystals.

Clearly the connection between acid sites and acceptor surface states is very close. It is not necessary that a material with strong acid sites will have acceptor ionic surface states, but if these strong acid sites arise from a strong Madelung potential, it can be anticipated that shallow ionic surface states will be present.

Before moving on to a discussion of the transition between ionic and covalent surfaces, the case of the heterogeneous surface should be recalled, and the influence of heterogeneity on Lewis sites and ionic surface states should be mentioned. Included as a source of heterogeneity is a random distribution of surface ions such as expected in mixed oxides. Heterogeneity will lead to sites with either or both of two special characteristics: (a) very high (or low) electrostatic potentials, and (b) orbitals oriented particularly favorably for reactions with adsorbates. The first characteristic will lead to strong Lewis activity, and also deep ionic surface states. The second characteristic may lead to strong Lewis activity or one-electron exchange with adsorbing species, but not necessarily to electron exchange with a bulk band.

The surface state behavior of crystals which are in the transition region between ionic and covalent is not entirely clear. As mentioned above (Section 4.2.1.1a), Yndurain and Falicov[23] have analyzed the surface states on the polar surfaces of GaAs, such a transition material, and concluded that both types of surface states, Tamm and Shockley, are simultaneously present. However, these particular results on GaAs may be related to the polar nature (either Ga atoms alone or As atoms alone) of the surfaces studied. From a qualitative point of view, it would seem more reasonable that as the bonding orbitals become less polarized, a gradual transition between ionic (Tamm) and hybrid covalent (Shockley) states would be expected.

The Shockley surface states on a covalent solid arise when the electron wave functions are highly directed covalent bonds. Then the rupture of a local bond during the formation of the surface results in a dangling bond and an unpaired electron. Chemically the surface atoms are equivalent to radicals, and the electrons are highly reactive.

With a clean surface, such reactive unpaired electrons represent a very high energy configuration, and a distortion (reconstruction) of the surface layer of atoms to permit some electron overlap between the dangling bonds may occur.[47,48] Such distortion to pair the dangling electrons will lead to a modest splitting of the dangling bond levels into bonding and antibonding bands of orbitals, with the associated lowering of the energy of the system. The probable connection in many cases between the observed reconstruction of the surface and such pairing was first pointed out by Schlier and Farnsworth.[49]

As with the ionic surface states, therefore, it is to be expected[50,51] that in general the dangling bond surface states on covalent solids will consist of two bands, one essentially occupied, one unoccupied. Such a surface state configuration shown schematically in Figure 4.5b will be expected to show up in the experimental measurements.

In the case of metals, where in the neighborhood of E_F allowed bulk energy levels overlap any levels localized at the surface, it is not usually fruitful to refer to surface states, but one must describe the "local density of states, LDS" associated with the surface layer, and to a great extent dismiss the atomic orbital picture. In the case of a transition metal, however, with its relatively narrow d band, an atomic orbital picture again becomes useful. Lobes associated with these atomic orbitals can be considered available for covalent bonding, and the model becomes similar in some ways to that of the covalent solid. Even with nontransition metals the interaction of adsorbing species with the metal surface can in many cases be described by an "induced covalent bond." In this case the bonding electrons become more localized than the LDS of the clean metal surface would indicate. The discussion of the LDS of such bonds will be continued in Chapter 5.

Measurements by the surface spectroscopies are expected to be suitable to show the LDS for a clean metal surface.

4.3. Measurements on Adsorbate-Free Ionic Solids

4.3.1. Reconstruction on Ionic Solids

On a clean ionic solid the tendency to reconstruct seems related to electrostatic effects. If, on the one hand, the surface plane is electrically neutral, such as the (110) surface of the zinc blende structure, then no reconstruction is normally observed by LEED measurements. Relaxation of a special kind has been observed[149]: ions of one charge move outward, ions of the other charge move inward. If, on the other hand, the surface plane of atoms has a net charge (a polar surface), it is usually found to be laterally reconstructed when clean. As reconstruction is more readily detected than relaxation, far more data is available to illustrate these cases. In general it is found that although reconstruction of polar surfaces is common, it is not inevitable. Chung and Farnsworth[52] found the (0001) surface of some II–VI compounds, ZnO, CdSe, ZnS, and CdS (a plane with only cations exposed) to be reconstructed, but the (000$\bar{1}$) plane (only anions) was more stable, and reconstruction depended on pretreatment. Others[53,54] observe no reconstruction for some of these polar surfaces. As summarized by Chang and Mark,[55] the polar surfaces of ZnO are not reconstructed if prepared below 600°C, but the (0001) face reconstructs at higher temperature. In the case of III–V compounds, again there are exceptions to the rule. MacRae and Gobeli[56] found the GaAs nonpolar surfaces to show integral order (presumably no reconstruction) LEED spots only, following the "rule," but Derrien et al.[57] found the GaP (111) surface unreconstructed, contrary to the rule.

Neutralization of excess surface charge is the reason for the reconstruction in the case of the polar surface. The system is in a lower energy state when reconstruction removes the strong electrostatic double layer.[58,59] A quantitative calculation by Nosker, Mark, and Levine[60] for the wurtzite and zinc blende polar surfaces showed that the surface can be stabilized and the surface potential eliminated if a surface charge of the opposite sign that is 0.25 times the charge on a bulk plane is developed. The desired surface charge can be produced either by reconstruction or by faceting (exposure of new planes).

Mark[61] has shown that the reconstruction observed on GaAs (111) surface is not inconsistent with the requirement of the 0.25 monolayer of oppositely charged ions. He proposes a lattice configuration which would be consistent with the observed LEED spots (a $\sqrt{19} \times \sqrt{19}$ pattern) and which would also be consistent with the electrostatic requirement. He also points out that the (2×2) superstructure observed on GaAs, GaSb, InSb, CdS, and ZnO, as well as the $(\sqrt{3} \times \sqrt{3})$ structure[62] on ZnO (0001) comply with the rule.

van Hove, Leysen, and co-workers[54,63,64] suggest that charge neutralization (at least on ZnO surfaces and doubtless the thesis applies to all polar surfaces) can be made not only by reconstruction, but also either by charged surface states or by impurities—a superstructure of appropriate electropositive or electronegative ions. In particular he found that etched faces of ZnO acquire a superstructure of Ca ions (3×3) which is very difficult to remove by sputtering. This group conclude that with impurity-free ZnO, the (0001) and (000$\bar{1}$) surfaces do not reconstruct, and suggest that here intrinsic surface states provide the necessary charge. Similarly Derrien et al.[57] suggest surface state capture of electrons stabilizes the surface. Chen,[13] on the other hand, suggests motion of Ga atoms to the ($\bar{1}\bar{1}\bar{1}$) surface of GaAs, when the sample is annealed, to neutralize the double layer.

4.3.2. Physical Measurements on Ionic Solids

The electrical measurement of surface orbitals on clean surfaces of ionic solids have naturally been made primarily on semiconductors because electrical measurements on insulators are more difficult and harder to interpret. However, the semiconductor measurements, both the electrical and the surface spectroscopy measurements, provide information about surface orbitals (acceptor and donor states) which can be carried over to the insulator cases qualitatively if not quantitatively. In this subsection we will first discuss those semiconductors of intermediate ionicity, then insulators of high ionicity, and conclude with GaAs, which is partially covalent.

Electrical measurements showing ionic surface states on ionic solids, and relating the band edges to the ionicity of the solid, were reported in a classic paper by Swank.[7] He studied the series of ionic solids: ZnO, ZnS, ZnTe, CdS, CdSe, CdTe. In each case the clean surface was prepared by cleavage. He was able to show behavior in accordance with the qualitative expectations of the theoretical models of the preceding sections.

First he was able to show that the position of the valence band at the surface, for a series of compounds with a common cation, is directly related to the ionicity of the anion. The more electronegative the anion, the deeper the valence band. The measurement was made by a photoelectric threshold study, where electrons from the valence band were excited to the free electron level. Figure 4.6, from the paper by Swank, shows the photoelectric threshold versus the electronegativity of the anion as calculated using Pauling's scale. A comparison with Figure 4.3 will indicate why a species X with a high electron affinity (high electronegativity) should lead to a deep valence band, $E_e - E_{vs}$ large, using the symbols of Figure 2.2. Harrison and Ciraci[65] have introduced a more detailed classical theory, relating the position of the valence band to covalent interaction as well as to the ionic interaction of the Madelung approach. They are able to show substantial agreement of theory with experimental results that include both Swank's data on ionic crystals and data on a covalent material such as silicon.

Second, Swank was able to show that ionic surface states were present in the band gap region on most of these compounds, as determined by a measure of the band bending V_s. The observation of band bending implies capture of bulk carriers by surface states. The band bending was determined by measuring the work function and the photoelectric threshold. The band gap E_G is known for these compounds, and the bulk Fermi energy is de-

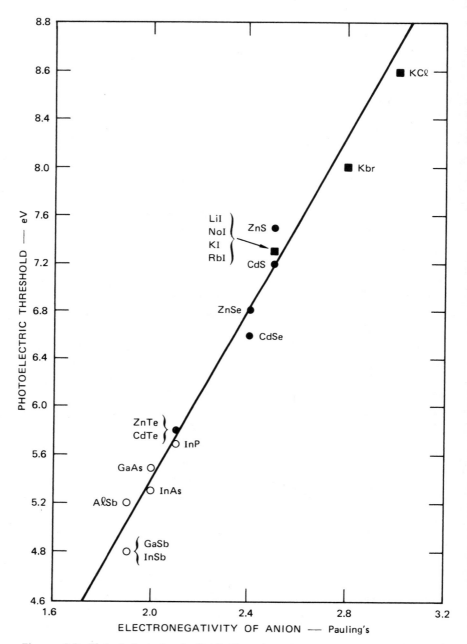

Figure 4.6. Plot of photoelectric threshold against electronegativity of the nonmetal constituent.

termined by the doping of the crystal. Then from Figure 2.2 it is observed that

$$E_e - E_{vs} = \phi - eV_s + E_g - \mu \qquad (4.21)$$

where the left-hand side is the photoelectric threshold, E_g is the band gap for the material, and so all terms are known except the band bending V_s.

From the value of V_s observed, it could be concluded that in all cases the majority carrier was captured at the surface. In the case of ZnO, however, the band bending is negligibly small. Thus acceptor states below the initial Fermi energy are present in all cases of n-type materials, although in negligible density for ZnO. Also for the one p-type material studied (ZnTe), it was found that donor states exist above the Fermi energy.

The results thus are not inconsistent with the theoretical expectations as summarized in Figure 4.3 and 4.5a. Unfortunately, other methods have not been so definite about the presence of ionic surface states on these solids of intermediate ionicity.

Marien, Leysen, and van Hove[66] suggest from field emission experiments that an unoccupied level may be 0.05 eV below the ZnO conduction band edge. However, they qualify this[67] to suggest that their observations may be associated with dissolved zinc which moves to the surface under the electric field rather than a surface state. Lüth and Heiland,[68] and later Lüth,[69] from photoconductivity measurements, observe optical absorption leading to photoconductance. They find steps in the photoconductance/wavelength response that suggest occupied states on ZnO at various energies including 0.18 and 0.45 eV below the conduction band edge. However, the results should be viewed with caution for, as discussed in Section 3.1.5, photoconductance and photovoltage measurements involving direct excitation from surface states can be difficult to interpret.

Brillson[70] has studied CdS cleaved surfaces, and found by photovoltage measurements no indication of surface state effects (no structure on the photovoltage/wavelength response that is not explained by bulk effects). Williams and McEvoy[71] find no surface states on gallium chalcogenides, another group of intermediate ionicity. Fischer[72] finds no donor states on p-type AlSb.

Electron spin resonance (ESR) methods also indicate that ionic surface states are not available in the band gap. This technique should give a signal if a reasonable fraction of such states are occupied by unpaired electrons. The ESR detection of unpaired electrons on crushed ionic solids has been explored by Haneman and his group.[73–75] No unpaired electron signals due to electrons in surface states were found with crushed GaAs,

AlSb, InAs, GaP, GaSb, CdS, CdSe, ZnSe, PhS, or PbTe. Haneman[75] suggests the negative results on the III–V compounds can be explained by preferential cleavage on the (110) planes, together with a predicted[76] slight dehybridization of the bonds at the surface. The preferential cleavage on the (110) means that a negligible fraction of polar surfaces will be exposed during crushing, and the dehybridization of the bonds at the surface will mean that even with the relatively covalent GaAs, for example, the surface will behave as an ionic solid. Electrons will be paired on the As atoms rather than occupying hybrid interatomic bonds which leave unpaired electrons when ruptured.

On highly insulating wide band gap ionic solids, on the other hand, a few measurements have been made that suggest ionic surface states that are charged. Hochstrasser and Antonini[77] have observed surface states on the SiO_2 surface, using a material crushed in ultrahigh vacuum, and using ESR to detect the presence of unpaired electrons on these surfaces. They conclude the surface states are associated with silicon surface atoms, and the density of unpaired electrons on these sites is the unexpectedly large value of $10^{12}/cm^2$ (about 4% of the surface silicon atoms). This is a high charge density, and may arise for surface neutralization reasons, as discussed in the last section (one of the mechanisms of van Hove *et al.*[54,64] for neutralization is capture on surface states). Alternatively it may represent a nonequilibrium electrostatic charge generated in the surface states during the crushing process, a charge that is quasistable on the insulator surface.

Nelson and Hale[78] found surface states spectroscopically on the alkaline earth oxides, MgO, CaO, and SrO. They detected a tail on the absorption edge which they attributed to absorption and detected fluorescence occurring at wavelength greater than 200 nm. Presumably electronic transitions between donor and acceptor surface states were being examined, the best indication being that adsorbed oxygen quenches the fluorescence. The quenching suggests the donor surface states must provide active oxygen adsorption sites.

Gallium arsenide is on the other end of the ionicity scale, on the borderline between an ionic and a covalent solid. At the present time it appears that there are surface states present in the band gap region on some of the crystal planes, but the exact situation is not clarified.

On the cleavage plane of GaAs, the (110) plane, Dinan, Galbraith and Fischer[79] conclude that any donor states must be below the band gap region. Specifically they find by measurements of photoemission, work function, and photovoltage [see the discussion of Equation (4.21) above] that V_s on p-type GaAs is zero, indicating that any donor states must be below the valence band edge. The model for donor states is supported by UPS studies of Eastman and Grobman[80] and of Spicer and his group,[81,82] who find any occupied donor band is entirely below the valence band edge.

On the other hand, Dinan et al.[79] find V_s on n-type GaAs is large, and the measurement can be interpreted as indicating a band of acceptor states 0.85 eV below the conduction band edge with a density of 2×10^{13} cm^{-2} eV^{-1}. Eastman and Freeouf[83] and Gregory and Spicer[81] also found evidence for such states, although in more recent studies[151] they no longer observe the levels. van Laar and Sheer,[84] with work function measurements, find $V_s = 0$ on both n- and p-type material, which suggests there are no surface states. ELS studies of GaAs (110) by Froitzheim and Ibach[85] also fail to show the presence of acceptor states as expected (no ELS energy absorption peaks). Huijser and van Laar[86] have made work function measurements on crystals where on the one hand the surface is damaged during cleavage and on the other hand the surface is undamaged. They find work function shifts only on the damaged material, which suggests that the acceptor states where observed may be associated with dislocations rather than with surface atoms. Thus by the observations of van Laar and his co-workers, the (110) surface of GaAs shows no ionic surface states in the band gap region. The admissibility of such an alternate model indicates the need for caution in interpreting work function measurements such as those of Swank as being definitive proof that intrinsic surface states are present.

On other surfaces of GaAs, surface states are found. Ranke and Jacobi[87] have investigated the polar (111) and ($\bar{1}\bar{1}\bar{1}$) surfaces. They find donor states overlapping the valence band edge and extending up to 0.25 (111) or 0.75 ($\bar{1}\bar{1}\bar{1}$) eV above E_v.

Ludeke and his co-workers[88,89] with ELS find a surface transition on the GaAs (100) face and attribute it to transitions between the valence band and the unoccupied surface states or between As donor states and Ga acceptor states. In these studies the Ga/As ratio at the surface is varied by an epitaxial deposition, so the coordination number of the atoms that show surface states is open to question. This point is discussed in detail by Jacobi,[90] who used the same technique with GaP and found both donor and acceptor states, depending on which species is in excess.

Thus at the present time the existence of intrinsic surface states on GaAs, particularly acceptor states on the (110) face, is under hot debate. As dislocations lead to states that can be confused with acceptor surface

Table 4.1. Intrinsic Acceptor Surface States on the (110) Face of III–V Compounds

Compound:	GaP	GaAs	GaSb	InAs	InSb
Energy gap E_g (eV)	2.2	1.4	0.7	0.4	0.2
Surface state, $E_c - E_t$ (eV)	1.2	0.9	0.6	0.5	0.4

states, experiments must be performed where it is shown that dislocations are not participating.

If we believe the surface states are actually present, a summary of the position of measured acceptor surface states on the cleavage surface of III–V compounds, by Freeouf and Eastman[91] is of interest, and is reproduced in Table 4.1. It is observed that with narrow gap (more covalent) materials the acceptor state is not found in the gap region.

To summarize, the state of knowledge of ionic surface states on solids in general is clearly far from satisfactory. The few measurements available on the very wide band gap ionic solids, using optical or ESR detection methods, suggest that for such materials (SiO_2, alkaline earth oxides), ionic surface state levels are found in the gap region. From the qualitative Madelung model it is understandable that for these highly ionic materials the surface state levels are more likely to be in the band gap.

At the other extreme is GaAs, a material which is almost in the covalent class. It also shows evidence of surface state bands in the forbidden gap, below the band gap for the donor states but in the band gap region for the acceptor states. In this case all the evidence points to a material in the transition region between ionic behavior and covalent behavior, and for the latter (see Section 4.4) surface states due to dangling bonds are clearly documented.

The few measurements available on materials (ZnO, CdS, etc.) of intermediate ionicity are not definitive with respect to whether the ionic surface states are in the band gap or outside the band gap region. The electrical measurements suggest levels in the band gap region; the few nonelectrical measurements show no strong evidence for such a picture. There may be several explanations. As discussed in Chapter 2, a very small density of surface states (the order of 10^{-3} monolayers or less) could account for Swank's results. Thus these electrical measurements could be accounted for by heterogeneity (flaws arising during cleavage) or the tail of surface state bands extending from the band region into the band gap region.

4.3.3. Chemical Measurements on Ionic Solids

4.3.3.1. Introduction. As described in Section 4.1.2, in general chemical measurements are made on powders. If the solid surface is relatively inert, the observations will reflect the properties of the clean surface. Thus our definition of "clean surface" is more relaxed in this case.

In addition, however, there is a further problem in definition. Many of the most interesting ionic solids from a surface chemistry point of view are oxides, prepared in the powder form as the hydroxide, the carbonate, or the oxalate, and then decomposed to form the oxide. There is little doubt from a theoretical or an experimental point of view that the surface characteristics determined on such material is a fair representation of the behavior of a clean surface of a mixture of oxide and, say, hydroxide. The characteristics of course change from the characteristics of a hydroxide to the characteristics of an oxide as the dehydration proceeds. Now with such a solid it is impossible to draw a definitive line that distinguishes between a clean sample of mixed hydroxide and oxide on the one side and a sample which has adsorbed water on its surface on the other. We will simply consider the case of mixed hydroxide/oxide samples or rehydrated samples all as clean surfaces and discuss them in this chapter on "clean surfaces." In other words in this section a clean surface on an ionic solid will be defined as a surface which has no adsorbates other than protons or hydroxide ions. An equivalent definition could apply to mixed oxides formed by decomposition of carbonates or oxalates, but we will only use results on hydrous surfaces as examples.

The surface activity pattern followed in such an experiment, starting with the hydroxide and heating, is common to many oxides. The surface covered with the hydroxide is reasonably inert, but as water leaves the surface, active sites are left, which show high acidity or basicity. This partially dehydrated solid is often the material of most interest. When the temperature is high enough for complete dehydration, it is usually high enough for reconstruction of the surface, and the active centers are passivated again by movement of surface ions to neutralize the strong electric fields.

In the following discussion of experimental results we will choose a few examples to illustrate the theoretical models of Section 4.2.3, and the order of the examples will follow the order of the theoretical models. Thus, first we will discuss the work of Boehm that illustrates the appearance of two types of centers on the oxide surface. Then we will select an example of an acid, a basic, and a neutral oxide (Al_2O_3, MgO, SiO_2) to illustrate various sites on a presumably uniform surface, in order to compare to the model of Tanaka and Tamaru. The appearance of exceptionally active centers on mixed oxides will be illustrated. Finally we will discuss experimental work, comparing acid centers and acceptor surface states.

4.3.3.2. Acid and Base Sites.

An excellent demonstration of the two competing (acid and base) sites on the solid surface has been provided by Boehm.[92] A group of oxides was examined: TiO_2 (both anatase and rutile), η-alumina, α-Fe_2O_3, CeO_2, and SnO_2. He found that in all cases, on surfaces with water adsorbed, the OH groups fell into two distinct classes, the one more acid than the other. In Figure 4.4 above, the origin of the two forms of OH groups is indicated schematically.

As an example of Boehm's results, consider the adsorption of various gases on hydrated anatase. By D_2O exchange the number of OH groups at the surface of his anatase was determined to be about 460 microequivalents per gram ($\mu eq/g$) of TiO_2. Adsorbates of intermediate acidity, such as acetic acid, were found to adsorb to the extent of 420 $\mu eq/g$. Adsorption of strong acids such as NO_2, however, occurred only to the extent of 210 $\mu eq/g$, showing that half the sites are strongly acidic and will not adsorb strong acids. Adsorption of bases such as ammonia, occurred also to the extent of the order of 230 $\mu eq/g$, showing that half the sites are basic. A test of particular interest was the adsorption (from solution) of NaOH:

$$(OH)_s^- + NaOH = (ONa)_s^- + H_2O \qquad (4.22)$$

This is a typical "ion exchange" reaction. Ion exchange reactions are of great technological importance, for example in exchanging Ca^{2+} for $2Na^+$ for the softening of water, or the exchange of heavy metal ions for Ca^{2+} in zeolites in catalyst preparation. We will not discuss ion exchange reactions in detail in this book, for to a good approximation such reactions can be considered in terms of a simple chemical equilibrium with equilibrium constant defined by the nature of the surface sites.

In the present case [Equation (4.22)], Boehm found that for very dilute solutions (low basic activity), the exchange of Na^+ saturated at about 200 $\mu eq/g$. Thus it is concluded that for the more acidic sites (the protons adsorbed on lattice oxygen ions as shown in Figure 4.4b), Equation (4.22) is driven to the right at sodium concentrations of only about 3×10^{-3} M. At high concentrations the sodium ion was found to replace all the available protons, adsorbing to a saturation at about 415 $\mu eq/g$. That is, for the more basic half of the sites (the OH^- groups adsorbed at and coordinated to a single cation in Figure 4.4b), it requires a sodium concentration greater than 0.1 M to drive Equation (4.22) to the right.

Such results show very clearly the simultaneous presence of two types of sites on this series of clean oxides, as would be expected from the various theories discussed in Section 4.2. Boehm's studies, of course, have to do with Brønsted activity, but, as discussed in the theoretical arguments and as suggested in Figure 4.4a, the same concepts should apply even more obviously to Lewis activity. Both Lewis acid and Lewis base sites should be present on the surface of a dehydroxylated ionic solid. At a gas/solid interface the acidity and the type of acid center are very sensitive to the hydration of the surface. A few examples will be described to illustrate the evolution of acid centers (with heat treatment, for example) at the gas/solid interface and to illustrate the strength of acid (or base) centers as a function of the parameters of the solid.

First, the characteristics of a typical solid acid, alumina, will be discussed. In this case the acid strength and the number of the acid centers is low[93] until the hydrated alumina is calcined at a temperature of at least 200°C and finally becomes low again if the alumina is calcined at a temperature above 900°C. Such behavior is common, as discussed above in Section 4.3.3.1.

Peri[94] in his studies of alumina has identified five different infrared absorption maxima due to OH⁻ groups after calcining at an intermediate temperature (800°C). Figure 4.7 shows his suggested interpretation for the various surface groups. The letters A to E identify the groups in accordance

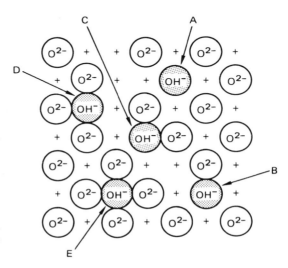

Figure 4.7. Sites of varying acid strength on an alumina surface that arise because of different configurations of surrounding ions. Site A is most acid (with four positive charges surrounding it), site E is most basic (with four oxide ions surrounding it). A $+$ denotes an Al^{3+} ion on the layer below the surface.

with decreasing acidity at the site at which the hydroxyl group is adsorbed. Site A has only cationic nearest neighbors, and so is the strongest acid (a proton is yielded easily); site E has four O^{2-} neighbors, and so is the strongest basic site (attractive to protons).

Further evidence that the dehydration of the surface of alumina/ aluminum hydroxide is still incomplete beyond 500°C was provided by Lee and Weller[95] in studies where the residual hydrogen was detected by exchanging it with deuterium gas: the appearance of hydrogen in the gas phase over the sample revealed the presence of the protons on the surface. Peri[94] suggests a 10% coverage by hydroxyl ions after his 800°C treatment. A detailed study of adsorbed water on alumina by Borello *et al.*[96] has demonstrated a high degree of complexity in its characteristics. In particular these authors show strong lateral interactions of the adsorbed water occur that may tend to make an otherwise heterogeneous (Figure 4.7) surface appear relatively uniform.

A convenient method of distinguishing the type and strength of acid centers on a solid is by the infrared absorption of pyridine on such centers. Parry[97] was able to show that a band from adsorbed pyridine at 1540 cm^{-1} is due to an $(N—H)^+$ group (adsorption on a Brønsted acid site); a band near 1440 cm^{-1} provides information about Lewis acid sites. If the wave number of the ir band is near 1440 cm^{-1}, the band is associated with a weakly adsorbed pyridine molecule (hydrogen bonded), but when this 1440 cm^{-1} band shifts to higher wave numbers, strong adsorption on a Lewis site is indicated. Thus the ir absorption of pyridine indicates distinctively the presence of Lewis or Brønsted sites.

Alumina shows no band at 1540 cm^{-1}, and the band at 1440 cm^{-1} is shifted substantially. Therefore, it is concluded that all strong acid sites on alumina are Lewis sites. In fact, in the case of alumina the presence of protons only leads to modest Brønsted acidity; water adsorption simply poisons the Lewis acid centers.[98] Thus site A of Figure 4.7 is a weak Brønsted acid site, due perhaps in part to the homogenizing effect observed by Borello *et al.*, but site A would be a strong Lewis acid when the hydroxyl ion is removed.

Alumina also shows strong basic sites. It has been proposed[99] that the strong basic centers on alumina are due to isolated oxygen ions beyond the surface plane of the alumina. However, anionic sites at site E of Figure 4.7 (when dehydrated) would be expected to show relatively strong basicity.

Peri[94] suggests that a site of particular interest is one he terms a β site, which combines a site of strong Lewis acidity near a strongly basic site. These are detected by their ability to dissociate NH_3 into NH_2^- and

H^+ even at 50°C. Peri suggests such acid/base combination sites are responsible for much of the chemical activity of alumina toward unsaturated hydrocarbons. He also discusses their appearance on mixed silica/alumina catalysts[153] arising as a result of a particular strained Al—O—Al group. Such active dual sites will be discussed further in Section 10.3.

A typical oxide on which basic sites dominate on the surface is MgO. As discussed in Section 4.2.3, the alkaline earth oxides, being highly ionic with a low Z cation, are basic, with electron donor surface states. The basic sites on MgO do not appear until dehydration is almost complete.[93] The experiments described in Reference 93 were made with Hammett indicators, titrated to determine the concentration of basic sites. Dehydration is almost complete at 400°C. Basic sites appear about 300°C; their number maximizes at about 600°C, and declines at higher temperatures of calcining. This behavior suggests that (as with alumina described above), the oxide ion at the surface provides the basic sites, but at high temperature reconstruction of the surface neutralizes the high potentials observed upon initial removal of the water.

A solid which shows neither strong acid or basic sites is silica.[93,100] Parry[97] with his pyridine test, described above, was able to show that the 1540 cm^{-1} line associated with Brønsted acidity does not appear on silica, and the 1440 cm^{-1} line was present but shifted only slightly, indicating weak hydrogen bonding of the pyridine, rather than strong bonding to Lewis acid sites. According to Boehm's review,[98] at low temperature the OH groups dominate at the silica surface, dehydration at about 400°C results in "strained siloxane groups" which can be rehydrated easily, but heating at temperatures > 500°C allows the strains to be relieved (reconstruction) yielding a passive surface. Hair and Hertl[101] and West et al.[102] suggest either (a) Si—OH or (b) Si—(OH)$_2$ groups or (c) one of these hydrogen bonded to a neighboring O^{2-} may be present.

Using the examples discussed above, we are in a position to compare the proposed direct analogy between the acid/base properties of the solid and the acid/base properties of the same compound in solution. The relation,[39] as discussed in Section 4.2.3, seems correct in its overall trend, but provides an incomplete picture. Consider the examples (MgO, Al$_2$O$_3$, SiO$_2$, and TiO$_2$) that have been discussed. The acid/base properties of these in solution[40] varies from strongly basic for MgO, through amphoteric for alumina, to a slight net acidity for TiO$_2$ and SiO$_2$. As a solid, MgO is dominantly basic, in agreement with the thesis, and the other three which in solution tend to be amphoteric show both acid and basic sites, again in agreement. It appears that while the analogy with the aqueous system suggests the expected "mean" acidity of the sites, it is difficult to predict the simultaneous appearance of the two types of sites of differing acidity. For example, it certainly does not predict that alumina will have both strong acid and strong basic sites[103,104] where with silica both types of sites are weak.

The very active sites on the clean surface of mixed ionic solids will be briefly mentioned as a final topic in this subsection. Of particular interest are the strong acid sites observed on oxides with mixed cations. Tanabe *et al.*[44] have reviewed a long list of mixtures of binary metal oxides and indicated whether the acidity of the mixture is significantly higher than the acidity that would be expected by just summing the contributions of the individual components. In 25 cases a significant increase is observed, in 6 cases not. These workers showed reasonable agreement with their theory that was described in Section 4.2.3, where the dominant parameter is the coordination number.

Shibata *et al.*[105] correlate the acid strength of a series of 22 mixed oxides with the average electronegativity of the cations and find some correlation. They find, of course, increasing acid strength with increasing electronegativity.

As an example of mixed oxide solid acids, we will discuss silica/alumina mixtures. Schwartz[106] has provided an interesting survey of the variation in acidity of SiO_2/Al_2O_3 with varying cation ratios. In particular, he has used the pyridine ir spectrum to distinguish between Lewis and Brønsted activity (see discussion of Parry's work, above). Schwartz finds that the Lewis acidity decreases monotonically as silica is added to alumina, reaching zero at 100% silica. Brønsted activity (zero for alumina) begins when the silica concentration is 25%, peaks at about 75%, and decreases to zero for 100% silica. This observation is not inconsistent with a suggestion by Thomas[107] that Brønsted acidity with this mixed solid arises by a substitution of $Al^{3+} + H^+$ for a Si^{4+} ion in the silica lattice structure. The proton is relatively loosely held, and one obtains strong Brønsted acidity.

4.3.3.3. Ionic Surface States. Ionic surface states, their presence and their activity, have also been examined by chemical tests. The object is to determine whether an ionic surface state will remove one electron from (or donate one electron to) a foreign species, leaving the foreign species as an ion with an unpaired electron. Tench and Nelson,[108] for example, have shown by electron spin resonance and optical absorption that electron transfer from donor states on MgO occurs to ionize certain nitro compounds. They suggest surface oxygen ions which are incompletely coordinated as the donors. As only about 15% of the surface becomes covered by the ions, each surface oxygen presumably does not provide a sufficiently strong donor site, and Tench and Nelson suggest that only those sites are active that have a particularly low coordination (a particularly low Madelung potential). Similar studies on alumina surfaces are reported.

Flockhart et al.[109] have made extensive studies of electron transfer between Al_2O_3 surfaces and adsorbed species. They suggest OH^- is the reducing species below 400°C, perhaps involving a chemical reaction, but oxide ions are the dominant donor above 400°C.

Many authors, as reviewed by Tanabe,[93] have shown similar one-electron transfer from various electron donors to the surface of a mixed silica/alumina surface, as determined by the electron spin resonance signal of the resulting unpaired electron.

An attempt to establish in more detail the connection between acid centers and ionic surface states has been made by comparing the acidity of a series of solid Lewis acids by chemical measurements to the energy of ionic surface states at their surface. The acidity was measured by NH_3 adsorption, the surface state energy by electrical measurements.[110] For these experiments the solid Lewis acids were deposited on the surface of a powdered semiconductor (TiO_2). Then any transfer of electrons from the conduction band of the TiO_2 to the ionic surface states on the Lewis acid surface could be monitored. The powder conductance method (Section 3.1.8) was used to measure the energy level of the surface states (all relative to the "conduction band" of TiO_2).* The ammonia adsorption (Section 3.3.3) was used to monitor the acidity of the deposited acids. A satisfactory correlation between acidity and the energy of ionic surface states was obtained, as indicated in Figure 4.8.

In the above discussion of chemical measurements of acidity and ionic surface states, a very few examples have been chosen for illustration. A broad background of literature is available on the behavior of the surfaces of ionic solids as measured by various chemical techniques covering most materials of interest. The importance of this work cannot be overstated in its role as background material for further understanding of surface effects and for comparison with the emerging theoretical models. However, clearly in the present overview we must restrict ourselves to enough examples to be representative and to show the general agreement of the results of these studies with the concepts emerging from other theories and measurements.

---

* As TiO_2 conducts by polarons, the term "conduction band" is inaccurate, but all that concerns us here is that the same reference energy is used in all cases.

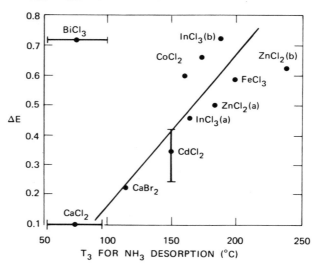

Figure 4.8. Comparison for a series of solid Lewis acids of the acceptor surface state energy ΔE associated with electron capture from the conduction band of TiO_2 with the acidity of sites on the surface, measured by the temperature T_3 required to desorb a gaseous base NH_3.

4.4. Measurements on Adsorbate-Free Covalent or Metallic Solids

4.4.1. Reconstruction on Covalent and Metallic Solids

As briefly discussed in Section 4.2.4, the primary driving force for reconstruction (movement of surface atoms parallel to the surface to form a new periodic structure) on covalent solids is considered to be an electron pairing or dimerization process. This contrasts with the case of ionic solids, where the driving force is the neutralization of electric fields. However, on the covalent solid other forces in addition to electron pairing may cause reconstruction. Appelbaum et al.[51] have performed a pseudopotential analysis of the silicon (100) surface (unreconstructed) and concluded that two surface states are present, the dangling bonds with lobes normal to the surface and a node at the surface atoms, and a second σ bonding state with lobes parallel to the surface, also with a node at each surface atom. These authors suggest that for this case it is most likely the strong interaction associated with the second type of surface state that causes the reconstruction of the surface, and pairing of the dangling bonds is not the motivating force. Harrison[152] reviews the theory of such an interaction and the resulting reconstruction in more detail. Tosatti and Anderson[50] consider the reconstruction to be associated with an excitonic state, stabilized by a longitudinal phonon, thus indicating why the pairing interaction has long-range order.

In any case silicon and germanium both show[49,111] reconstruction. Diamond does not.[112] On the cleaved Si(111) face, a (1×2) reconstruction occurs, which becomes a (7×7) reconstruction when annealed at an intermediate temperature. Above 840°C reconstruction is not observed.[113] Germanium also is dominated by a (2×1) reconstruction at room temperature, converting to a (8×8) and then to a (12×12) pattern beyond a temperature region between about 470 and 570°C.

Auer and Mönch[114] in simultaneous electrical, work function, and LEED studies of the Si(111) face, found that during the transformation of the pattern from (2×1) to (7×7) at about 350°C a decrease of both surface conductance and work function occurs, followed by a substantial restoration of both as the (7×7) pattern develops. The changes in the electrical properties during the transformation were ascribed to nonuniformities in the interatomic spacing arising from the increased disorder. More important, they found that the temperature of the change in reconstruction is a function of the density of steps at the surface. The more numerous were such surface imperfections, the more easily the surface transformed to its high-temperature structure.

It has been observed with silicon[113] that a very small adsorbate concentration (0.04 monolayers of chlorine) suffices to remove the evidence of reconstruction.

On metallic solids reconstruction is not common.[115] For example, Kesmodel and Somorjai[116] find no reconstruction on the Pt(111) surface to an experimental accuracy of 5%. Aberdan et al.[117] find no reconstruction on the (100) or (110) planes of aluminum. On the other hand, as discussed in Section 7.3.2, there is evidence that the Pt(110) surface shows a (1×2) structure when clean.

Aberdan et al. find relaxation of the surface (the movement of the surface plane toward the crystal) of 3–5% for the Al(110), although none on the (100). Appelbaum and Hamann[51] suggest that such relaxation must also occur on silicon, and argue that the Si(111) surface should relax by 0.33 Å. Phillips[118] analyzes such distortion and relates it to the covalent radii of compounds.

In summary, there is good reason to believe that the chemical and electronic surface behavior of Group IV semiconductors is intimately bound up with the reconstruction and relaxation on the surface. Without such a process the surface would be much more reactive chemically, and the intrinsic surface states would not be separated into two bands, one occupied and one unoccupied, but remain as a half-filled band of states associated with the dangling unpaired electrons.

4.4.2. Electrical Measurements of Intrinsic Surface States on Covalent Solids

Most of the very early work on clean surfaces was done on germanium surfaces that were either argon-bombarded or simply outgassed at high temperature. A high density of acceptor surface states was found close to the valence band; the surface was always p type independent of the bulk doping. The first work on cleaved germanium surfaces[119,120] confirmed this. The method used was the Statz channel method (Section 3.1.7), which permitted the occupancy of surface states to be determined while the Fermi energy was varied from a value near the valence band to about 0.15 eV above the valence band. It turned out that varying the Fermi energy over this span showed no change in the surface charge, which indicated that the density of states in that energy span is negligible. Thus all the occupied acceptor surface states were found to be below the valence band edge, and a gap of at least 0.1 eV was determined to be present between the occupied and any higher-energy (unoccupied) surface states.

The density of these acceptor surface states below the valence band edge was found to decrease as a function of oxygen adsorption. With oxygen exposure to about 10^{-6} Torr min, the density changes negligibly, but with further exposure the density decreased. Such a decrease has been found typical of covalent solids—in fact in more recent work the disappearance of states upon adsorption is often used to distinguish surface from bulk states in the surface spectroscopy measurements. A plot of the density of surface states[120] versus the oxygen exposure is shown in Figure 4.9. Samples PNP-3 and PNP-4 of Figure 4.9 were not doped sufficiently to shift the Fermi energy, but samples PNP-5 and PNP-6 were heavily doped, and it could be shown as discussed above that over a span of 0.1 eV above E_v no acceptor surface states were present before or after oxygen absorption. Margoninski[121] suggests the region of change ($>10^{-6}$ Torr min) corresponds to a second layer of oxygen. If this is true, the acceptor surface states measured are relatively unreactive with oxygen.

Subsequent studies, discussed below, suggest that the initial surface state density observed in this work was too low, probably due to an initial exposure of the sample to gases during the cleavage process. If this is the

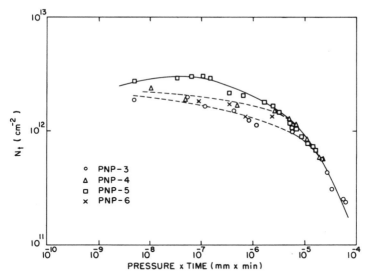

Figure 4.9. Density of acceptor states below E_v on the cleaned germanium surface (111). For samples PNP-5 and PNP-6 it was shown a gap with no surface states at least 0.1 eV wide was present above E_v.

correct interpretation, contrary to these observations, there is a possibility of surface states above the valence band (in the region 0.05–0.15 eV above the valence band) for the entirely clean surface. However, for consistency, if these were present, they would have to be the first surface states to react and they would have to be highly reactive. Unfortunately these measurements using the Statz channel method, where the density of states can be determined, have not been repeated, and so there is no other direct electrical information about the density distribution of states on a clean surface. Other electrical methods, to be discussed below, can only provide the density of states when a model is assumed, and the spectroscopic methods are too insensitive to determine state density with the required energy resolution.

Allen and Gobeli[122] measured the effect of surface states by examining the work function ϕ and photoelectric threshold $E_e - E_{vs}$ as the bulk Fermi energy was varied. This technique when used on germanium is unrewarding except to confirm that E_F is close to the valence band upon cleavage. So many surface states are present near the valence band that the Fermi energy is pinned at the surface near the valence band independent of the doping of the crystal. Thus the work function and the photoelectric threshold (the energy necessary to excite an electron from the valence band to the vacuum level) are found identical for all doping.

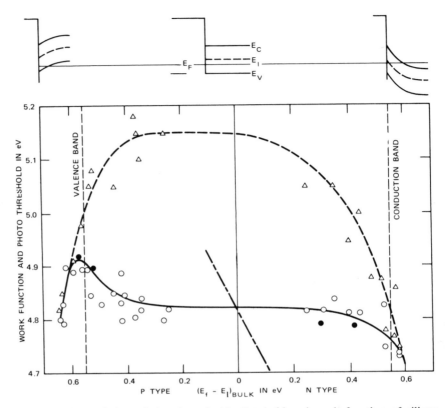

Figure 4.10. Variation of the photoelectric threshold and work function of silicon (111) with bulk Fermi energy, as determined by Allen and Gobeli (Ref. 122). The dotted line is the photoelectric threshold, the solid line is the work function. E_I is the midgap energy.

The technique applied to silicon surfaces is more rewarding. The photoelectric threshold and the work function, determined by Allen and Gobeli as a function of bulk doping of a series of cleaved silicon samples, is shown in Figure 4.10. The symbols used in the figure are those of the authors: E_w is work function, ϕ_{AP} is the photoelectric threshold. Equation (4.21) is used to obtain the band bending V_s over the region where the photoelectric threshold is $E_e - E_{vs}$. The value of μ is known from the bulk properties of course (Figure 2.2). Over a wide range of doping it is observed in Figure 4.10 that neither the work function nor the photoelectric threshold vary, so

$$(E_e - E_{vs}) - \phi = \text{const}$$

and from Equation (4.21) it is concluded that V_s increases linearly with decreasing bulk Fermi energy μ over a wide range. This will occur if a large density of surface states is present such that they become charged to a given level and "pin" the surface Fermi energy independent of the bulk Fermi energy. [See the discussion following Equation (2.14).] Since the pinning occurs by experiment in a region somewhat below the midgap position, it is concluded that a high density of partially occupied surface states must be located in this region.

Conductance measurements were used by Handler[123] to determine the value of the bulk Fermi energy which would lead to no surface conductance (flat bands) upon cleavage of silicon. They found this occurs when the bulk Fermi energy is 0.23 eV below the midgap position. This result agrees with the midpoint of the surface states as determined by the Allen and Gobeli method. Henzler[124] and Henzler and Heiland[125] measured the field effect and computed the total density of states assuming a model for their distribution. As reinterpreted by Davison and Levine,[1] the results could be fitted to a distribution with two discrete levels. With this model the total density becomes $2 \times 10^{14}/cm^2$, and the position of the Fermi energy when there is no charge on the surface is $E_0 = 0.25$ eV below the midgap position. Thus a satisfactory picture of the Si(111) surface is emerging.

Many other workers have contributed to the study of the surface states on the clean surface using electrical measurements. Most of these are reviewed in the study by Davison and Levine.[1] The objective here is simply to show that such electrical measurements can provide information of the surface states associated with the clean surface, and, in fact, can provide a reasonably clear picture of the distribution. But the important contribution, when compared to the spectroscopic techniques below, is their ability to provide surface state energies under certain conditions to the nearest tenth or even hundredth of an electron volt.

Haneman[75] has provided an excellent review and analysis of the use of electron spin resonance (ESR) to investigate the presence of unpaired electrons in dangling bonds on various materials. Silicon and germanium produced ESR signals probably associated with the surface. It is found that crushing silicon at room temperature in various gases or vacuum leads to about $2 \times 10^{14}/cm^2$ unpaired electrons, while germanium crushed in vacuum at 77°K has a factor of 10 fewer unpaired electrons. In each case exposure to oxygen irreversibly affects the signal. With germanium the signal disappears at 300°K, but is only desaturated, not removed, by exposure to 10^2 Torr oxygen at 77°K. With silicon, extremely high oxygen exposure is required to affect the signal. Such behavior is difficult to reconcile with the expected chemical activity of unpaired electrons, and the source of the signals is not clear. Thus ESR measurements show unpaired electrons are present on covalent semiconductors, but the source of the electrons detected is not known. In fact, the technique provides new questions about the behavior of these materials that were thought to be reasonably well understood.

4.4.3. Measurement by the Surface Spectroscopies

As discussed earlier, the surface spectroscopies can be used to detect surface levels and orbitals far from the Fermi energy. Because of its technical importance in semiconductor devices, silicon has been studied in the most detail. Occupied surface states are revealed both by FEM and UPS but are hard to distinguish from valence band states. FEM measurements on the (111) surface by Lewis and Fischer[126] suggest a band of levels below E_F and overlapping E_v. Many authors[80,127-129] have reported a shoulder on the UPS spectrum where, of course, the main peak in the spectrum arises from emission of valence electrons. The shoulder is identified as a surface state just at or below the Fermi energy using the standard criterion[130,131]—the peak disappears as oxygen gas is adsorbed. However, this means of identification is not entirely trustworthy. As Lewis and Fischer comment,[126] fading of a UPS peak assigned to the bulk d bands on nickel also occurs, so the fading of a peak upon adsorption may not be sufficient as proof that the peak is a surface state.

Sebenne et al.[132] made UPS measurements on silicon with a high-resolution monochromator and electron energy analyzer. They concentrate on the photoelectron spectrum in the region of the valence band energy, studying silicon with varying bulk doping. For a reference material they examined a heavily doped p-type sample for which the occupancy of any surface states above the valence band would be negligible. The extra photoelectron emission for samples with lower p-type doping and for n-type material was ascribed to electrons in the surface states below the Fermi energy in each case. They conclude there is a band of energy levels starting at about 0.27 eV above the valence band edge and extending into the valence band. This seems the best evidence to date showing the position and shape of the occupied dangling bond surface state on Si(111).

As discussed in the theoretical arguments above (Figure 4.5), there is reason to believe that another band of surface states that is normally empty should be present at some energy region above those occupied states. Chiarotti et al.[133] summarize the evidence for such a band of empty levels. The results of Rowe et al.[134] using ELS, of Müller and Mönch[135] using photoconductivity, and Chiarotti et al. using infrared absorption all suggest a band of levels above the occupied surface states. In particular the referenced article by Chiarotti et al. compares an infrared peak at 0.45 eV to an ELS peak at 0.52 eV and concludes they are identical. Both peaks disappear when the silicon reconstruction changes from (2×1) to (7×7) (see Section 4.4.1). Thus we conclude that the Si(111) face with (2×1) reconstruction has (as in Figure 4.5) a split set of surface state bands: one that is normally occupied overlapping the valence band, and one that is normally unoccupied about 0.5 eV higher. From the electrical measurements we conclude that with no charge on the surface the Fermi energy is between these bands, at 0.23 eV below the midgap position.

The geometrical orientation of the lower (occupied) surface state orbital has been explored by Traum, Rowe, and Smith.[136] They have investigated the angular distribution of the photoelectrons emitted from the surface state. If the surface state were a simple dangling bond surface state at the (111) surface, oriented normal to the surface, there should be no preferred azimuthal angle. In other words, the electrons emitted at an angle of 30° from the normal should show no intensity dependence upon the azimuthal angle. However, it turns out that there are three directions of high-intensity emission: the $(\bar{2}11)$, the $(1\bar{2}1)$, and the $(11\bar{2})$ directions. These are the directions, as viewed from the top, at which the surface atoms bond to the first substrate layer of atoms. Thus it is concluded that the observed occupied surface state is not a pure dangling bond surface state, but contains a substantial contribution from the bonds between the first and second layers. Strangely enough, the part that disappears upon adsorption of residual gases is not the isotropic part of the contribution which may represent the dangling bond, but that part with strong angular dependence. It is suggested as a possibility that the directional characteristics are associated with diffraction effects rather than actual orbital shape. In any case, there is an awkward inconsistency here which must be resolved. By reasonable expectation the isotropic dangling bond orbital should be the orbital most affected by the adsorbed gases.

In summary, we have a fairly complete picture of the silicon bonding orbitals, and in later discussions we can note the reactions between these orbitals and the orbitals of impinging gas molecules. Less information from the surface spectroscopies is available for germanium. Surface states on germanium as measured by UPS have been reported by Murotani et al.[129] and by Eastman and Grobman[80] with the occupied levels so detected lying just below the Fermi energy. Shepherd and Peria[137] report FEM measurements suggesting an occupied level about 0.18 eV above the valence band on the (100) face, of density about $6 \times 10^{12}/cm^2$. Ernst and Block[138] report the presence of unoccupied states as observed by FIM on both the (111) and (100) faces. These results are in disagreement with the electrical studies. On the other hand, the ELS studies of Ludeke and Koma[139] agree more closely with the electrical measurements, suggesting the acceptor levels are at the valence band edge.

In the case of metals, when one is interested in available bonding sites, not only surface but also bulk orbitals play an extremely important role (see the next chapter). However, the description of these bulk orbitals will be postponed until Section 5.2, preparatory to discussing the actual bonding configurations of adsorbates on metals. In the present section we will restrict the discussion to "surface states" on the clean metals, energy levels localized at the surface and arising because of the presence of the surface discontinuity. As discussed in connection with Figure 4.3, such surface states can be considered to be moderate perturbations of the local density of states (LDS) at the metal surface. Such surface state perturbations may cause a higher density of states with orbitals directed out from the surface and so affect the bonding character.

The tungsten surface has been studied most intensively. Using UPS, Murotani et al.[129] measured the spectrum on polycrystalline W and found a peak due to a surface state at 0.6 eV below the Fermi energy. It is considered a surface state since it disappears with time (adsorption). The position of this surface state peak and the position of the center of the bulk d-band peak are claimed to be in satisfactory agreement with the calculations of Petroff and Viswanathan[140] that predicted a surface state peak at 0.6 eV and predicted the bulk d-band peak at 1.6 eV below the Fermi energy. Other workers observe this surface state plus another deeper level. Waclawski et al.[141] in studies of the W(100) surface report some angular dependence in the two peaks they identify as surface states (at 0.4 and 2 eV below the Fermi energy), but the behavior is complex and is not analyzed. Feuerbacher and Christensen[142] and Feuerbacher[143] describe more detailed UPS investigations on the (100), (110), and (111) faces of tungsten. They only monitored electron emission normal to the surface, which made the identification of surface states easier, because a larger fraction of electrons emitted in this direction come from surface states. They report a strong surface state emission on the (100) and (110) faces at an energy near the Fermi energy, and on the (111) face a similar though smaller peak plus larger surface state peak at about 2–3 eV below the Fermi energy.

Plummer and Gadzuk[144] report the same two surface states on clean tungsten (100) using the FEM technique. Again the peak at 0.4 eV or so disappears with a small adsorption of gas, including CO, H_2, N_2, O_2, and even Kr.

It should be noted that the appearance of these two peaks is in agreement with the predictions of Forstman and his co-workers[34-36] and many others as discussed in Section 4.2.1.2c. These models predict a surface state associated with the d band but appearing at certain crossover points of the s and d bands.

Other materials have not been studied so thoroughly. Kasowski[145] reviews work that shows that copper surfaces show no surface states, in agreement with his and other models. Murotani et al.[129] report that surface states on molybdenum are analogous to those on tungsten, with both the bulk d band and the measured surface state higher (closer to the Fermi energy) by 0.2 eV.

As clear from the above discussions, experimental information on surface states on covalent and metallic solids is not abundant, but is sufficient to provide a general picture of the surface and to provide confidence that the theoretical models are able to present a fairly good description of the surface orbitals on a clean surface. However, as Feibelman and Eastman[146] point out, caution must be exercised regarding a quantitative comparison with theory. Not only are there approximations in the theory, but also there are experimental problems. Polarization or screening effects may affect the apparent ionization energy if comparison with the ionization energy of a free atom is of interest, and absorption of energy by plasmons may also affect the observed energy distribution. These problems are discussed in more detail in connection with adsorbate surface states (Section 5.4.1), for in that case the problems are even more serious.

5 | Bonding of Foreign Species at the Solid Surface

In this and all subsequent chapters, we will be concerned with foreign species at the surface of a solid and how these foreign species affect the chemical and electrical properties of the surface. Here and in the next chapter the emphasis will be on describing the energy levels associated with the steady state system, rather than on describing rate processes. In Chapters 7–10 the emphasis will be switched to the kinetics of adsorption and desorption, of electron transfer, and of chemical reaction.

In the present chapter the objective is to describe the various ways a foreign species can interact (bond) with the solid surface. The discussions will be dominated by the theory of the chemical bond as modified to include the presence of the solid. But in addition other factors, such as the positioning of surface atoms (both adsorbate and substrate), will be discussed, as will be the influence of Franck–Condon effects and the influence of coadsorbed polar molecules. Each of these factors affects the surface state energy level associated with local adsorbate/solid chemical bonds. In fact, it will become evident that any local bonding interactions substantially change the energy, capture cross section, and the distribution of surface states. Thus knowledge of the local bonding of adsorbates to a solid is needed not only in describing the chemical reactions of the adsorbates but also in describing the influence of the adsorbates on the electrical properties of the solid.

5.1. Reconstruction and Relocation in Bonding

In order to understand the interaction between a particular adsorbate and a solid surface, it is desirable to know where the adsorbate atoms and host lattice atoms are located relative to each other. For example, with the help of valence theory such information will provide an indication of the type of bonds formed. Again changes in the configuration (with temperature, amount adsorbed, etc.) will indicate corresponding changes in the surface state energy levels associated with the bond.

The location and relocation of a fractional monolayer of surface atoms is determined with the aid of the surface spectroscopies, in particular, LEED and recently ESD. Unfortunately, although the generalization can be made for many systems that both adsorbate and host atoms occupy other than normal lattice positions, detailed knowledge is much harder to obtain. LEED can immediately provide information about superstructure, that is, about arrangements that show long-range order of periodicity different from that of the host surface. However, it is much more difficult to extract information from LEED about (a) adsorbate position or movement at low coverage before ordering can occur, (b) host atom relocation normal to the surface, or even (c) exchanges of host and adsorbate atoms that leave the periodicity of the surface unaltered. Detailed analysis of peak heights and intensities has been used successfully to provide information in the latter two cases, but the analysis is difficult.[1,2]

Although it is necessary for understanding of surface phenomena to recognize that displacement of surface atoms from lattice positions during adsorption is common, it is still unrewarding to attempt a general classification for predictive purposes. We still have too few cases that have been completely analyzed. Thus in this section we will merely present data somewhat randomly to indicate the generality of the phenomenon.

First, consider a few examples showing the relocation of substrate atoms. May[3] provides an excellent review of relocation processes on adsorption. He points out for example that usually on covalent semiconductors such as silicon or germanium, adsorption relieves the reconstruction characteristic of the clean surface and restores the substrate atoms to their normal position. Henzler[4] has shown that about 0.1 monolayer of oxygen on germanium suffices to remove the reconstruction. Another example is the movement of the host nickel atom upon the adsorption of oxygen. Here the LEED pattern moves toward that of NiO, indicating reconstruction. The adsorption of CO on tungsten also causes reconstruction, as known from LEED, FEM, and FIM work. There is no chemical reaction associated with the reconstruction: the CO desorbs intact. Sachtler and his group[5,6] report extensive work showing the movement of metal ions at surfaces upon adsorption of foreign species. Riwan et al.[7] show from measurements that in the initial phases of oxygen adsorption on Mo(100) the work function decreases, indicating the oxygen becomes located below the surface Mo atoms.

Another process of importance is the positioning (superstructure) of the adsorbate atoms themselves. For example, Demuth et al.[8] have thoroughly analyzed S adsorbed on Ni and show the S occupies highly coordinated sites in hollows of the Ni surface, and, in fact the Ni–S bond lengths are shorter than those of the compound NiS. As an example of relocation of an adsorbate with pretreatment, in Section 5.4.3 we will discuss in detail the changes of position of adsorbed CO on W with changes in temperature and amount adsorbed. As an example of the use of ESD to determine sorbate position, we have the work by Yates' group, who have shown[9] (see Section 3.2.9) that oxygen adsorbate ions on tungsten occupy different locations depending on the temperature and the amount adsorbed. With minimal adsorption the oxygen ion occupies an interstitial position between the W atoms on a (100) surface. With more adsorption the oxygen ions bond to one W surface atom, but depending on the temperature pretreatment, the position can be directly over a W atom or bonded at an angle indicating the bonding orbital is not normal to the surface.

The apparent ease during and after adsorption with which relocation and reconstruction occurs has several implications. In the experimental studies to be described it will be shown that the surface state energy levels are sensitive to the position of the adsorbate relative to substrate atoms. It will be shown that changes in the surface state energy level spectrum can be correlated with shifts in the LEED pattern of the adsorbate on the surface. With adsorbate location so easily changed, surface state energy levels cannot be considered stable. Finally, the ease of atom movement becomes of major concern in the discussion of Franck–Condon effects in Section 5.5.1 below.

5.2. The Semiclassical Model of Bonding: The Surface Molecule

5.2.1. Surface Molecule Versus Rigid Band Model

As described in Chapter 1, there has been through the years a continuing controversy between scientists who have preferred a strictly localized "surface molecule" description of an adsorbed species and those who have preferred the so-called "rigid band model." In the pure surface molecule model, the adsorbate is assumed to interact with one or a few substrate atoms, forming bonds according to a molecular orbital theory or other chemical bonding theory. The interaction is described without invoking the band characteristics at all. In the rigid band model, all local interaction is neglected, and the adsorbate is viewed as a simple electron acceptor or donor which removes or donates an electron to the bands of the solid. Neither the electronic energy levels of the adsorbate nor the energy level distribution in the bands of the solid are considered to be altered in the adsorption process. This rigid band model is relatively easy to visualize, whereas the surface molecule model has all the variability of bonding theories in chemistry. A convenient model to help visualize the surface molecule model, one that is used in some cases as the starting point for quantum calculations is: (a) a host atom is removed from the surface, (b) a molecule is formed composed of the host atom and the adsorbate atom, (c) the "molecule" is reattached to the surface.

There are good arguments that particular cases can best be described by one approximation or the other. As an example suggesting a rigid band model, Butler et al.[10] found that a saturated amalgam of platinum in mercury showed no excess activity in hydrogen ion reduction (electrochemically) over mercury alone. But the rate should have been 10^5 higher if the platinum atoms were acting as localized centers for proton adsorption. Platinum is an excellent electrocatalyst for this reaction, mercury is not. Thus they conclude that the activity cannot be associated with individual atoms and local bonding. Similarly Rand and Woods[11] find that Pd/Au and Pt/Rh alloys have properties at the solid/liquid interface as if one homogeneous phase were present. Ponec,[12] on the other hand, studying catalysis at the gas/solid interface, favors a surface molecule approximation for alloy effects. This view is supported by the behavior of Na/Ag alloys or of Pt/noble metal alloys where he observes that the alloy is more active than the sum of the components. He ascribes this exceptionally high activity to certain geometrical configurations of the catalyst atoms that develop at the surface when the catalyst metal is diluted by an inert metal. Sachtler and van der Planck[13] have studied Cu/Ni alloys. They note that with an 80 Cu/20 Ni alloy the d bands of Ni should be filled so adsorbing hydrogen cannot interact with d electrons. But experimentally the alloy still adsorbs hydrogen. They suggest the hydrogen must be adsorbing on individual nickel atoms and hence claim this as proof of the "surface molecule" approximation. Further evidence is offered by Soma and Sachtler[14] who find the desorption energy and the ir absorption frequency of CO on Pd/Ag alloys is the same as observed on pure Pd, and conclude that a Pd—CO local bond is indicated.

As will be described in the next sections, the best model is a complex compromise between the two extremes, with more or less emphasis on the one or the other depending on the details of the sorbate/sorbent system. If the sorbate/solid interaction is very weak, the quantum calculations will suggest that the rigid band model is best. If the sorbate/solid interaction is very strong, the quantum calculations will suggest the surface molecule model is preferred. Thus when weak interactions are encountered at the solid/liquid interface, the rigid band model provides the best description, whereas for the reactions studied at the solid/gas interface involving energetic adsorption, the surface molecule model is best.

Based upon such interaction energy concepts, a few generalizations regarding the choice of model can be made. Both in the case of adsorption on clean covalent (or metallic) solids and in the case of acid/base reactions on ionic solids it is generally best to describe the system in terms of the surface molecule approximation, and then introduce the effects of the bands of the solid. In the case of clean covalent (or metallic) solids, the high reactivity of the "dangling bonds," as described further below, almost inevitably leads to strong interaction. In the case of acid/base bonding, there has been no discussion in the literature. It has been implicitly assumed without discussion that bonding is best represented by a pure surface molecule picture. If a basic molecule interacts with a solid acid, donating two electrons to the cationic orbital, the resulting bonding orbital is tacitly assumed to be far below the conduction band, so that any interaction with the band can be neglected. The equivalent is assumed for adsorption of acids on solid bases. Such a shift of the bonding orbitals was discussed in the preceding chapter. We must await further quantum calculations before we can determine when, if ever, the surface molecule approximation fails for acid/base bonding.

On the other hand, in the case of ionosorption, either on ionic solids or on "real" covalent solids, the rigid band model has been universally used. On an ionic solid the local interaction of an electron donor or acceptor (a reducing or oxidizing agent) is expected to be weak, so the rigid band model is not unreasonable, at least as the first approximation. On covalent "real" solids, that is, on covalent solids that have been exposed to room air, again the approximation may be reasonable, for if the surface of the solid is reactive, it will have oxidized and hydrated and hence the adsorption of interest will be occurring on an ionic solid, the oxide layer. But it must be emphasized that even in these cases the rigid band model is only a first approximation, and some local bonding is almost inevitable, whether the bond formed is electrostatic, crystal field, hydrogen bonding, covalent, or acid/base in nature. The ionosorption case, as described by the rigid band model with local bonding treated as a perturbation is discussed more in Section 5.5.1.

In the present section we will discuss the surface molecule model and provide examples of how one extrapolates from chemical valence theories to describe bonding to a surface. If it is recognized that the models are oversimplified—that for an accurate model the perturbation of band formation must be included—these models will provide an excellent first approximation for the discussion of strong sorbate/sorbent bonding.

5.2.2. Adsorbate Bonding to Covalent or Metallic Solids

As described above, the bonding of atoms to covalent solids has all the diversity of chemical valence theory, plus a few added complications. In the present section, we will attempt to (a) point out some of the "added complications" that are associated with the fact that the orbitals (dangling) available at a surface are different from those of a free atom, (b) introduce the concept of π bonding to a surface which will be of major concern in later discussions of adsorption and catalysis, (c) indicate how electro-negativity differences affect the bonding, and (d) review semiclassical calculations based on simple bonding concepts that have been used for adsorbate/solid interaction. In essence we are trying to pick out topics in bonding that are somewhat unique to the adsorbate/surface bond and avoid topics that are equally well treated in terms of molecular valence theory.[15,16]

5.2.2.1. Orbitals Available on the Solid. Bonding to dangling hybrid s–p orbitals on covalent solids with the formation of a simple covalent bond between the adsorbate and a surface atom on the crystal is undoubtedly the most common form of bonding on nontransition covalent solids such as silicon and germanium. Here orbitals with essentially unpaired electrons are available at the surface. The concept is a direct carryover from the concepts of Section 1.2 and needs no further elaboration. For example, Ludeke and Koma[17] conclude on the basis of ELS that on germanium and silicon the adsorption of oxygen leads to a simple double-bonded oxygen atom (Figure 1.2b). Bootsma and Meyer[18] measured room temperature adsorption of HCl, HBr, H_2S, H_2Se, NH_3, and PH_3 on Ge and Si and found (by ellipsometry) saturation after a monolayer, which indicated simple bonds with negligible relocation of host atoms. Fahrenfort, van Rijen, and Sachtler[19] concluded that upon adsorption of formic acid, HCOOH, on metals the adsorbate has all the characteristics of the formate compound of the metal involved (as measured, say, by ir). In some cases with transition metals also the concepts are similar. For example, Bond[20] describes hydrogen adsorption on transition metals in terms of simple covalent bonding between hydrogen orbitals and d orbitals perpendicular to the surface.

Bonding to d orbitals, however, is in general more complex, and because transition metals play such an important role in surface chemistry the topic warrants more detailed discussion.[21] In order to understand better the bonding of adsorbates to d orbitals, it is desirable to determine what d orbital lobes will be available at the surface of the metals for bond formation. These are identified by simple crystallography. From a knowledge of the crystal structure and the crystal face under study and assuming no relocation of atoms, it is conceptually straightforward to determine for a surface atom the direction and occupancy of the "dangling orbitals." Johnson[22] has classified metal catalysts based on this approach. He considers the surface usually exposed to be the (111) face on cubic close-packed and hexagonal close-packed structures, and the (110) face on body-centered cubic crystals. Then:

Class I: Mo, W have a vacant d orbital perpendicular to surface

Class II: Rh, Ir, Ru, Os, Tc, Re have a vacant d orbital 36–45° to surface

Class III: Fe, Co, Ni, Pd, Pt have partially occupied orbitals, 30–36° to surface

Class IV: Zn, Ga, Cd, In, Ge, Sn, Pb show asymmetry in the d shell

Johnson relates the high catalytic activity of the Class III group to the ability of these metal surfaces, with the indicated orbitals available, to form bonds with olefins (unsaturated hydrocarbons).

5.2.2.2. π Electrons on the Sorbate and Their Interaction with a Solid.

A particularly important class of surface bonding to transition metals is the bonding of the π electrons of unsaturated hydrocarbons (olefins or aromatics) to the surface of transition metal catalysts. Such bonding is critical to hydrogenation, dehydrogenation, or isomerization reactions (the addition or removal of hydrogen from the molecule or the movement of hydrogen from one carbon atom to another), or to oxidation reactions of hydrocarbons.[23-26] To provide a basis for future discussions we will develop in this subsection the Hückel representation of π electrons on the simplest olefin, ethylene, and will discuss its π bonding to platinum.

Ethylene has the formula

Three of the orbitals on each carbon atom are coplanar: the two that overlap the hydrogen atom orbitals and one of the two bonding orbitals in the double bond between the carbon atoms. These all form σ bonds. Only one of the orbitals extends in a direction normal to the plane of the paper, the orbital that participates in the second of the carbon/carbon double bonds. The two electrons in this bond, one from each carbon, are the π electrons.

To provide a description of the molecular orbital of the π electrons we will use the Hückel formalism introduced in Chapter 4, Equations (4.1)–(4.7). The parameter α is still the Coulomb integral, and the parameter β is the resonance integral describing the overlap interaction between the two π electrons. The summation in Equation (4.3) now is only over two atoms, so from Equations (4.3)–(4.7) we have immediately the two equations

$$c_1(\alpha - E) + c_2\beta = 0 \tag{5.1}$$

$$c_1\beta + c_2(\alpha - E) = 0 \tag{5.2}$$

The solution to these equations is

$$E = \alpha \pm \beta \tag{5.3}$$

$$c_1 = \pm c_2 \tag{5.4}$$

The normalizing condition

$$\int \psi\psi^* \, d\tau = 1 \tag{5.5}$$

leads to

$$c_1 = c_2 = 2^{-1/2} \tag{5.6}$$

and from Equations (4.1) and (5.3) the two wave functions and the associated energies are

$$\psi_1 = 2^{-1/2}(\phi_1 + \phi_2) \qquad E = \alpha + \beta \qquad (5.7)$$

and

$$\psi_2 = 2^{-1/2}(\phi_1 - \phi_2) \qquad E = \alpha - \beta \qquad (5.8)$$

The atomic orbitals ϕ_1 and ϕ_2 are shown in Figure 5.1a, the molecular orbitals ψ_1 and ψ_2 arising from Equations (5.7) and (5.8) are shown in Figure 5.1b. As β is negative, the MO ψ_1 has the lower energy E and is the bonding orbital occupied by the two electrons, one contributed from each of the carbon atoms. This occupied bonding orbital is represented by the solid lines in Figure 5.1b. The unoccupied antibonding orbital ψ_2 is represented by the dotted lines.

In Figure 5.1c is shown how this configuration of orbitals can match with a platinum atom at the surface of a metal. Two hybrid dsp orbitals of the platinum are shown, one of which, the $5dsp$ orbital, is occupied; the other, the $5d6s6p^2$, is unoccupied. Again solid lines indicate occupied orbitals, dotted lines unoccupied orbitals. The point is that for all these orbitals there is positive overlap. Thus the electrons from the bonding orbital in the ethylene can be shared with the unoccupied orbital of the platinum, the electrons from the occupied orbitals of the platinum can "back-bond" and be shared with the unoccupied orbitals from the ethylene. Thus with such d orbitals available on the surface, the platinum metal provides an adsorption site which forms a very strong bond with this particular gas molecule. In fact, as observed experimentally and quantitatively rationalized by the BEBO model (Section 5.2.2.4) the bonding is so strong that the ethylene will dissociate.[27]

Transition metals with their d electron orbitals are particularly suitable in orientation, in occupancy, and in energy, to accommodate adsorbate π orbitals that originate from $2p$ electrons.

Another class of π orbitals on unsaturated hydrocarbons are the π orbitals on the aromatic hydrocarbons. Gland and Somorjai[28] suggest such aromatics always tend to adsorb on transition metals by π bonding. Benzene, C_6H_6, provides the simplest and most common example:

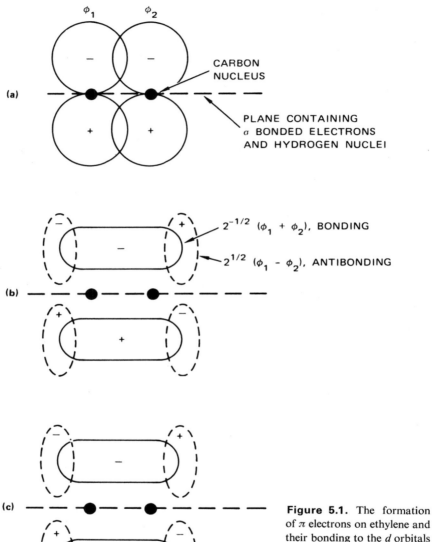

Figure 5.1. The formation of π electrons on ethylene and their bonding to the d orbitals of a transition metal: (a) atomic orbitals of π electrons in ethylene, (b) molecular orbitals of π electrons in ethylene, and (c) overlap leading to bonding between π electrons of ethylene and d electrons of platinum.

Again each carbon atom takes part in three σ bonds, one to a hydrogen atom and one to each of the neighboring carbon atoms. The fourth orbital on each carbon extends perpendicular to the plane of the paper, and the overlap between these orbitals provides π bonding around the ring. In the formal description above three alternating double bonds are shown, but actually the ring is entirely symmetric, and the better formula is

where the positions of the carbon atoms and the hydrogen atoms are understood, and the circle in the center represents a bonding orbital for the six π electrons. The bonding orbital is smeared (by a Hückel model) in the same way as the bonding orbital of the ethylene π electrons was smeared in Figure 5.1b. When the benzene molecule, or an equivalent aromatic (unsaturated ring compound) lies flat on a metal, the π orbitals may overlap the dangling bonds of several surface metal atoms.[24,25] Catalytic reactions involving these aromatics on the surface of metals are visualized as first π bonding of the molecule, then perhaps the molecule can flip and form a σ bond with a particular metal atom. Hydrogen atoms can be acquired or lost, or carbon/carbon bonds can be broken, while the molecule is alternating between π and σ bonding.

Another variation on π bonding is that suggested for CO adsorption on a transition metal.[29,30] By infrared spectroscopic studies it is concluded that the CO is linked to the transition metal surface by a σ bond, with back donation from a filled d orbital to the vacant π orbital on the CO molecule. Queau and Poilblanc[29] observe that the presence of a Lewis base at the surface increases the electron availability in the metal and strengthens the back π bonding.

5.2.2.3. Electronegativity. The tendency of adsorbates to donate or accept (polarize) electrons in a bond, the electronegativity, provides another important variable in the bonding pattern. Sargood *et al.*[31] have examined this factor in detail, particularly for the adsorption of a series of adsorbates on tungsten. In particular they studied the surface double layer due to the adsorption and found a linear relation between the electronegativity of the adsorbate (Cs, Ba, U, I, Br, Cl, O, H_2O) and the resulting change in the work function. The radius of the adsorbate ion and the polarizability of the adsorbate had to be factored out to obtain the relation. Such a relation shows the expected tendency of the electrons in the bonding orbitals to shift toward the more electronegative species.

Naturally the strength of the bond and the partial charge on the atoms in the "surface molecule" are strongly dependent on the electronegativity difference between the host and adsorbate atoms. The analysis is equivalent to molecular valence theory.

5.2.2.4. BEBO Calculations. Weinberg, Merrill, and Deans[27,32,33] have utilized the surface molecule approach to provide quantitative calculations of adsorption characteristics, using what they term the BEBO (bond energy-bond order) approach. They use a semiempirical method: estimating for a site on the solid the orbitals as predicted by crystal field theory, estimating the energy of the bonds likely to form to a given adsorbate as observed in analogous bulk compounds, and estimating the variation in this energy with bond order (single, double, triple bonding). For, example Weinberg[33] discusses the dissociation of nitrogen on platinum. This requires the severance of the $N \equiv N$ triple bond, forming eventually three bonds in an $N \equiv Pt_2$ complex. They are able to show that the energy of the system increases rapidly during the disruption of the first $N \equiv N$ bonds. If the energy is plotted as a function of the bond order of the Pt—N bonds, there is a maximum in system energy at a bond order between 1 and 2, with the energy of the system decreasing again to less than the energy for physisorbed $N \equiv N$ when the $N \equiv Pt_2$ configuration is reached.

Bond[34] has offered a similar semiempirical calculation for hydrogen bonding to tungsten using the Pauling relations for bond strength and found reasonable agreement. However with the emergence of the cluster and other quantum approaches to be described, these semiempirical methods are less necessary for simple systems.

5.2.3. Adsorbate Bonding to Ionic Solids

5.2.3.1. Types of Bonds to Ionic Solids. In adsorbate bonding to ionic solids the most energetic local bonds will normally be the acid/base bonds. However weaker interactions are possible, for example, electrostatic effects and covalent bonding. Finally, here as with the covalent solids, bonding can occur that is best described as an incipient new phase, where the formal valence state of the host atom can be considered to have changed upon adsorption.

Acid/base bonding was discussed in some detail in Sections 4.2.3 and 4.3.3, where the presence of Lewis active centers on a clean surface was reviewed. The best way to test for Lewis active centers is to observe the formation of an electron pair bond with an adsorbate, so it was necessary to describe acid/base interactions at that point.

The electrostatic potential (Madelung) contribution to the energy of adsorption of an ion on an insulating ionic solid has been calculated for a smooth plane by Mark.[35] The difference in potential above a surface cation from that above a surface anion was not found to be very large. Thus strong electrostatic bonding of a charged species such as O_2^- is not expected. Of course at flaws such as steps and dislocations, differences in potential are expected to become significant again for such monopole adsorption. Such flaws would provide sites intermediate between the strongly interacting sites in the outermost plane (Section 4.2.2) and the weakly interacting (according to Mark) sites beyond the outermost plane.

On the other hand, the electrostatic (Madelung) contribution to the adsorption of a dipolar species such as NaCl is expected to be much stronger. When NaCl adsorbs, and the Cl^- occupies a site over a cation, the Madelung potential at the site over the neighboring lattice anion will be much lower, attracting the Na^+ with strong electrostatic forces. Quantitative Madelung calculations for a dipolar adsorbate interacting with a lattice anion/cation pair have not been made as yet.

An example of a combination of the above two mechanisms (acid/base and dipolar Madelung) is provided by the adsorption of water. In many cases of water adsorption, hydrogen bonding and (for transition metal adsorbent) ligand field effects will provide even further interaction between the OH^- group and the solid. James and Healey[36] provide cogent arguments comparing water adsorption to hydrolysis of ions in solution.

Covalent interaction, orbital overlap where the sorbate and the solid each contribute an unpaired electron, may also occur on ionic solids. The covalent interaction must induce hybridization of one of the orbitals of the solid, and so depends on the availability of a potential partly occupied hybrid orbital at an appropriate energy. For example, as suggested by van Laar and Scheer[37] (see Section 4.1.1), the "ionic solid" GaAs may best be described as ionic with a clean surface. However, it will hybridize its surface orbitals either to provide covalent bonding with adsorbing metals as indicated by the covalent characteristics at the metal/semiconductor interface, or to provide local covalent bonding between the surface As atom and adsorbing oxygen.[38,106] Effectively, such an interaction amounts to covalent interaction between an adsorbate and an ionic surface state as described in Chapter 4.

If modest relocation of surface atoms occurs, we can have cases where a familiar chemical group forms during adsorption. Obviously, formation of such groups is a most common occurrence with covalent materials or metals, where for example adsorption of oxygen, with metal atom relocation, is best described as incipient oxide formation. With ionic solids the possibility of such processes is often overlooked in discussions of adsorption. When the interaction to form the new surface group involves bonding primarily with only one of the lattice ions it can usually be described best as a change in the oxidation state of that ion. An example might be the adsorption of oxygen on Cu_2O, forming a CuO "molecule" at the surface, or the adsorption of oxygen on CdS, forming an incipient SO_4^{2-} group at the surface. A well-studied example[39] is the adsorption of CO and oxygen on various oxide surfaces to form carbonate groups.

5.2.3.2. The Adsorption of Water on Silica. An interesting example of strong local interaction on ionic solids that illustrates acid/base bonding and relocation of substrate atoms is water on silica. As discussed in the preceding chapter, impurity-free silica is neither strongly acid nor strongly basic. This means that special sites of very high electrostatic potential such as those postulated for alumina (Figure 4.7) are uncommon. It also means that when water is adsorbed the tendency to give up protons and to give up hydroxide ions is similar. This does not imply that the bonds are weak, it simply implies that both fragments of the water molecule are held equally firmly.

Figure 5.2 shows the suggested model for water adsorption on silica. In Figure 5.2c siloxane groups at the surface of the dehydrated material are illustrated. Boehm[40] suggests the siloxane groups may be more polarized than shown, with the oxygen ion unequally shared. The presence of chemisorbed water, (Figure 5.2b) leads to a coverage of the surface by silanol groups, and Figure 5.2a shows "physically adsorbed" water. If the system is heated, the physically adsorbed water desorbs at a temperature of about

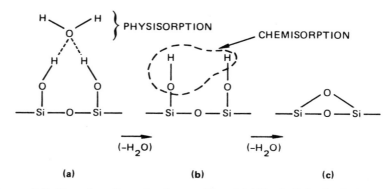

Figure 5.2. The adsorption of water on silica. (a) "Physically" adsorbed water; (b) "chemically" adsorbed water; (c) dehydrated surface.

100°C or less. The desorption of chemisorbed water occurs at temperatures between 180 and 400°C. This implies strong bonding. If the sample is maintained at a temperature below about 500°C, so-called "strained" siloxane groups are present. Boehm[40] suggests the strained groups may be similar to those of Figure 5.2c but more polarized, with the oxygen bonded to a particular Si ion. Readsorption of water on such strained siloxane groups is rapid. If the sample is heated above 500°C, relocation of the surface ions occurs and a passive surface results, presumably of the form of Figure 5.2c. Readsorption of water on such a surface is slow.

In summary, although silica does not show a net acidic or basic character, the high desorption temperature implies that a strong acid/base bond with water is formed. The lack of a net acidic or basic character suggests that both the proton and the hydroxide group of the water become bonded equally firmly to the respective surface site. The well-documented elimination by relocation at $T > 500°C$ of adsorption sites for water is particularly instructive.

5.2.4. Multilayer Adsorption: The Development of a New Phase

A clear distinction between the various degrees of adsorbate interaction on a surface has always been difficult. The interaction can vary from very weak physisorption, to chemisorption, to an incipient new phase, and finally to a well-defined new phase. As a matter of fact, there probably is no clear distinction; there are broad gray areas between each class of interaction which will frustrate any attempts at a definitive classification. However, improved observations of reconstruction and relocation as the adsorption progresses are providing new insight into the problem.

In the present volume, we are attempting to describe primarily chemisorbed species, but for clarity the discussion must be extended to the case where there is an incipient new phase. We will only discuss the adsorption of oxygen, as this case is by far the most studied. However, the concepts can be applied to any adsorbate where compound formation is possible.

In the rest of this subsection we will describe many examples of oxygen interaction on solid surfaces. There will be examples where oxygen adsorbs and interacts with dangling bond surface states. There will be examples where oxygen adsorbs and the result can best be described as a valence change of a surface atom or ion. There will be examples suggesting an incipient new phase, and results suggesting an extensive new phase. In general the best description of oxygen adsorption requires a combination of the above processes.

We have discussed in the preceding chapter cases of oxygen adsorption (or adsorption of most other gases) removing intrinsic surface states from a clean surface. The so-called "removal" of course means a shift of the state to a new bonding orbital energy where it is not detected by the particular technique used.

The strong local bonding of oxygen on elemental semiconductors can be summarized by reference to a few results. Henzler[4] shows that the first fraction (0.1) of a monolayer of oxygen adsorption on germanium causes a reconstruction, changing the surface structure from the (2×1) surface structure characteristic of the clean surface to the (1×1) structure associated with the normal lattice. Further adsorption removes intrinsic surface states. Burshtein et al.[41] find that when the stoichiometry Ge_s—O_2 is reached (where Ge_s is a surface atom), the surface is passivated and no more oxide will form until water vapor is provided to catalyze oxide growth.

On GaAs also there is evidence that oxygen forms a local bond with the surface, in this case with the As surface atoms.[38,108] Along the same line, but more surprising, is the observation of Russell and Haneman[42] that even at 77°K, enough O_2^- adsorbs to provide a LEED pattern. This is much more than the 10^{-3} or 10^{-2} monolayers permitted for ionosorption (the Weisz limitation, Chapter 2), and the results suggest local electron transfer from surface As atoms (oxidation of the arsenic).

Such formal valence changes may be more easily understood on the chalcogenides. The sulfur or selenium atom commonly can exhibit a valence of -2, 0, $+4$, or $+6$. Thus it is to be anticipated that a surface sulfur ion on a material like CdS can be oxidized by oxygen adsorption. For example, the sulfur could take on a formal valence of $+4$ or $+6$ with the formation of a group at the surface resembling a sulfite or sulfate group. And indeed there are indications[43] of large oxygen effects on such chalcogenides. Guesne et al.[44] find monolayer oxygen adsorption on CdSe, again much more than can be accounted for by ionosorption with the Weisz limitation. The bond can be considered a local covalent bond or an incipient new phase, according to preference.

Oxygen adsorption on molybdenum provides an instructive example showing extensive relocation during adsorption and oxidation, as is necessary on metals. Studies using LEED, AES, work function, and ELS by Riwan et al.[7] illustrate the development of the surface from a small fraction of a monolayer to a surface oxide. They observe three stages in oxygen adsorption. In the first stage, up to $\frac{1}{2}$ monolayer, a (2×2) LEED superstructure is present. The work function decreases as oxygen is adsorbed, so relocation of the Mo must be occurring—the oxygen atoms are becoming buried in the Mo lattice interstices. Between $\frac{1}{2}$ and 1 monolayer further relocation occurs—the surface becomes faceted. Streaks in the LEED pattern suggest steps of three or four lattice parameters deep, exposing (110) facets. Finally with exposure of more than a monolayer, the surface flattens out again, and the LEED patterns are identified as representing an oxide film.

There are many tests other than LEED that provide information about the various changes between adsorption and a new phase. Morgan and King[45] measured the sticking coefficient of oxygen on nickel (the probability that a molecule when it strikes the surface will adsorb) and found variations with coverage corresponding to changes in the LEED superstructure. Hall and Mee[46] have measured work function changes with the adsorption of oxygen on iron, cobalt, and manganese. They distinguish three regions: adsorption (work function increase), the burial of oxygen (work function decrease), and oxide formation (stability), in that order. Others also have found work function to be a useful test.[7,47]

Benninghoven and Wiedmann[48] have followed the oxidation of the alkaline earth metals using secondary ion mass spectroscopy (SIMS, sputtering) to distinguish between adsorbed oxygen and oxide formation. The test is simple. If the sputtered metal comes off as an ion, it is considered to have been oxidized. It is not clear whether this test will be useful in describing other differences between adsorbed and absorbed species, but it is an interesting approach. Their results suggest that 1 Langmuir (10^{-6} Torr sec) suffices to form a monolayer of oxygen on these metals and an oxide phase appears by their criterion only after the monolayer adsorption.

Another technique to distinguish adsorbed and absorbed oxygen is XPS. Robert et al.[49] find XPS peaks at 530.4 eV that they ascribe to lattice oxygen, 531.4 eV that they ascribe to oxygen in an adsorbed hydroxide group, and 532.2 eV that they ascribe to chemisorbed oxygen. The identification of the peaks is made by using various treatments of the sample which should enhance the one form or the other.

In Section 5.4.3 below, where the surface state energy levels associated with absorbates on metals as measured by UPS are discussed, the variation observed as the oxygen levels progress from the (relatively) narrow band of adsorbed oxygen to the broad band of the metal oxide is described. Such surface spectroscopy measurements are providing further insight into distinguishing between adsorption and the development of a surface phase.

5.3. Quantum Models of the Adsorbate/Solid Bond

In the preceding section the semiclassical surface molecule models of adsorbate bonding have been reviewed. In this approach little consideration is given to the fact that the atom from the substrate that participates in the bond is actually a part of a solid, and that in the solid many of the energy levels are best described as a broad band rather than the localized atom levels assumed in the surface molecule picture. In the present section we look to quantum theories to show how realistic the surface molecule approximation was, and how it must be modified to present a more realistic picture, and finally to identify any cases where the surface molecule approach fails completely. Such questions will be reviewed in Section 5.3.5 following a description of the methods and results of the quantum analyses.

Three general approaches have been used to describe in quantum calculations the energy and orbital characteristics of the bond between an adsorbed species and a solid. The first is the "semi-infinite crystal" approach, where nonlocalized wave functions are assumed, corresponding to electrons in the periodic solid. In this approach bands associated with the solid are calculated, and the adsorbate is treated as a perturbation on the potential. This turns out to provide a qualitative picture, as it is difficult to introduce the surface perturbation in a quantitative way. The second is the opposite extreme, where quantitative molecular orbital (MO) calculations are made of the adsorbed species plus a few substrate atoms that form a "cluster," which can be analyzed by computer methods. The third, the "interacting surface molecule," or "model Hamiltonian" approach, is a compromise between the solid state and the MO approaches, where the single atom adsorbate wave functions are joined to the wave functions of the semi-infinite crystal to provide a semiquantitative model. This model is highly valuable in developing an intuitive picture of the behavior of adatoms.[50,51] Both the cluster and interacting surface molecule models can include the directional characteristics of bonds at the surface as found so important in the last subsection. As Shrieffer and Soven[52] point out, the next step forward will be an "interacting cluster" model, which when developed should provide the most quantitative results of all.

As was the case with the quantum theories of the clean surface (Section 4.2.1), we will make no attempt to develop the mathematical methods used in the various approaches, giving only general descriptions. The objective of this section is only to provide some insight as to the results and the models emerging from the quantum approaches.

5.3.1. Solid State Theories: The Semi-infinite Crystal

The semi-infinite crystal calculations were discussed briefly for the clean surface, and there is very little to add which contributes to a quantitative or even a qualitative understanding of the surface states associated with adsorbates. Hückel and extended Hückel calculations have been performed.[53-57] The latter are Hückel calculations as described in Section 4.2.1.1 where better descriptions for the Coulomb, resonance, and overlap ($S_{\mu\nu}$) functions are calculated. Approaching the problem from this viewpoint, where the band model is considered dominant and the variations necessary to describe the surface are introduced as perturbations, must of course in principle result in a quantitative description. However the same problems, only more complex, arise in describing the surface perturbation as were discussed with respect to clean surfaces (Section 4.2.1.1). At the present time it appears that the approaches starting with the MO model of the adsorbate are both more easily visualized and more quantitative.

5.3.2. Cluster Models

The cluster models are made possible by good approximations to the Schrödinger equation and by the development of rapid computers. Basically they analyze in detail a molecule of up to 15 or 20 atoms. With such a number of atoms in the "molecule" the result shows characteristics of a large crystal, including "band" formation. There are only 15 or 20 energy levels in the band, but these energy levels are grouped much as the solid state bands are grouped. Such a cluster can include foreign atoms at its surface. Thus one can solve for the energy levels of a cluster comprising several host atoms and a foreign adsorbate.

The greatest advantage of the cluster models over the interacting surface molecule approach below is the detailed picture one obtains of the shape of the individual orbitals of the surface atoms and how they are modified by the presence of many other atoms. The disadvantage, according to Schrieffer and Soven[52] and van der Avoid et al.[58] is that it is not clear how closely the cluster analysis actually approaches the description of the solid. Also, the large surface to volume ratio of the cluster causes shifts in the spectrum of energy levels.

Slater and Johnson[59] present a simplified review of one approach to the analysis of a cluster, the "self-consistent field X–alpha scattered wave method" (SCF-Xα-SW). This is a descendant from a "muffin tin"* approx-

---

* The term "muffin tin" means that the potential near an atom in the crystal is not allowed to go to $-\infty$, but is cut off at a minimum value.

imation from solid state theory that has the advantage of rapid convergence of the computation as compared to LCAO (tight binding) theories. The "self-consistent" part of the description means, as discussed earlier, that the calculated wave functions are fed back into an iteration of the Schrö-dinger analysis, and the iterations are repeated until the wave functions resulting from an iteration are the same as the wave functions introduced into that iteration.

Some of the SCF-Xα-SW analyses are of particular interest. For example, the clusters Ni$_8$ and Cu$_8$, among others, have been analyzed.[59,110] A group of levels that Slater terms a "band" of d levels (16 levels) appears for copper, with hybrid spd levels split off above and below the d levels. Slater regards these hybrid levels as the precursor of the sp band in bulk crystalline copper. The Cu$_8$ cluster is described further in Section 6.3.

Batra and Robaux[60] studied the chemisorption of oxygen on nickel, using a cluster of the form O(Ni)$_5$, and obtained good agreement with UPS experimental results. Batra and Bagus[61] studied CO on Ni. They were able to conclude that CO adsorbs by electron donation from the highest filled CO molecular orbital and that back donation occurred from the solid into the empty antibonding $2\pi^*$ orbital of the CO. Further, they concluded that the CO 5σ and 1π orbitals are badly perturbed when CO bonds to the metal, and are changed substantially in relative energy from the corresponding levels in the free molecule. Thus they were able to suggest that the assignment of the UPS peaks by Eastman and Cashion[62] was in error. The accepted model at present suggests that with CO adsorbed on Ni, the 5σ and 1π levels are approximately at the same energy.

Rosch and Rhodin[63] have used a small cluster in analysis of ethylene adsorbed on nickel. The "cluster" of nickel actually comprised only two nickel atoms, so the model was close to a surface molecule model. Using the SCF-Xα-SW analysis they examined two orientations:

$$
\begin{array}{c}
\mathrm{H_2C{-}CH_2} \\
\mid \quad\ \mid \\
\mathrm{Ni{-}Ni}
\end{array}
\qquad\qquad (a)
$$

which resembled an adsorbed ethylene bonded with two σ bonds (a di-σ bond) to the "nickel surface," and

$$
\begin{array}{c}
\mathrm{H_2C{=}CH_2} \\
\vdots \\
\mathrm{Ni} \\
\mid \\
\mathrm{Ni}
\end{array}
\qquad\qquad (b)
$$

which resembled an adsorbed ethylene molecule π bonding to the "nickel surface." In the above models the plane of the the ethylene molecule is perpendicular to the plane of the paper. They compared the results of these calculations to the experimental results by UPS of Demuth and Eastman,[64] to be described below in Section 5.4.1. They found the energy shift of the ethylene π electrons as calculated from model (b) is greater than that calculated from model (a) and is closer to the experimentally observed value of 0.9 eV. Thus their cluster calculations indicate that the earlier molecular models described in Section 5.2 are realistic, that "true" π bonding occurs, as opposed to the di-σ bond. They also determined (as is important in catalysis) that if, with a modest expenditure of energy, the π bond switches to a di-σ bond, the carbon/carbon bond in the ethylene molecule becomes much weaker. They therefore suggest the di-σ bond may be an active intermediate in the dissociation or in other reactions of ethylene. The reason the carbon/carbon bond becomes weaker is that there is substantially more back donation of electrons from the nickel to the antibonding π orbitals of the ethylene in the di-σ configuration than in the π configuration. Thus by analysis of this simple little cluster, with a qualitative comparison to experiment, Rosch and Rhodin[63] were able to provide a substantial insight into the behavior of adsorbed hydrocarbons. More recently, Demuth and Eastman[109] have developed a more detailed theory of such olefin bonding to nickel, using a tight-binding theory, and reach similar conclusions about the prevalence of π bonding.

5.3.3. The Interacting Surface Molecule (the Model Hamiltonian Analysis)

This analysis[52,65−68] is similar to the surface molecule theories (Section 5.2) in that the adsorbate species is presumed to interact with a single atom or small group of atoms on the surface. Thus one obtains bonding and antibonding levels, reminiscent of the simple valence theory of diatomic molecules. However, instead of the single atom or group of atoms at the surface being represented by atomic wave functions, in this model it is represented by the orbitals and the local density of states (LDS) that arises from the surface state models of Section 4.2. Thus, for example,[52] the binding energy of an adsorbed species may be described in terms of the energy of three processes: (1) the energy to decouple a surface atom from the solid substrate (with its LDS maintained unchanged during the decoupling), (2) the energy resulting from the interaction of the adsorbate atom and the decoupled substrate atom to form an unusual complex, and finally (3) the energy obtained by reattaching this complex back to the substrate.

Now the point is that the LDS of the substrate atom does not resemble at all (particularly in the valence band energy region) the atomic energy levels used for the substrate atom in the pure surface molecule approach. Rather than monoenergetic molecular energy levels the LDS is a broad distribution of energy levels extending over the energy span of the valence band. Further there is no requirement that this LDS be occupied by an integral number of electrons. When the complex is formed, between this surface atom with its strange "electronic orbital" and an adsorbate, it resembles the surface molecule approach in that bonding and antibonding "orbitals" are formed, separated in energy more and more as the orbital overlap increases. But in the present model the bonding and antibonding "orbitals" are now each a distribution of energy levels.

One of the major contributions of this model to an understanding of surface bonding is the fact that the model provides a semiquantitative picture describing when the rigid band model is applicable and when the surface molecule model is applicable. The development of bonding and antibonding states with increasing interaction energy is sketched in Figure 5.3. The figure represents the case of a simple interaction between say an *s* orbital of hydrogen and an uncomplicated LDS of a metal surface atom. If the "hopping" interaction (overlap interaction) is very weak, then as would be expected, the levels of the adsorbate are perturbed little on the energy level picture (Figures 5.3a and b). Then the rigid band model is the best first approximation. If on the other hand, the hopping interaction is very strong, the bonding and antibonding levels develop and become narrow, widely spaced bands, located, respectively, near the bottom and top of the valence band of the solid (Figure 5.3d). The energy levels become reminiscent of a direct atom/atom interaction, and the surface molecule picture becomes the best first approximation.

Variations in the adsorbate position and in the directionality of the bonding orbitals extending from the surface are handled in an obvious way in this mathematical approach. For adsorption at a site between atoms, a group orbital must be used to describe the LDS, and, as described in Chapter 4 with reference to Figure 4.2, information regarding the directionality of the LDS is obtained from the clean surface analysis.

Correlation effects are more difficult to introduce quantitatively, yet they have a large effect on the energies. By correlation effects one means effect of one electron in an orbital on the average position of another. Such effects are not included in analyses based on one-electron models.

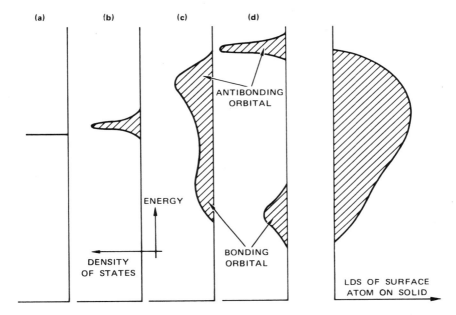

Figure 5.3. The development of bonding and antibonding states from a state on an adsorbate atom (extreme left) interacting with the local density of states (LDS) of a surface atom of a solid (extreme right): (a) $d = \infty$, no interaction, (b) weak interaction, (c) intermediate interaction, and (d) strong interaction. The integrated density of states is not normalized in these sketches.

Schrieffer and Soven[52] describe ways in which such effects can be artificially introduced into the result. Hyman[69] has been able to introduce iteration for self-consistency into this approach, using a pseudopotential model for the solid.

Lyo and Gomer[70] have used the general approach in the analysis of hydrogen adsorption on the tungsten (100) surface. The calculated density of states is shown in Figure 5.4. The s band of the solid overlaps the d band, but the interaction of the hydrogen orbital is primarily with the d band of the solid. In this calculation the LDS for the d band of W is assumed to be composed of two peaks with a deep minimum in the center. The bonding orbital between the H atom and the low-energy d peak can be seen developing at the bottom of the d band as the overlap, s, increases. The antibonding orbital between the H atom and the high-energy d peak can be seen developing (with increasing overlap) near $E = 1.3\,\text{eV}$. The peak at $E \approx 0$ is associated with the minimum between the d bands and is not a simple bonding or antibonding orbital.

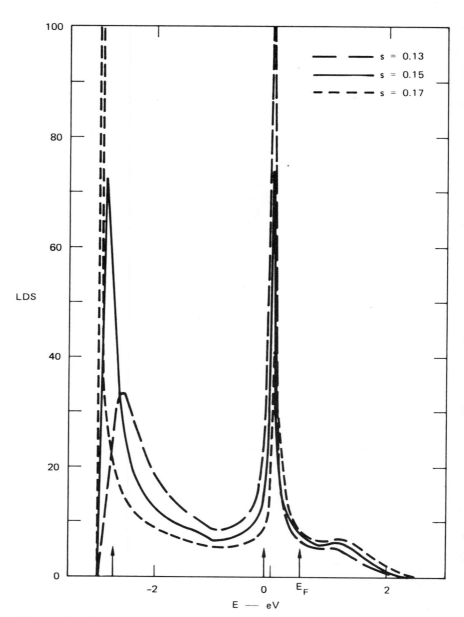

Figure 5.4. The local density of states (LDS) calculated for the adsorption of hydrogen on tungsten. The zero on the energy scale is the center of the d band of W. With the adsorption, the LDS increases near the bottom and top of the unoccupied levels of the d band. These are the "bonding" and "antibonding" states of the complex.

Davison and Huang[111] apply this general approach to determine the heat of adsorption of oxygen on narrow-band semiconductors, and conclude the heat should depend on E_g and ϕ. Foo and Davison[112] have used a self-consistent Green function method of this class to describe hydrogen and halogens on silicon and germanium.

5.3.4. Other Quantum Models

Other variations on the possible approximations have been utilized which have advantages in particular cases. For example van der Avoird, Liebmann, Fassaert, and Verbeeck[58,71,72] have emphasized a compromise between the cluster model and the interacting surface molecule model that they term the "resolvent method." It is based on a semi-infinite crystal approach, but is applicable to finite crystals. Larger finite crystals can be treated than is possible by the cluster method, because some of the LCAO methods such as a periodic boundary can be introduced to simplify the analysis.

Einstein[73] has introduced a model that he calls the "short chain model" where only four atoms in a chain are analyzed. Here the important advantage is that the calculations are greatly simplified to the extent that they can be done approximately by hand, thus providing rapid semi-quantitative results.

Ying et al.[74] and Huntington et al.[75] have introduced electrostatic self-consistency but to do so have used the simpler "jellium model" for the solid, which assumes the solid has no discrete charges, and so the model loses much detail in the LDS calculated. In addition a simplified potential for the adatom is used, so the approach resembles the semi-infinite crystal model discussed in the preceding section. However, with the simpler model the effect of such variables as adsorbate/solid separation can be more easily explored.

5.3.5. Remarks

The applicability of the pure surface molecule and the rigid band models, as described in Section 5.2, can now be evaluated by comparing their predictions to the results of the quantum analyses. Reference to Figure 5.3 is of particular value. From the quantum analyses it is clear that the best model for the bonding of an adsorbate to a solid is actually intermediate between the two semiclassical models. Realistically when the interaction energy is high, one should incline toward the surface molecule model. The shape of dangling orbitals at the surface will affect the bonding in a way qualitatively as anticipated by the surface molecule model. The cluster calculations suggest the surface molecule model provides a good picture of π bonding and back bonding. The errors in the pure surface molecule model are (a) the estimated energy levels of the bonding orbitals are at best qualitative and (b) the broadening of the bond energy levels into bands is not predicted.

It must be recognized that the quantum analyses above applied to local covalent bonding. Special effects in acid/base bonding have not been analyzed. Even strongly heteropolar covalent bonding has not been analyzed in detail. However, the surface molecule concept will undoubtedly be best also for other bonding mechanisms where the local interaction is strong.

A quantum analysis of acid/base bonding is overdue. There is little doubt that it will reflect more the "surface molecule" configuration even than covalent bonding, for the acid/base interaction seems highly local. But it would be very enlightening to have a fundamental analysis showing how ionic surface state bands transform into localized acid centers upon adsorption of a gaseous base.

As the interaction becomes weaker, the rigid band model becomes more appropriate. If there is no local interaction, the "pure" rigid band model is clearly preferred.

In most cases, there is a broad overlap where either model can be used and the model is best chosen depending on the characteristics of interest. If electron transfer from a semiconductor band to a surface species is of interest, then the rigid band model is clearly the most useful picture, as it describes the source of the electrons. However, if at the same time there is a strong local interaction, the effect of this interaction shifting and broadening the LDS of the surface states must be recognized and somehow fed into the model. If only local chemical interaction is expected and of interest, the surface molecule model is most useful. However, the effect of the bands of the solid on the LDS of the bonding orbitals must be recognized and used to modify the picture.

5.4. Measurement of Adsorbate Surface States on Covalent or Metallic Solids

5.4.1. Screening Shifts and Other Inaccuracies in Measurement

In this section the energy levels of surface states associated with covalent sorbate/surface bonds are described and compared to predictions of the quantum models.

The greater fraction of the work to be discussed was done with the tools of the surface spectroscopies. In particular, UPS has played a dominant role in the elucidation of the energy levels associated with bonds. As described briefly in Section 3.2.1, UPS provides information about bond energy and direction. However, there are problems in interpretation, which will be discussed in this section. The particularly interesting work of Demuth and Eastman on hydrocarbon adsorption will be introduced in this section because, in addition to providing valuable data on bonding energy levels, it illustrates some of the interpretation problems and how they can be overcome.

Demuth and Eastman[64] examined (with UPS) the energy levels that developed when ethylene, acetylene, or benzene adsorbed on the Ni(111) surface. Now it would appear that these molecules would be much more difficult to analyze than a simpler adsorbate like oxygen, but actually the greater complexity of the adsorbate permitted much more accurate analysis of the bonding. It was found that some of the energy levels of the molecule (the C—H σ bonds) that should not participate appreciably in the bonding showed nonetheless substantial displacement in energy from the values for the free molecule. Such nonbonding levels should not shift so much, and this points out the problem. In Section 3.2.1 it was mentioned that such apparent shifts occur, and are due to unwanted "relaxation" effects—when the electron is removed from the bonding orbital, the electrons in the solid move to screen the resulting positive charge, and this movement is called relaxation. The relaxation energy is given to the photoelectron. Unfortunately the screening and hence the relaxation energy will be different for an adsorbed atom than for a bulk atom, so if a bulk energy level (say, the Fermi energy) is used as a reference, a correction is needed to account for the difference in relaxation energy between the bulk and surface levels. Now in the case under discussion, where the molecule is complex enough to have nonbonding electrons with measurable energy levels, the shift in the energy levels of these nonbonding electrons can be determined and used as calibration to correct the peak energies for the bonding electrons.

Figure 5.5 shows a bar diagram of the energy levels obtained by Demuth and Eastman for the ethylene/nickel system. The UPS spectrum of ethylene is measured in three forms: gaseous, physically adsorbed (little interaction), and chemisorbed. The four σ levels should be unchanged as the molecule is adsorbed, so the levels are all normalized to these energies. In other words, the σ bond energies of the sorbate are set equal to the σ bond energies of the gas-phase molecule to correct for relaxation differences in the UPS. Comparing in Figure 5.5 the energy levels of the free molecules with the

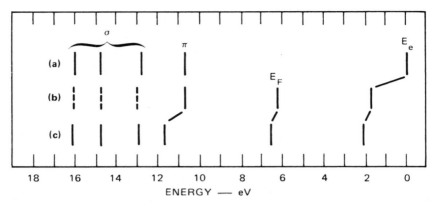

Figure 5.5. Energy of UPS peaks observed associated with ethylene (C_2H_4) on the Ni (110) face. (a) Photoelectron peaks from gas-phase ethylene, (b) UPS peaks from physisorbed ethylene, (c) UPS peaks from chemisorbed ethylene. The energies are normalized so the σ levels are not affected by the adsorption processes.

energy levels for the physisorbed species while the σ orbital energies are made constant, we note that the π electron energy does not change upon physisorption. However, the apparent free-electron energy changes substantially. This is the relaxation energy associated with adsorption. Upon chemisorption, the relaxation energy is unchanged—the screening effect is insensitive to whether the molecule is physisorbed or chemisorbed. However, the energy of the π electrons from the ethylene changes. This shift between the chemisorbed and physisorbed value can be associated with the π bonding. Also the Fermi energy in the solid at the surface changes, as the LDS of the surface atoms has changed.

This experiment is thus not only another convincing demonstration of π bonding (see the analysis by Rosch and Rhodin in the preceding subsection) but is also an excellent demonstration of the difficulty of obtaining quantitatively accurate energy levels when such effects as relaxation occur and methods to correct for the effects are not used.

Gadzuk[76] discusses the relaxation effect in detail. Feibelman and Eastman[77] and Menzel[78] discuss this plus other problems in quantitative interpretation of UPS peaks as surface states. They point out there are problems even in qualitative analysis, such as satellite lines that can be misinterpreted as surface state peaks. Brundle[79] provides another excellent review, both of UPS and XPS, discussing the various causes of satellite lines.

Further evidence of the shift in energy due to relaxation is offered by Yu *et al.*,[80] who studied adsorption of a series of molecules at 77°K on an inert substrate (the cleavage plane of MoS_2) and found the shift was about the same for any molecule; the levels shifted upward by about 1–1.65 eV. This suggests that a given substrate may be "calibrated" with respect to its ability to screen the holes produced on adsorbates, the calibration accomplished by any convenient physisorbed gas. However, the need for making the measurement at low temperature ($77°K$) may be a problem.

A difficulty in the interpretation of ELS measurements has been studied by Sickafus and Steinrisser.[81] At times it is found that the ELS peaks depend on the energy of the primary beam, and of course (Section 3.2.2) the simple model of ELS predicts no such dependence. Sickafus and Steinrisser were able to demonstrate that an ordered adsorbate that forms a LEED superstructure may cause problems. The electron beam may be diffracted before the energy loss event occurs, and as the energy loss event to be expected depends upon the direction of the beam, such diffraction effects can cause complications.

In the following discussions reference to such problems will not usually be made because the author of the referenced article has not clarified whether a correction has been made. And independent of whether the correction, say, for relaxation effects in UPS has been made, the results provide an extremely valuable insight into the nature of bonding orbitals of foreign adsorbates.

5.4.2. Bond Angles

In earlier sections (Section 5.1, for example) we discussed relocation of atoms on the surface. The direction of bonding orbitals can be inferred from measurements such as LEED or ESD that show the position of the sorbate relative to the lattice atoms. In the present section we will illustrate with a few examples the feasibility of measurements of bond angles where the electron in the orbital is observed directly.

In the UPS measurement the photoemitted electrons to a first approximation are concentrated along the bond directions. Liebsch and Plummer[82,83] have analyzed in detail the expected angular dependence of the photoelectrons as a function of the shape of the orbitals and the symmetry of the adsorption site. Gadzuk[84] has developed a similar model. Unfortunately detailed experiments of angular dependence to date have been primarily on clean surfaces, as, for example, the work of Traum, Rowe, and Smith,[85] which was described in the last chapter, and that of Traum, Smith, and Di Salvo,[86] which included angular photoemission studies on the cleavage plane of the layer crystal $TaSe_2$. The method is successful, however. In the latter study, for example, the authors were able to show that the d orbitals of the tantalum in the sublayer were directed between rather than toward the Se atoms in the surface layer. They rationalized this observation by noting that d orbitals should be nonbonding. Anderson and Lapeyre[87] observed a shift in the angular dependence of photoemission upon H adsorption on W. It was concluded however that the results were deceptive, in the sense that the direction of the H—W bond was not being detected, but rather the changes in photoemission angles were related to reconstruction of the substrate lattice upon adsorption.

Another concept based on UPS that was intended to provide information regarding directionality of bondings was the use of polarized light. Unfortunately Feuerbacher and Fitton[88] and Rowe and Christman[89] have shown that both for bulk and adsorbate states the photoemission depends only on the intensity of light polarized normal to the surface. Thus, little structural information is obtained.

Waclawski et al.[90] studied the UPS angular distribution for oxygen adsorbed on the tungsten (100) plane. They found a maximum emission normal to the sample for an energy level at 5.2 eV below the Fermi energy, which suggested that this surface state has a p_z symmetry. Another surface state at -6.3 eV shows a minimum for electron emission in the normal direction. It is concluded that this orbital must be involved in bridged bonding of the oxygen atom to two tungsten atoms. More studies on the energy levels of oxygen on tungsten will be reported below. Such an orientation was partially confirmed by Yates' group, Madey et al.[9] In their studies of ESD as discussed in Section 5.1, the latter authors were able to show indeed that at low coverage the oxygen occupies the interstices between tungsten atoms causing a (1×4) LEED pattern. At higher concentrations another state dominates, with the oxygen on top of the W atoms, but with a bond angle not necessarily normal to the surface.

5.4.3. Surface State Energy Levels of Sorbate/Sorbent Bonds

The adsorption of CO on W has been thoroughly reviewed by Gomer,[91] and a summary of his conclusions provides an excellent example of the analysis of surface state energy levels. When CO is adsorbed at low temperature: (a) the work function increases, (b) there is TPD peak at about 200–400°K, (c) the coverage is about one CO molecule per W atom, and (d) ESD measurements show the desorbed ion to be CO^+. Thus it is concluded that the low-temperature form is molecular CO. From the change in work function, and by analogy with the molecule $W(CO)_6$, it is concluded that a bond forms between the carbon and the tungsten, with the negative oxygen in the outermost layer. A probable bond is a σ bond between an sp carbon orbital and the d orbital (d_{z^2}) of the tungsten that extends normal to the surface. In addition, back bonding occurs with electron density from the tungsten d_{xz} and d_{yz} orbitals contributed to the CO antibonding π orbitals (see Figure 5.1 for an analogous example).

The UPS data are not well defined. Feuerbacher[92] finds the UPS structures for CO on an adsorbed layer of intermediate treatment are "featureless." Menzel,[78] on the other hand, in his review states that the CO levels on tungsten are very similar to the levels of CO adsorbed on Ni, Ru, Pd, Pt, of CO gas, and of $W(CO)_6$. In each of these cases he concludes that two CO peaks are resolved, one at -7.5 to -9 eV (varying somewhat for the different metals), the other at -11 to -12 eV below the Fermi energy. Gomer[91] reports results showing up to 4 peaks [on the (100) surface at room temperature], all below -8 eV. The reader can recognize that the UPS spectrum is somewhat ill defined and sensitive to pretreatment.

An FEM resonance tunneling peak is found by Young and Gomer[93] at about 2 eV below the Fermi energy.

The assignment of these peaks to particular bonding orbitals is most probably straightforward with the help of the analysis of Batra and Bagus[61] (see Section 5.3.2). The peak at -11 to -12 eV below the Fermi energy, is most likely the σ $2p$ bond, the peak at -7.5 to -9 eV is the π bonding orbital, while for CO on W the resonance detected with FEM at 2 eV below the Fermi energy is most likely the π antibonding orbital. Note that Batra and Bagus conclude the σ and π bonds are interchanged in energy from their values on the free molecule, due to the strong interaction with the surface.

The above represents the picture of CO adsorbed at low temperature. If the CO is adsorbed at a higher temperature, several different bonding structures can be present. For example, with a monolayer of CO adsorbed at low temperature, increasing the temperature in vacuum causes the CO to desorb between 200 and $400°K$. However $\frac{1}{2}$ monolayer is desorbed; the other half is retained. The latter $\frac{1}{2}$ monolayer is of an intermediate transition form which desorbs by ESD as O^+. With ever-increasing temperature this CO transforms into the "β phase" at temperatures above about $800°K$, and the β phase does not desorb easily under electron bombardment. TPD shows a peak for the β phase at about $1100°K$. Another type of adsorption, the α phase, occurs if the low-temperature phase is desorbed; then after the β phase is formed, more CO is adsorbed. It is understandable how the form of CO on tungsten can become confused. The transition and the β phases presumably correspond to the CO molecule lying parallel to the surface. The UPS spectrum of this phase resembles the sum of the C and O spectra.

We will not describe in detail measurements of the higher-temperature forms of CO on tungsten, but refer the reader to the cited reviews. The summary of measurements on the low-temperature form provides an excellent example of the use of the various tools to explore the energy levels of a species on a surface. The summary of the behavior at high temperature demonstrates that the surface bonding varies dramatically with relocation during pretreatment, and that one must be cautious about accepting energy level values unless it is clear from various measurements that a single bonding mechanism is under study.

Hydrogen adsorbed on tungsten provides an interesting system, primarily because the system can be compared to the quantum models which predict a bonding/antibonding split of the W/H levels (see Figure 5.4). Feuerbacher[92] in his review suggests two surface states associated with different sites termed β_1 and β_2 hydrogen. The β_2 hydrogen which adsorbs first to the extent of half a monolayer, causes a LEED (2×2) superstructure on the (100) face.[94,95] Further adsorption is in the form of β_1 hydrogen, with a lower binding energy. Two UPS peaks are observed, for which the relative intensity varies with coverage. The energy of the peaks varies with crystal face. These two peaks can be associated with the β_1 and β_2 hydrogen, which, as stated, arise because there are two sites, thus cannot be related to the bonding/antibonding peaks of Figures 5.3 and 5.4. The peaks observed are the order of 2.5–5.5 eV below the Fermi energy, depending on the crystal face exposed. On the same scale the d band extends 3–5.5 eV below the Fermi energy. Thus both these levels could be considered "near" the bottom of the d band. Feuerbacher and Willis[113] in angle-resolved photoemission find that these levels, and a third occupied level at 12 eV below the Fermi energy, show angular dependence. They suggest qualitative agreement with a model such as Figure 5.4, but conclude sp hybrids must be involved in generating the bonding and antibonding levels.

There is evidence of a higher orbital, unoccupied and thus not detectable by UPS. Anderson et $al.$,[96] using a highly sensitive optical absorption measurement with which they claim to resolve reflectance changes of 0.01%, have observed reflectance changes as a function of wavelength as H_2 (and other gases) are absorbed on the tungsten (100) surface. They observe an absorption peak at about 2–3.5 eV, which suggests a transition either to an antibonding level or to the empty levels of the solid above the Fermi energy.

Further evidence is provided by Plummer and Bell,[97] who measured a FEM tunneling resonance with H_2 on W(100) and (110). They find that at half a monolayer (β_2 state) they observe a level 0.9 eV below the Fermi energy. This could be associated with one of the upper levels in Figure 5.4.

A system that is especially interesting is the adsorption of oxygen on various metals. As would be expected, the adsorption of oxygen is complex, since adsorption, incipient oxidation, and oxidation can all occur (Section 5.2.4). Gadzuk and Plummer,[98] in their review of FEM studies, describe the variation in the characteristics of the FEM signal from tungsten with oxygen exposure in three such phases. First there is the removal of the band of intrinsic surface states near the Fermi energy; then at about 0.1 L exposure a broad FEM peak develops [on the (112) surface] at an energy about 1.2 eV below the Fermi energy; and finally peaks associated with oxides appear at about 5–20 L exposure as the 1.2 eV peak diminishes. By 50 L exposure the oxide peaks are the only structure left. All the peaks are poorly resolved, which might suggest heterogeneity.

In general, UPS data[78,99] show a dominating peak due to adsorbed oxygen at 5–7 eV below the Fermi energy. Such an energy level is in agreement with cluster theory[60] as discussed in Section 5.3.2. The studies of Waclawski et al.[90] of the angular dependance of the emission, as described in the preceding subsection, show that this peak is composed of two subpeaks associated with oxygen on different sites.

Kress and Lapeyre[100] show on Ba and Sr that the narrow UPS oxygen peak from adsorbed oxygen develops into the broad valence band peak characteristic of the oxides if oxygen adsorption is continued. Helms and Spicer[101] note that the metallic structure (on Sr) is unchanged by the heavy oxygen adsorption, so postulate the oxygen atoms are absorbed into the lattice. Similar broadening of the oxygen peak into the oxide peak is observed with oxygen on nickel.[62]

In the discussions above we have examined strong interactions such as observed when oxygen and hydrogen adsorb on metals and moderately strong interactions such as observed when CO and C_2H_4 (Section 5.4.1) are adsorbed on metals. A few studies are available of weak interactions. However, it turns out that the uncertainties in the experimental techniques make it unrewarding to use the surface spectroscopies in such studies, for the small shifts in the energy level position of the sorbate upon bonding cannot be resolved. For example, Norton and Richards[102] have examined CO_2 adsorbed on Pt with UPS. They find the peaks associated with adsorbed CO_2 are essentially the same as the photoelectron peaks from gaseous

CO_2, but are all shifted in energy. Sources of such shifts, discussed in Section 5.4.1, presumably have little to do with bonding energy. Another example, that of Yu *et al.*,[80] who showed the same result for a long series of molecules adsorbed on a relatively inert substrate (the cleavage plane of MoS_2), was described in Section 5.4.1.

We have chosen only a few examples from the expanding literature on measurements of the energy levels associated with adsorbate bonding orbitals. Any attempt to cover the literature would become unrewarding as far as the development of new concepts is concerned, and the intent in this book is to describe concepts rather than to review data. The few examples have been selected to show the general character of the emerging results, how they fit into the general models, and what difficulties are involved in defining a surface under study with respect to the bonding type. From the examples chosen it is clear that the bonding of a single adsorbate can be very complex because many different bonding configurations are simultaneously possible. In order to obtain a complete picture about the adsorption of a species, one measurement technique is usually insufficient. LEED and/or TPD and/or ESD and perhaps others must be utilized to determine, if possible, when adsorbate/solid interaction is reasonably homogeneous. Then and only then do the measured surface state energy levels permit easy interpretation.

However, with suitable precautions and with some inaccuracies, the surface state energy levels associated with sorbate/sorbent bonds are being measured by the surface spectroscopies. The measurements are confirming many of the models in detail that have been applied in surface chemistry for many years, as well as providing quantitative values to use in these models.

5.5. The Chemistry of Surface States

In the above sections we have been describing the surface state energy level associated with a bond between an adsorbed species and a solid. In the present section we address the problem of how that surface state energy level changes upon electron (or hole) transfer between the bands of the solid and the surface state. That is, we are interested in comparing the oxidized and reduced form of the adsorbate—how the surface state energy level changes depending on the oxidation state of the sorbate.

There are several ways in which the surface state energy can be affected: in the first section below, we will discuss the effects of shifts in adsorbate bonding; in the second we will explore the effect of the reorientation of polar molecules; and in the third we will investigate the behavior if an unstable oxidation state results upon electron transfer.

5.5.1. Change of Surface State Energy Associated with Bonding

In general for any surface species involved in electron transfer, we must consider the possibility that the local bonding differs for the two oxidation states (valences) of the surface species, resulting in a different surface state depending upon whether the level is occupied or unoccupied by an electron.

As an example, consider the shift in energy level between the neutral and reduced (O_2^-) form of an adsorbed oxygen molecule on an ionic semiconductor. For the neutral molecule it will be assumed there is little bonding interaction with the surface. Before electron transfer the acceptor energy level on the oxygen (relative to the free electron energy E_e) will be the electron affinity of the free oxygen molecule, modified by the Madelung potential at its equilibrium position as a physisorbed molecule. However, following the electron transfer (a) the resulting negative ion will undoubtedly move to a position closer to a cation (relocation) and (b) in this new position there may be local covalent interaction leading to a lower energy level for the electron. Thus, neither the adsorption site nor the surface state energy level are the same for the neutral as for the ionized adsorbate. This behavior is a manifestation of Franck–Condon behavior. Because of the ease with which a surface adsorbate can relocate a few tenths of an angstrom, such Franck–Condon effects can be expected in general in electron transfer processes.

The relation between the shift in energy level between the occupied and unoccupied surface state and the shift in local bonding energies is easily derived.[103] Consider the adsorbate A^{n-} with the preferred bonding site S_1 that forms the surface complex $S_1 \cdot A^{n-}$. The valence state $n-$ represents the oxidized form of the adsorbate when the surface state is unoccupied. The corresponding energy level of the surface state, when the atoms are in the configuration represented by S_1, will be denoted by E_{t1}. Then the reaction

$$S_1 \cdot A^{n-} + e = S_1 \cdot A^{(n+1)-} \qquad (5.9)$$

represents the exchange of electrons between the state E_{cs} (say the conduction band of the solid) and the surface state E_{t1}. During this rapid electron transfer process the system is maintained in the configuration represented

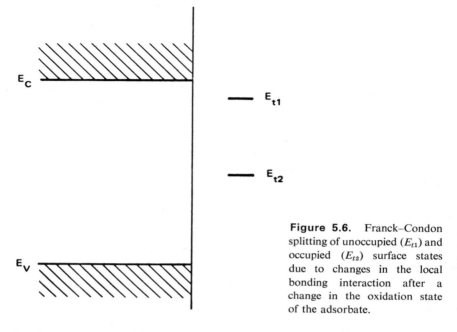

Figure 5.6. Franck–Condon splitting of unoccupied (E_{t1}) and occupied (E_{t2}) surface states due to changes in the local bonding interaction after a change in the oxidation state of the adsorbate.

by S_1. The energy $(E_{cs} - E_{t1})$ will be released by the electron transfer. Figure 5.6 shows the energy level E_{t1} on a band diagram.

In the oxidation state $-(n + 1)$ the adsorbate A will probably form a different bond with the surface. In particular, suppose the ion relocates after election capture according to Equation (5.9) and the system shifts to a lower-energy configuration denoted by S_2. The relocation can be described by

$$S_1 \cdot A^{(n+1)-} = S_2 \cdot A^{(n+1)-} \tag{5.10}$$

which represents a chemical equilibrium between the complex $S_1 \cdot A^{(n+1)-}$ and $S_2 \cdot A^{(n+1)-}$. There will be a free energy change $\Delta G''$ associated with the transformation. Another equation can be written describing the equilibrium electron transfer process while the atomic configuration is maintained as S_2:

$$S_2 \cdot A^{(n+1)-} = e + S_2 \cdot A^{n-} \tag{5.11}$$

with the electron moving from the surface state E_{t2} to the conduction band, releasing the energy $E_{t2} - E_{cs}$. Finally another chemical transformation equation can be written describing the relocation of the system to its lowest-energy configuration S_1 while it is in the oxidation state $(-n)$:

$$S_2 \cdot A^{n-} = S_1 \cdot A^{n-} \tag{5.12}$$

and we will denote the free energy change in this case as $\Delta G'$. The total energy change as the system progresses around the cycle represented by Equations (5.9)–(5.12) is of course zero, whence

$$E_{t1} - E_{t2} = -(\Delta G'' + \Delta G') \tag{5.13}$$

where it is noted that the ΔG's as defined above are negative, so $E_{t1} - E_{t2}$ is positive, as indicated in Figure 5.6. Thus the splitting of the surface state energy levels by the Franck–Condon behavior is equal to the gain in free energy realized by the atom movement, summed over the values for the two relevant oxidation states.

The ratio of occupied to unoccupied surface states, that is (equivalently) the concentration ratio of reduced to oxidized species at the surface, can be calculated as a function of E_F. In the two electron transfer expressions (5.9) and (5.11) the occupancy must be described at equilibrium by the Fermi distribution function [Equation (2.32)], and in the two chemical transformation expressions (5.10) and (5.12) the equilibrium is described thermodynamically. Then we have from Equation (5.9) and the Fermi function:

$$\frac{[S_1 \cdot A^{(n+1)-}]}{[S_1 \cdot A^{n-}]} = \exp\left(-\frac{E_{t1} - E_F}{kT}\right) \tag{5.14}$$

Here the square brackets on the left-hand side indicate density, so the left-hand side represents the ratio of occupied to unoccupied states. From Equation (5.10) we have

$$\frac{[S_2 \cdot A^{(n+1)-}]}{[S_1 \cdot A^{(n+1)-}]} = \exp\left(-\frac{\Delta G''}{kT}\right) \tag{5.15}$$

where thermodynamic equilibrium is assumed. Two equivalent equations obtain from Equations (5.11) and (5.12). From these four equations the ratio of the concentration of $S_1 \cdot A^{n-}$ to $S_2 \cdot A^{(n+1)-}$, the two stable forms, in terms of the Fermi energy can be immediately obtained. The result is

$$E_F = \tfrac{1}{2}(E_{t2} + E_{t1}) + \tfrac{1}{2}(\Delta G'' - \Delta G') + kT \ln\left(\frac{[S_2 \cdot A^{(n+1)-}]}{[S_1 \cdot A^{n-}]}\right) \tag{5.16}$$

(written for convenience as an expression for E_F). The occupancy at equilibrium (the ratio of reduced to oxidized species) is thus the same as if there were a single energy level, not at E_{t1} or E_{t2}, but at an intermediate energy E_m given by

$$E_m = \tfrac{1}{2}(E_{t1} + E_{t2}) + \tfrac{1}{2}(\Delta G'' - \Delta G') \tag{5.17}$$

Thus at equilibrium the system can be described as a single energy level halfway between E_{t1} and E_{t2} but modified by the difference in relocation energies for the two oxidation states of the adsorbate.

In summary a description of electron transfer to or from surface states must account for chemical bonding at two stages of the theory. First, in order to predict the surface state energy level of the adsorbate/solid bond before electron transfer, the local chemical interaction must be taken into account as was described in Section 5.3. Second, in order to describe the occupancy of the surface state (as determined by electron transfer), the difference in the local bonding for the two oxidation states must be accounted for. The first consideration is usually neglected when discussing electron transfer by the rigid band model, using the philosophy that it is not necessary to know the origin of the surface levels. The implications of the second consideration are much more complex, and unfortunately cannot be neglected in general. As the energy differences associated with bonding shifts (the ΔG's) become large, by Equation (5.13) the levels split substantially, and the surface molecule approximation may become preferred. Even with less splitting, the bonding shifts must be treated differently for different surface state measurements. For example the UPS measurement will give the energy level E_{t2} of Figure 5.6, whereas an equilibrium surface conductance experiment will measure the level as E_m of Equation (5.17). In general, kinetic (nonequilibrium) measurements will yield E_{t1} or E_{t2}, while equilibrium measurements will yield E_m.

5.5.2. The Influence of a Polar Medium or Coadsorbate on the Surface State Energy

In Section 2.3.3 the Gerischer model for the rate of electron transfer to a surface state in a polar medium was described. In that discussion it was pointed out that the "reorganization energy" λ represented the change in the energy of the polar medium due to reorientation of dipoles following the process of electron transfer to or from the surface. It was also pointed out that this dipole reorientation, depending on the occupancy of the surface state, causes a shift in energy levels, a Franck–Condon splitting of the mean occupied from the mean unoccupied energy level. This Franck–Condon splitting due to polar effects, to be described in this section, is entirely analogous to the splitting due to bond shift described in the last section.

Consider an unoccupied surface state associated with a foreign ion in the presence of a polar species. Its energy level is affected by the presence of the polar species, particularly if the foreign ion is charged, for then the surrounding dipoles have become oriented in response to this charge. Now let an electron be transferred from the solid to the surface species. The electron transfer is "instantaneous" in accordance with the Franck–Condon principle. Following this electron transfer, the surrounding dipoles will reorient in response to the new charge on the surface state. During the reorientation the electric potential at the surface state changes, so the surface state energy level changes. When the system reaches equilibrium with the surface state occupied, the electron is at an energy level different from its energy level at the instant of transfer.

Mathematically, the analysis of the reorganization of the dipoles near the surface state is identical to the analysis of shifts in bonding described in Section 5.5.1 above. Thus Figures 5.6 and 2.5 are entirely analogous insofar as the shift of the mean energy of the surface states is concerned. Of course in Figure 2.5 a large amplitude of temporal fluctuations around the mean surface state energy is indicated, because, as was described in Section 2.3.3, the presence of a highly polar species causes the surface state energy level to fluctuate and calculations show these fluctuations are substantial. With Franck–Condon shifts due to shifts in bonding (Figure 5.6), such temporal fluctuations are not so easily calculable.

The magnitude of the Franck–Condon splitting in the case of a polar medium is obtained directly from Equations (2.48) and (5.13). From Equation (2.48) the energy ΔE_p is the chemical energy shift, related to the ΔG's of Equations (5.10) and (5.12). With $\gamma = 1$, $\Delta E_p = \lambda$ is the energy change in the polarization field that follows the removal or addition of one electron. Further, we can say that $\Delta E_p = \Delta G'' = \Delta G' = \lambda$, for in each case [Equations (5.10) and (5.12)] the G's represents the reorganization energy after the electron transfer, in one case electron capture and in the other electron injection. Thus from Equation (5.13) we have immediately

$$E_{ox} - E_{red} = 2\lambda \tag{5.18}$$

and the Franck–Condon splitting is as shown in Figure 5.6.

For convenience we have returned to the symbols for surface state energies that were used in Section 2.3.3 and in other sections where the occupied surface states and unoccupied surface states are intentionally added (as chemical reducing agents and oxidizing agents, respectively). Equation (5.18) will be referred to primarily when discussing charge transfer to the surface at the solid/liquid interface, because such polar effects are most obvious there and thus all studies to date have been made on this system. Thus the notation with "red" indicating a reducing agent and "ox" indicating an oxidizing agent is convenient.

However, it should be pointed out, as was done also in Section 2.3.3, that the analysis of polar effects is not restricted at all to movement of charge to species in liquids. Coadsorbed polar molecules such as water will reorient and cause such effects, as will the dipoles in polar semiconductors such as TiO_2. The only question in such cases is the derivation of an effective λ. With coadsorbates in particular the analysis for λ is not as simple as in the case of ions in a polar liquid.

5.5.3. Surface States due to Multiequivalent Foreign Adsorbates

In the above models showing the influence of chemical interaction on the energy levels of surface states associated with sorbate/sorbent complexes, it was assumed always that the system was in its ground state. Yet some of the most important chemistry is associated with species in the form of radicals, that is in an unstable valence state. And the production of radicals at a surface is not uncommon; in fact it is to be expected. Transfer of a hole or an electron to a species adsorbed on the surface of a solid obviously changes its charge by one equivalent, and many if not most chemical species are two-equivalent: a change in valence by one results in an unstable species (a radical) and it requires a change in valence by two equivalents to reach the next stable valence state. Thus transfer of a hole or electron to the surface will result in an unstable radical that must either capture or give up an electron to regain a stable valence. Radical generation by minority carrier capture is particularly important, as with the limited availability of the carriers the radical so produced may be long lived.

In the following we will first describe qualitatively the model of the energy level diagram that represents[104] the two-equivalent surface species (using the rigid band model, and ignoring local chemical interaction). Then a few examples of radical generation and reaction will be cited to illustrate the practical implications both in the chemistry and physics of surfaces.

The surface state scheme associated with a two-equivalent adsorbate is indicated in Figure 5.7. The example of an arsenic ion adsorbed on an n-type semiconductor is chosen for illustration; the arsenic ion has stable valences $+3$ and $+5$. If the adsorbed ion is in the $+3$ oxidation state, the energy level of its most energetic electron is indicated in Figure 5.7 as the level As^{+3}/As^{+4}. If the level is occupied, the species is As^{+3}; if the level is unoccupied, the species is As^{+4}.

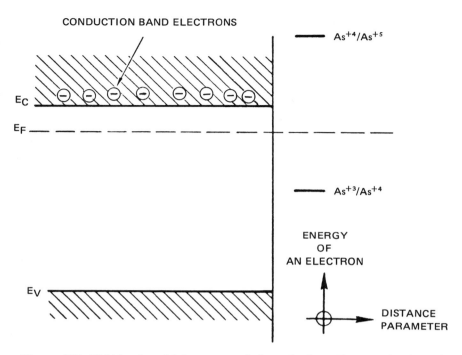

Figure 5.7. Rigid band model for a two-equivalent adsorbate. The example of arsenic is used, where $+4$ is the unstable valence. The semiconductor could be ZnO, for example (see Section 8.3.2).

Now as mentioned, the intermediate radical species As^{+4} is unstable. If it is formed, it is energetically favorable to recapture an electron and restore the stable As^{+3} form so the As^{+3}/As^{+4} level is as indicated at a low energy level. However, the unstable As^{+4} radical alternately can reach stability by losing its unpaired electron, forming the stable As^{+5}. This process also can be exothermic. That is, the unpaired electron on the As^{+4} is at a high energy, and its energy level (As^{+4}/As^{+5}) can thus be as indicated above the As^{+3}/As^{+4} level in Figure 5.7.

In summary when the ion is in the valence state $+4$, the topmost level is occupied and the lower level is unoccupied, and as is clear on the figure, it is energetically favorable either to lose an electron forming As^{+5}, or to gain an electron forming As^{+3}. Obviously there is no position of the Fermi energy that will lead to a high density of As^{+4}.

The analysis of the equilibrium occupancy of the two states[104] is a straightforward application of Fermi statistics [Equation (2.32)] to each of the energy levels, and the algebraic elimination of the concentration of the radical species to obtain the relative concentration of the two stable species as a function of the Fermi energy. Thus with E_1 the energy level of the lower level (As^{+3}/As^{+4}) and E_2 the energy level of the upper level (As^{+4}/As^{+5}), we have from the Fermi function, Equation (2.32):

$$\frac{[A^{(n+2)-}]}{[A^{(n+1)-}]} = \exp\left(-\frac{E_1 - E_F}{kT}\right) \tag{5.19}$$

$$\frac{[A^{(n+1)-}]}{[A^{n-}]} = \exp\left(-\frac{E_2 - E_F}{kT}\right) \tag{5.20}$$

where for generality we use the symbol A and let A^{n-} be the oxidized species, for which the next stable oxidation state is $-(n + 2)$. Eliminating the concentration of the unstable radical $[A^{(n+1)-}]$ from the equations, we find

$$E_F = \tfrac{1}{2}(E_2 + E_1) + \tfrac{1}{2}kT \ln \frac{[A^{(n+2)-}]}{[A^{n-}]} \tag{5.21}$$

Thus if both species A^{n-} and $A^{(n+2)-}$ are at the surface at equilibrium, the Fermi energy will be between the two energy levels, modified by the ratio of $[A^{(n+1)-}]$ to $[A^{n-}]$.

Experimentally, the best evidence for radical generation effects by electron (or hole) transfer comes from studies of optical effects at a semiconductor surface.

For example, it has been shown in ESR measurements[105] that when the semiconductor TiO_2 is illuminated, holes are produced which react with the surface species OH^- forming the radical OH. This radical generation is a cause for concern in paints pigmented with TiO_2, for when the paint is illuminated by solar ultraviolet, the resulting radicals lead to chemical reactions in the paint binder and cause chalking of the paint. The same generation is a cause for optimism that TiO_2 can be used to photolyze water using solar uv,[106,107] thus converting solar energy to chemical energy in the form of hydrogen.

Other more quantitative studies of radical production at the surface of a semiconductor have been made at the semiconductor/electrolyte interface, and will be reported in Section 8.3.2. In most of these studies the radical is formed by minority carrier capture and subsequently injects a majority carrier into the appropriate band of the semiconductor, so that in the end the species changes its oxidation state by two units. Included in such studies are measurements of the competition for the reactive radicals between the solid and other surface species. In other words, the radical once formed can react by injecting a carrier into the solid or by reacting with a coadsorbate.

To conclude the discussion we will mention qualitatively how two-equivalent effects lead to overvoltages in electrochemical reactions and to trapping in semiconductor photoprocesses. For a more quantitative analysis see Ref. 104. Consider the reduction of As^{+5} to As^{+3} at an electrode, either a semiconductor or metal electrode, so that Figure 5.7 can be used for illustration. For such reduction to occur, by a strict rigid band model, neglecting local bonding, an electron must first transfer to the upper level. When this transfer is accomplished, the second transfer, to the As^{+3}/As^{+4} surface state, is highly exothermic. Thus the overall free energy of the reduction may be favorable, but the first step provides a large activation energy for the process, a large overvoltage for the electrochemical reduction. The band model illustrates the source of the overvoltage very clearly.

If one is concerned with the electrical effects of surface states, it should be noted that such a set of energy levels can lead to carrier trapping. Again consider the example of Figure 5.7. If a photoproduced hole is captured at the energy level As^{+3}/As^{+4}, the resulting unpaired electron on the As^{+4}/As^{+5} level will be immediately injected into the semiconductor. Essentially two holes are trapped at the surface. The return of the ion to the As^{+3} form, that is the neutralization of the holes by electrons, requires promotion of an electron to the upper level, a highly activated process.

5.6. The Formation of Surface State Bands

The broadening of the single-valued molecular orbitals of an adsorbing species into bands or apparent bands can occur through several mechanisms, each of which has been noted previously, but not emphasized. Four of the most important are heterogeneity of the surface sites, overlap of the orbitals of the adsorbate, temporal fluctuations due to the presence of polar species, and finally, the interaction of the molecular orbitals of the adsorbate with the bands of the solid. This section is to summarize these effects that lead to apparent band formation. We will distinguish between "real" bands, broadened by orbital overlap effects, where such effects as impurity band conductivity can be expected, and "apparent" bands, due to a variation in surface state energy with time or distance. In the case of real bands, orbitals can be fractionally occupied, whereas in the other cases each surface state must be occupied by an integral number of electrons.

Heterogeneity of the surface sites will broaden the observed surface state distribution into apparent bands if the measurement is averaged over an area of surface. The various forms of heterogeneity described in Section

1.4 will all contribute, for the bonding and hence the location of the surface state will differ in general depending on the details of the bonding site. In addition, variability in bonding site can occur due to relocation processes even with a homogeneous surface. An example of this is the bonding of CO to W, discussed in Section 5.4.3. In that case partial reconstruction or low-temperature adsorption on a reconstructed surface was shown to give a multiplicity of surface state energy levels.

Franck–Condon splitting as described in Section 5.5.1 may have the effect of broadening the surface state energy levels into a distribution, particularly if the splitting is associated with heterogeneity of sites.

Temporal fluctuations due to polar effects will effectively broaden the surface states into an apparent band if the measurement is averaged either over time or over an area of the surface (Figure 2.5). In the case of a surface state when coadsorbed polar molecules are present, the number of adsorbed polar species associated with each surface state will not be constant over the surface and this will provide further "heterogeneity" and so apparent band formation.

Overlap of the orbitals of the adsorbate will cause real surface state band formation as the surface coverage approaches a monolayer. The band formation is based on the same quantum effects that cause band formation in all condensed phases. As will be discussed in Chapter 7, adsorption often occurs in "islands," and in this case band formation will occur at very low coverage.

Finally the surface states can be broadened into bands if there is a strong interaction between the orbital of the adsorbate molecule and a particular band of the solid (Section 5.3.3). With no interaction there is of course no broadening of the orbital of the adsorbate molecule. As the interaction increases, a surface state band forms as illustrated in Figure 5.3.

6 | Nonvolatile Foreign Additives on the Solid Surface

6.1. General

In this chapter we will discuss foreign species on the solid surface which, at the temperatures of interest, are nonvolatile, and which, under the conditions of the experiment, are not removed by chemical interaction. Such species can be accidental impurities or can be intentionally added to provide some control or function on the surface. They can be in the form of a low concentration of atoms present in less than a monolayer on the surface, perhaps on particularly active surface sites, or they can be in the form of a thin film of a second phase, but most often they are in the form of small crystallites deposited out of solution.

Such additives on the surface are of dominating importance in many surface phenomena. We can cite examples of unintentional and undesirable additives: The deposition of copper on germanium is one such case, the copper appearing during etching and washing of the transistor material even when there is much less than a part per million copper in the de-ionized wash water.[1-3] Such copper decreases the minority carrier lifetime of the germanium wafers[1] and also provides a source of acceptors which diffuse into the germanium if it is subsequently heated above 500°C. In the field of chemistry we can cite the example of many poisons that deposit during catalytic processes. On the other hand, we can cite examples of the use of intentionally added nonvolatile species to improve the surface properties of solids: silica on silicon to passivate the surface with respect to surface recombination, platinum on silica or silica/alumina to provide a high activity for various catalytic reactions, and corrosion inhibitors applied to a metal surface to prevent oxidation, are a few cases that are well beyond the research stage and in industrial use. As our understanding of the behavior of such additives increases, there is no doubt that their use in the control of surfaces, either for passivation or activation, will expand. Thus although there is to date little fundamental theory of the behavior of such additives, it is important to describe what is known, for the approach represents an important phase of surface science.

Foreign additives on the surface can have two functions. One is to control the surface properties of the substrate. The other is to act as an independent or semi-independent material. The additives copper on germanium, silica on silicon, or corrosion inhibitors on metals, control the properties of the substrate material. The additive platinum on silica or alumina acts as an independent material. In the latter case the substrate is termed the "support" and is usually of secondary interest. However, the support is important, for it permits large surface areas of platinum to be exposed, which would be impossible with a pure metal.

In most cases the additives are deposited on the surface from the liquid phase. As will be discussed in the next subsection, deposits from solution usually precipitate in the form of crystallites rather than monolayers on the surface. To form a monolayer of additive from solution one needs a very strong interaction between the substrate and the additive, and usually the interaction is relatively weak. Thus one of the first problems in using a surface additive to control the surface properties of a solid is to obtain a high dispersion of the additive over the solid surface.

6.2. Dispersion of Additives

6.2.1. Techniques for Dispersing Additives

Normally when depositing additives on a surface the objective is to make the dispersion D, defined as the ratio of the number of atoms exposed to the number of atoms added, equal to unity. A high dispersion means that the additive is being used effectively. If one is concerned with cost of the additive, as with platinum, it is clear that the surface area (and hence the catalytic activity) per gram of platinum is directly related to the dispersion. If one is concerned with control of the surface properties of the substrate, a high dispersion is needed so that the electrical or chemical effects will be uniform over the surface. If the additive forms large crystallites with broad expanses of additive-free area between, it will be ineffective. In this section we will describe dispersion techniques, measurement of the dispersion, and sintering (coalescing) of the dispersed additive.

There has been little work on dispersing additives on the surface of single crystals. In the few studies in the author's laboratory[4,5] the additive was sprayed on with an atomizer with the sample heated to evaporate the solvent (water) as the droplets settled. The rate of deposition was calibrated by placing a container of de-ionized water at the sample location, and observing the conductivity of the water increase as the droplets reached the surface. Anderson *et al.*[6] have successfully utilized an evaporation technique to obtain metal particles as small as 20 Å in diameter on a flat surface.

On the other hand reasonably sophisticated techniques have been developed for depositing additives on powders.[7] In this case the motivation has been strong: such structures are needed in industrial catalysis.

Perhaps the least sophisticated technique for dispersion on powders, and yet probably the one most used in research studies where maximum dispersion is not required, is simple "impregnation" where the support powder is impregnated with a solution of the additive, and the solvent is evaporated to dryness. This has several disadvantages, for example, during the solvent evaporation much of the additive moves to the surface of the powder residue, so that the concentration is not uniform.

If the suspension (including the solid) is sprayed from an atomizer[8] such that each support particle is coated with a film of solution and is deposited on a warm surface, better uniformity is obtained. The particle size of the additive can be 70–100 Å using these techniques.

A preferred technique is the precipitation of the additive onto the support.[9] The reason for the difference is discussed in Section 7.3.1. For example, careful precipitation of chloroplatinic acid onto alumina, effected by increasing the pH until the platinum precipitates as the hydroxide, permits particle sizes averaging 20 Å, compared to the hundred angstrom range by simple impregnation. Nickel precipitated from the nitrate yields 20–40 Å particles. Another technique that results in high platinum dispersion is the use of H_2S to precipitate platinum as a sulfide.[10] A colloidal suspension of the platinum sulfide and the support is oxidized by bubbling air through the liquid, and the resulting supported catalyst is dried and calcined. Heard and Herder[11] improved further on this sulfide approach by introducing ammonium sulfide to permit much higher concentrations of platinum sulfide to be maintained in the solution without agglomeration.

Several authors[12–15] have used an ion exchange method (Section

4.3.3.1) of dispersing metal ions, particularly dispersing nickel and palladium on a silica/alumina support. Morikawa *et al.*[15] used a two-step process, where first they exchange the protons normally present (Section 4.3.3.1) on the washed surface of the silica/alumina for ammonium ions. This first exchange is effected by soaking the support in ammonia. They find the resulting adsorbed ammonium ion can exchange readily in a second exchange step with metal cations, whereas the original adsorbed protons do not. By such a double exchange they obtain, for example, palladium in what they estimate must be close to atomic dispersion. Others[9] consider 10 Å particles more realistic.

Yermakov *et al.*[16] have developed an attractive method of obtaining a high dispersion and maintaining it. To provide a strong bond between the metal additive and the ionic solid supporting it, they introduce an intermediate species that will bond to both. The intermediate species is an irreducible (under the conditions of interest) transition metal ion deposited first onto the surface. The ion on the one hand forms an ionic bond to the ionic solid, and on the other hand with its remaining d electrons forms a covalent bond to the metal atom. Thus for example Mo^{+2} is deposited using an organometallic molybdenum molecule, then Pt or Pd is deposited on top of the Mo^{+2}. The reduction process is chosen such that the noble metal is reduced but the Mo^{+2} is not. An atomic dispersion is apparently attained if a moderately low temperature of reduction is used in the process.

6.2.2. Measurement of Dispersion

The determination of the particle size of small metal crystallites once the additive is dispersed can be made by many methods. Whyte in his review[17] suggests chemisorption, x-ray line broadening, electron microscopy, small-angle x-ray scattering, and magnetic methods.

Measurement of the adsorption of a gas that chemisorbs only on the additive yields the surface area of additive, and a "surface averaged" particle diameter is obtained from the formula

$$d = 5/\varrho A_s$$

where ϱ is the density and A_s is the measured surface area per gram of additive metal. This formula for diameter is obtained simply by assuming the additive metal is in the form of cubes with one of the six faces inaccessible to the adsorbing gas. The key problem in this method is to have a gas that chemisorbs in a known way on the additive but not on the support. Table

Table 6.1. Chemisorption of Gases on Metals[a]

Metal	N_2	H_2	O_2	CO	C_2H_4
W, Mo, Zr, Fe	+	+	+	+	+
Ni, Pt, Rh, Pd	—	+	+	+	+
Zn, Cd, Sn, Pd, Ag	—	—	+	—	—
Au	—	—	—	+	+

[a] + indicates chemisorption, − indicates no chemisorption.

6.1 from Whyte,[17] indicates which gases form chemisorbed layers on various metals and provides a starting point for developing a chemisorption-based surface area measurement.

Other related methods are titration methods. An increase in sensitivity and in reliability can be obtained if one first adsorbs oxygen on the metal, then reacts it with hydrogen,[18-22] or with carbon monoxide.[23,24] By measuring the original adsorption (of O_2), the adsorption of the reducing agent (H_2 or CO), and finally desorption of the product (H_2O or CO_2), one obtains several crosschecks that the adsorption is proceeding according to the theory. Also, the amount of the reducing agent adsorbed is increased when the titration rather than the simple adsorption measurement is used. Enough H_2, for example, is consumed to form the H_2O as well as to form an adsorbed monolayer of hydrogen, and this increased consumption increases the sensitivity.

The use of x-ray line broadening to obtain the average particle size provides a different averaging process than the surface area measurement, an averaging process that weights the larger particles more. Electron microscopy is the most reliable method for moderate size particles,[17,25] but is a tedious technique.

6.2.3. Sintering of Dispersed Particles: Surface Diffusion of Adsorbates

To complete the discussion of the dispersion of additives, it should be pointed out that maintaining a highly dispersed additive is not straight-forward. The term "sintering" is used in this field to mean the decrease in surface area of the additive with time arising from a general increase in particle size. The atoms from the small particles become absorbed into the large particles. Sintering of a supported additive can occur either by gas-phase transport, diffusion of atoms as an adsorbed species, or even motion of crystallites across the surface of the support.[26]

In general, sintering is most rapid when the supported material is in the zero-valent state, although other factors also modify the sintering rate. For example, in catalytic work platinum that has been formed by impregnation of chloroplatinic acid must usually be reduced to platinum metal to provide the required catalytic activity. The reduction in hydrogen will require an elevated temperature, and during or following reduction the platinum metal atoms may sinter into much larger clusters.[27] Baker *et al*.[28] have studied the sintering of platinum by electron microscope observation (minimum particle size observable $= 25$ Å). They examined platinum prepared by vacuum evaporation, by spraying an ammoniated platinum hydroxide, and by spraying chloroplatinic acid. In all cases the initial particle size was below the observable limit. However, metal particles originally deposited by the evaporation and the hydroxide techniques nucleate at 500°C to form particles visible in the microscope. The metal particles prepared using the chloroplatinic acid technique require 650°C before sufficient sintering occurs.

The mechanism of sintering is still under heated discussion, probably because there are so many mechanisms possible and the mechanism for any particular system depends on many details of the system used (the metal, the support, and the ambient gas and temperature). In general one can say that during a sufficiently high-temperature pretreatment or during a high-temperature catalytic reaction sintering can and will occur for the very small clusters under discussion.

The diffusion of adsorbed species is of interest not only in sintering but in adsorption (where motion to form certain surface structures is observed, as discussed in Chapter 7) and in catalysis (where reactants may diffuse together to react). We will not review in detail the mathematics of two-dimensional diffusion that describes the possible movement of adsorbates, but will digress for a few paragraphs to discuss the qualitative features.

The activation energy for surface diffusion is related to the strength and localization of the bonding of the sorbate to the surface. If the bonding is strong and highly localized, a strong bond must be broken and re-formed as the adsorbate moves from adsorption site to adsorption site, and the energy of the activated state (where the sorbate is between adsorption sites) is expected to be high. If the bond is weak or nonlocalized, the energy of the activated state is not much greater than the energy on an adsorption site.

Thus, the activation energy for diffusion, the difference of these energies, is low and diffusion rapid.

For example, a neutral atom on the surface of an ionic solid may in many cases move relatively freely,[29] as there is no strong bond. Thus in the above discussion of sintering of platinum metal particles on silica, rapid sintering at reasonably low temperatures (here the order of 500°C) was expected and found. Similarly, spillover by hydrogen atoms diffusing rapidly onto an ionic support is understandable (Section 6.4.2.3). On the other hand, diffusion of an ion on an ionic solid, particularly if, say, a strong acid/base bond is formed with the solid surface, is expected to be much less rapid, because bonds between ions are usually highly localized. Thus sintering of the platinum on silica is not a problem until the platinum is reduced to the metal. In the case of covalent bonding there is more variation both in the localization and in the strength of the bonds, and thus more difficulty in defining the energy of the activated state. It is harder therefore to classify groups that should move easily. In some cases, although the heat of adsorption is substantial, the activation energy for diffusion can be low. We discuss in Section 7.1.2.1 that adsorption on metals may occur preferentially through adsorption on active sites and then diffusion to cover the surface. Klein[30] and Gomer et al.[31] report an instructive study using an FEM tip to monitor diffusion on metals. They deposit several monolayers of adsorbate on one side of their tip using a molecular beam so that the other side is in the shadow. After depositing the adsorbate at low temperature, they raise the temperature and they observe, by the associated FEM signal, that the adsorbate diffuses at reasonably low temperature to the other side of the tip. It is found that the diffusion is interrupted as the moving boundary reaches a corner or edge where there are active sites. However, the sites become saturated and the diffusion proceeds. Such processes are reviewed by Bonzel[89] and by Bassett.[90]

Heterogeneity of the surface will affect surface diffusion because adsorbate species will be trapped at highly favorable sites. Such adsorbate atoms must not only overcome the activation energy between identical surface sites in order to move, as is the case with a uniform surface, but they must make the initial jump from their low energy position. Thus, for example, Tanaka and Miyahara[32] in studies of O_2^- adsorbed on ZnO, first adsorbed $^{16}O_2$, then $^{18}O_2$. Upon thermal desorption the $^{18}O_2$ desorbed first, indicating the $^{16}O_2$ occupied the active sites, and there was no subsequent mixing by diffusion between the more active and less active sites.

6.3. The Cluster, the Transition between a Molecule and a Solid

In addition to the obvious properties (high surface area per unit weight, effective coverage of the substrate) associated with a highly dispersed supported additive, these clusters of atoms, with from 2 to 100 atoms in a cluster, are expected to have some very unique bulk and surface properties not attainable in any other form of a solid. Such properties will be reviewed briefly in this section. We will first describe the theoretical models and then discuss the experimental approaches that have been used to test them. With the tremendous technical interest in high surface area catalysts, most of the incentive for studies of such clusters has stemmed from catalytic chemistry. Fortunately for those with other specialities, catalytic behavior seems to be related to practically every property of matter, so most of the unique properties of such clusters have been mentioned in the catalytic literature.

As the number of atoms in a cluster increases from one to several hundred, the atomic or molecular levels become transformed into a band of levels. The energy level diagram of intermediate size clusters is a subject of theoretical and experimental interest. We have already discussed the transition very briefly in Section 5.3.2, where we described the quantum mechanical cluster theories that showed that as the number of atoms in a cluster increases, the energy levels split, multiply, and develop into incipient bands. Figure 6.1 shows the results of cluster theory of calculations reported by Slater and Johnson[33] on a cluster of eight copper atoms. The emerging d band is represented as shown by a dense group of energy levels; the $a_{1g} - t_{1u}$ interval can be considered the precursor of the sp band in bulk copper. Jones et al.[34] report calculations on a 13 atom cluster (of Fe and Ni as well as Cu) where the influence of surface steps is suggested. The importance of surface steps in these clusters is discussed below. One of the most interesting factors with respect to the "bulk" properties of the cluster, apparent in Figure 6.1, is the large energy gap between the highest occupied level (here the t_{1u} level) and the lowest unoccupied level (here the t_{2g} level). Such a broad gap will not occur in clusters of transition metal atoms, of course, since for those cases the Fermi energy is in the d-band region where the levels are close together. Baetzold[35] reports calculations showing the "band gap" for Ag changes from a few electron volts for a single atom to 0.5 eV with 20 atoms. With semiconductors or insulators, or even carbon, large band gaps persist for larger clusters of up to 35 atoms. Baetzold[35] reviews other unique bulk effects in clusters. For example,

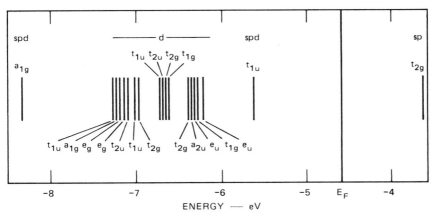

Figure 6.1. Energy levels for a Cu_8 cluster as calculated by $SCF_x\alpha$ analysis. The t and e symbols indicate the symmetry of the energy levels, and the s, p, and d symbols indicate hybridization of the levels. The Fermi energy E_F lies midway between the uppermost occupied level and the lowest unoccupied level.

there will be effects on the bonding energy of the atoms in a cluster—the bond energy will differ depending on whether there is an odd or even number of atoms in the particle. Similarly, the spin susceptibility depends on whether there is an odd or even number of particles. The ionization potential varies: the highest occupied orbital determines the ionization potential and is intermediate between the ionization potential of an atom and the work function of a solid. According to semiempirical molecular orbital theories, the ionization potential decreases from 7.6 eV for an Ag atom to 5.5 eV for Ag_{27}, approaching the work function of silver of 4.5 eV.

Another unique characteristic of these clusters is the number and type of active sites on the surface. Thus in addition to the rapid variation in the band structure between say 40 Å and 5 Å clusters, an equally important variation that must be taken into account is the variation in the coordination of surface atoms. A cluster of 4 atoms means every atom is a corner atom, and every atom is poorly coordinated. A cluster of up to about 64 atoms still must have more atoms as edge atoms than as atoms in a plane, and such edge atoms are also relatively poorly coordinated (have too few nearest neighbors) and so are active sites. As pointed out by Hardeveld and Hartog,[9] it is reasonably straightforward, assuming symmetric clusters, to estimate how many corner atoms, how many edge atoms, and how many planar surface atoms are to be expected. However, in real clusters the clusters are not expected to be symmetric. Actually, they seldom will have the necessary number of atoms to be symmetric. We will not go into the symmetric models. For analysis of such models see, for example, Schlosser[36] or Romanowski.[37] It should be sufficient for the interpretation of experiments to recognize the qualitative feature—the number of poorly coordinated surface species will increase rapidly as the cluster size decreases from about 40 Å.

Theoretical cluster calculations have been made by Jones et al.[34] on 13 atom clusters assembled to demonstrate the orbitals at steps. They find pronounced changes in the lobes at the steps. The lobes have little resemblance to atomic orbitals, and their shape suggests enhanced catalytic activity.

As a final comment on the unique theoretical characteristics of clusters, it is of interest to mention the suggestion of Ruckenstein[38] and Sinfelt[39] that clusters may be expected to form homogeneous alloys even for compositions where the bulk materials would segregate into phases. Ruckenstein suggests that greater lattice strains can be accomodated in the very small crystallites.

Experimentally a reasonable amount of work has been done on the effects of clusters, but seldom is it clear whether the effects can be definitely ascribed to "bulk" band effects (the transition from a band to separated energy levels) or to coordination effects (the increase in the number of poorly coordinated surface atoms). One observation which appears to stem from band effects is that of Tanner and Sievers,[40] who made optical absorption measurements in the far infrared, and have concluded that they were detecting transitions between the highest occupied and lowest unoccupied levels in the clusters. However the lack of uniformity in particle size makes quantitative comparison between experiment and theory difficult.

An experiment which appears on the other hand to measure coordination effects is the work of Hardeveld and his co-workers.[41,9] They found that nitrogen adsorption on nickel at room temperature only occurred if the nickel particle size was between 15 and 70 Å. From a semiquantitative analysis they concluded that the special sites involved were sites where one of the nitrogen atoms in the molecule could coordinate simultaneously with five nickel atoms, "B_5 sites." On the (111) or (100) planes that are normally exposed this configuration is not possible. But small clusters cannot be formed with only (111) and (100) planes, and here (110) and (113) facets appear, and these have ridges where B_5 sites are available. For a series of samples of varying particle size estimated by electron microscope measurement, models were used to estimate the number of B_5 sites. Infrared absorption was used to measure the density of adsorbed N_2 on these samples. The adsorption followed the expected site density as a function of particle size.

Other effects described by Hardeveld and Hartog in their review[9] are not so clearly assignable to band or to surface site effects. One observation is a shift of wave number for the infrared absorption of adsorbed CO molecules that appears when CO adsorbs on small clusters. Another is a decrease in the hydrogen/deuterium exchange rate in benzene as the particle size decreases. Curiously although the H–D exchange is sensitive to particle size, the hydrogenation rate of benzene is roughly independent of particle size.

Other examples can be quoted of effects that appear to be associated with very high activity of small clusters. Ostermaier *et al.*[42] studied the oxidation of ammonia over platinum clusters and found the larger crystallites showed much higher activity per unit area. They concluded that the oxygen adsorbed on the small particles was more strongly bound and hence less active than oxygen adsorbed on the large particles. Wu and Harriott[43] have shown that a silver catalyst, for oxidation of C_2H_4 to C_2H_4O, begins to oxidize the C_2H_4 to CO_2 and water if the particle size becomes of the order of 20 Å. They conclude edge or corner sites that bond oxygen strongly produce CO_2, and the oxygen weakly bonded to the planar regions adds to the ethylene and forms the desired ethylene oxide.

It should be pointed out that particle size does not always have a strong effect on the activity (per unit area) of a catalyst. Reactions such as those where the presence of surface steps, corners, etc., are assumed to alter the rate significantly are termed "demanding" reactions, in that very specific catalyst properties are required. Reactions that are not demanding in this sense are often termed "facile." Schlosser[36] reviews various catalytic processes with respect to such sensitivity to particle size.

As more and more attention is being given to the behavior of clusters, there is no doubt that our understanding of their properties will increase rapidly in the coming years. And the topic is certainly a most fascinating one from a very fundamental point of view as well as being such an important one from a technical point of view, both in the established field of catalysis, and, as will be described, in the less established field of surface control by nonvolatile additives.

6.4. The Control of Surface Properties with Additives

6.4.1. Theoretical Discussion

There are two ways in which a solid surface can be controlled so that moderate changes in the ambient gas surrounding the surface no longer affect its properties. The first is to lay an inert protective film over the surface. Thus, for example, a thermally grown oxide on silicon provides a beautifully stable surface with excellent properties for the semiconductor.

The second method of controlling the electronic properties of a surface is to deposit a chemically active species on the surface in sufficient surface density that it will determine the surface properties and maintain these surface properties, resisting the influence of adsorbing gases. The additive species must of course be nonvolatile, and its reactions with gases and with holes or electrons from the solid must not lead to volatile products. This section is to a great extent devoted to examining the state of our knowledge about the use of such nonvolatile additives to control the chemical and electrical properties of solids. In the present subsection a few of the simpler and more general theoretical models will be described, and in the next subsection a few illustrations of the use of surface additives will be discussed.

6.4.1.1. Theory of Fermi Energy Pinning by Redox Couples.
On a semiconductor or insulator, in particular, surface state additives can be used, intentionally or accidentally, to control the surface barrier (pin the Fermi energy). The theory of their use in surface control is straightforward,[44] although the realization at times is difficult. Suppose $10^{18}/m^2$ (about 0.1 monolayer) of each valence state of one-equivalent (differing in oxidation state by 1) redox couple is deposited on the surface of an n-type semiconductor. For example, $FeCl_3$ and $FeCl_2$ could be deposited, with Fe^{+3} providing the acceptor state. In Chapter 2 it was shown that by the Weisz limitation the maximum number of electronic charges that could be transferred to the semiconductor surface was the order of $10^{16}/m^2$. Electrons from the conduction band would be captured at the surface, but a negligible fraction, only 1% of the states, would change their electron occupancy. Thus the surface state corresponding to this redox couple would be half-occupied (within 1%). Now the Fermi energy is given from Equation (2.32):

$$E_F = E_t + kT \ln \frac{[Red]}{[Ox]} \qquad (6.1)$$

where we have denoted the concentration of occupied states n_t by the surface density of reducing agent [Red] and have denoted the concentration of unoccupied states p_t by the surface density of oxidizing agent [Ox]. Clearly if the concentrations of reducing and oxidizing additives are about equal, the Fermi energy is firmly pinned (anchored)[44,45] to E_t, the surface state energy. If the concentrations are unequal, the Fermi level is still pinned close to E_t, as the last term in Equation (6.1) is small.

As will be discussed below, it is relatively simple to identify a redox couple with a suitable value of E_t to pin the Fermi energy at a desired position. The difficulty in the approach arises because such redox couples cannot maintain their stability in the presence of atmospheric gases. We will first show that a high dispersion is not usually needed, then discuss the choice of redox couple to provide a given surface configuration. Finally we will discuss the electronic and chemical difficulties associated with the approach.

The use of this approach to pin the Fermi energy at the surface of a semiconductor requires a reasonable dispersion of the additive surface states. However, atomic dispersion is not required because the distance between additive clusters can approach in magnitude the thickness of the double layer. With a depletion layer, this means the distance between additive clusters can be up to 1000–2000 Å (Figure 2.3) and the one-dimensional approximation to the double layer will still be adequate. It should be noted that although in the above discussion the term "surface states" was used, with a cluster of atoms including reducing and oxidizing agents of the same species (say, Fe^{+2} and Fe^{+3}), the levels are best described as an incipient band, a half-occupied band if Fe^{+2} and Fe^{+3} are added in equal amounts. In this case the Fermi energy of the solid at equilibrium is pinned at the Fermi energy in the cluster, and, with its band half-occupied, the Fermi energy of the cluster is near the midpoint of the band.

In order to apply these concepts, appropriate additives must be selected that will pin the Fermi energy at the desired value. Thus energy levels associated with various redox couples must be determined. For a first approximation to the expected surface state energy E_t (or E_F of the cluster) it has been found of value to relate the expected energy level of one-equivalent redox couples to their chemical behavior. It was indicated in Chapter 1 (Figure 1.1) that there should be a relation between the chemical activity of a species as a reducing agent (its tendency to give up electrons) and the surface state energy level of that species. As a measure of the tendency of an ion to give up electrons, one can begin with its ionization potential,[46] or some equivalent physical parameter, or with a chemical parameter such as its redox potential. The ionization potential is unsatisfactory for an ion such as Fe^{2+} because the surface state energy would be calculated from the difference between two very large numbers, and have a corresponding error. The redox potential of the additive couple is a more direct measure of chemical activity, and to some extent the environment of the ion in solution has some resemblance to that of the ion at the solid surface. The arguments are discussed in more detail in Section 8.4, but it seems that the best indicator of the expected surface state energy of a one-equivalent redox couple is the redox potential of the couple.[47] If the redox potential is large and negative, the couple has a strong tendency to give up electrons, and the surface state energy level will be expected to be correspondingly high in the band diagram. Since this model neglects energetic factors such as local surface bonding of the adsorbate at the gas/solid interface, it is clearly only a very rough first approximation.

With this approach the more fundamental electrochemical measurements can be used, as discussed in Section 8.3.3.4, to suggest where the energy levels of various redox couples will lie relative to the band edge of the semiconductor of interest. Then each redox couple that appears suitable can be tested quantitatively by methods suitable for the solid/gas interface (for example, the powder conductance measurement) as described in Chapter 3.

Surface state control of the Fermi energy, using the models above, with Equation (6.1) appears straightforward. Unfortunately, there are both electronic and chemical problems that must be considered. A problem in control of the electrical properties of the semiconductor arises if not only the surface Fermi energy but also the surface recombination velocity (Chapter 9) must be minimized, as is the case for most semiconductor devices. The same states that pin the Fermi energy will be expected to act to recombine excess electron/hole pairs. For example as shown by Jackson et al.,[48] surface state control of silicon by amines as electron donors or by cyano groups as electron acceptors is feasible. Such additives, however, will seldom be used in practice because the passive thermally grown silica layer is so much preferable: it not only protects the silicon against adsorbed gases but also provides a surface with a very low surface recombination velocity.

A problem associated with chemical interaction arises if the ambient medium contains a material that will react with the redox couple. For example, the redox couple Cr^{+2}/Cr^{+3} probably will not be useful in contact with air, for the atmospheric oxygen will react with the Cr^{+2}, oxidizing it to Cr^{+3}. Then the adsorption/desorption of oxygen will "control" the surface Fermi energy. Thus although surface state control by Equation (6.1) should be effective by theoretical arguments, in many cases practical problems must be recognized.

Another approach to the use of additives as surface states to control the surface properties is the use of two equivalent species as trapping centers at the surface. The expected effect of such additives as traps has been mentioned in the literature,[44] and was discussed in the last chapter, but controlled experiments have not been reported. Their use, for example, in photosensors may be of particular interest, for minority carrier trapping enhances photosignals from semiconductors.

6.4.1.2. Theory of Additives to Induce Acid or Base Centers.

Acidity control is an important form of control on ionic solids that has been developed to the point of industrial use (in catalysis primarily). Specifically, if the acidity of a surface is to be maximized, a treatment with a strongly electronegative anion such as the fluoride ion is used. If the acidity is to be lowered, a treatment with a strongly electropositive cation (of low valence) such as sodium or calcium is effective.

The effect of the electronegative anion in increasing the acidity is usually rationalized primarily by an induction mechanism as follows. The fluoride, say, exchanges with an OH^- group at the surface:

$$HF + OH_s^- = H_2O + F_s^- \tag{6.2}$$

where the subscript s indicates an adsorbed species. With Equation (6.2) shifted to the right, the effect is to remove basic sites, and this represents one step toward a more acid surface. Further, the high electronegativity of the fluoride atom draws charge from a neighboring cation (the partial charge on the fluoride is highly negative, inducing an unusually high partial charge on the neighboring cation). A proton on a site beyond the cation is repelled by this induced positive charge (the Madelung potential at the site of the proton is higher) so the proton is more easily given up—the solid is a stronger Brønsted acid. Similar induced charge arguments can be made to explain an increased acidity of Lewis sites.

The action of an electropositive cation to make the surface more basic is entirely analogous. As the first step, the cation, say, Na^+ from NaOH, exchanges with the protons that are most easily given up at the surface, thus decreasing the Brønsted acidity. As the second step, the sodium ion, now bonded to a lattice oxide ion, induces an increased negative charge on the oxygen and decreases the Madelung potential at the neighboring site where an OH^- ion is bonded to the lattice cation. The OH^- ion can now be released more easily, so the solid is more basic.

A third type of additive for acidity control (enhancement) is the use of foreign cations which are expected to exhibit valence and coordination different from those of the cations of the host lattice. This approach is equivalent to the use of mixed oxides to cause strong acid sites. The theory behind the approach was described in Section 4.2.3, the theory being that of Tanabe et al.[49]

Such approaches to the control of acidity or basicity are in common use,[50-53] but, as is the case with much acid/base theory, quantitative analysis of such effects has not been developed.

In the next section a few examples will be described of the use of additives in the control of the surface properties of solids. Some of these examples can be rationalized in terms of the above models. Some can be rationalized in terms of other models entirely (as, for example, the galvanizing of iron by zinc to protect against electrochemical corrosion). But in many cases the use of additives for surface control must be classified somewhere between an art and black magic. The elucidation of these phenomena is one of the most important challenges of surface science.

6.4.2. Observations of Additive Effects

The practical utilization of surface additives to control the surface properties of solids is of course extensive. The most common class of additives is the inert protective coating on a surface, but that is of interest here. We are more interested in the addition of active species that either change the reactivity of a surface (including the passivation of surface states present on the surface) or use the surface as a support to permit the additive itself to exhibit certain properties (such as a supported catalyst). Both these classes of additives are utilized to develop certain electrical or chemical properties of solid surfaces. A few examples will be described in the present section illustrating their use. The objectives, however, are so varied, and their action so varied, that it is difficult to break down the classification further.

6.4.2.1. Surface Passivation by Additives.

The passivation of a surface, either desensitizing surface electrical effects or desensitizing surface chemical activity in catalysis or corrosion, is a problem that is of continuing interest in technology. A few examples will illustrate the use of surface additives for such purposes.

The effect of surface states as recombination centers is a continual problem in the semiconductor device industry. The great success of silicon device technology stems from the fortunate observation that a thermally grown SiO_2 layer almost completely neutralizes the intrinsic surface recombination centers and provides almost perfect passivation. It is not absolutely perfect, of course. Due to imperfect crystallographic registry a large density of oxygen ion vacancies remain in the silica layer near the silicon interface, and these can at times act as traps. Impurity ions such as sodium can diffuse through the silica layer and cause problems. But the major problem in the use of the semiconductor silicon is solved by the thermally grown silica. Unfortunately this solution is not so effective on other materials. Germanium, for example, is much harder to passivate. It too has an oxide layer, but presumably the registry is much poorer, and the recombination centers are not passivated. It is found that humidity control is needed in germanium transistors to maintain a low density of recombination centers when the temperature varies. Although it has been shown that water vapor control maintains a low surface recombination velocity,[3] the exact mechanism of how a low controlled humidity interacts with the oxide to eliminate surface states is not clear. But it is clear from the discussions of the effect of water on energy levels in oxides (Section 4.3) that hydration of the germania will be expected to have a marked effect on its electrical properties.

The varied classes of corrosion inhibitors can be used to illustrate the many ways chemical reactivity can be minimized. There is a class of additives based on simple electrochemical action such as zinc deposited on iron. There is a class of sacrificial additives, introduced to scavenge oxygen or to scavenge free radicals from a coating before such active species can attack the metal, and of course there is a class of inert materials that simply blocks access of the active species to the surface. But in addition there is a broad class of additives that play an active role to change the surface properties of the solid, and thus are of particular interest in the present discussion.

We will mention a few surface active corrosion inhibitors as examples. One type of inhibition is that of some polymers, species where (a) the monomer bonds to the metal surface, (b) in its bonding to the metal surface, the monomer acts as a Lewis base, and a positive charged layer results which electrostatically excludes protons and further inhibits corrosion and (c) polymerization occurs forming a moderately thick coherent protective layer. An interesting example of this form of corrosion inhibitor is discussed by Tedeschi.[54] This is a group of acetylenic inhibitors, which act as above, with the bonding of the molecules to the surface and to each other somewhat more complex than indicated above. The action of these acetylenic inhibitors can be improved by the introduction of amines to the basic alkynol to strengthen the Lewis base action and accentuate the double-layer rejection of corrosive protons. Arbogast[55] suggests the action of amines as a Lewis base is to donate electrons to Lewis centers on the incipient oxide, rather than to metal surface atoms, and this interpretation is more consistent with our earlier discussions of acid/base activity. However the suggestion provides only a minor modification of the overall picture.

A second interesting case is the use of benzotriazole or naphtotriazole[56] as a particularly effective corrosion inhibitor for copper. Here again, Walker[56] suggests that the action of the benzotriazole is to form a complex with the thin layer of cuprous oxide present on the copper surface and to stabilize this oxide film. Presumably the final film, of about 50 Å thickness, is a mixed layer of cuprous oxide and cuprous benzotriazole.

As a final example of additives which inhibit corrosion, we can cite the study of Fugassi and Haney,[57] who examined the effect of Lewis bases in preventing the corrosion of titanium by hydrochloric acid. Assuming that the corrosion was initiated by chloride ions attacking Lewis acid sites on the titanium oxide film, they added Lewis bases to deactivate the Lewis acid sites. Indeed such bases were found to inhibit the corrosion. Corrosion by HCl is not always rate limited by chloride ion adsorption of course, and where the mechanism is different, the Lewis base additive is ineffective. Thus in later work it was shown that Lewis bases provided little protection of steel against HCl, and it was concluded that the iron oxide is not a strong Lewis acid, so the mechanism of the corrosion by HCl is different. But the titanium corrosion study, where corrosion is inhibited by the passivation of the acid sites with Lewis base additives is most instructive.

Passivation of acid sites on a catalyst by use of an alkali or alkaline earth wash is actually quite common in catalysis technology. The point is that in a reaction not depending on acidity, stray highly active acid sites can cause unwanted reactions, lowering the selectivity. There seems little quantitative in the literature regarding this passivation technique.

Additives on paint pigments provide a very interesting class of passivating additives. In this case the reaction that must be prevented (with TiO_2 pigmented paint, for example) is a photocatalytic action where electrons and holes are created in the TiO_2 semiconductor by uv (from the sun), and these hole/electron pairs catalyze the oxidation of the organic binder of the paint. This process (discussed more in Section 9.4) leads to chalking of the paint, for the oxidized resin will no longer adhere to the TiO_2 particles. To prevent the reaction, the holes and electrons must be prevented from reaching the binder. One way that has proven highly successful employs an alumina additive on the pigment particle surfaces. It is possible that the alumina additive is effective simply as an inert barrier to the holes and electrons, but it is unlikely because usually, with the modest (averaging about 15 Å) thickness of coating used, there are large uncovered areas on each pigment particle. Thus it is more likely that the additive has a positive action—perhaps the acidity of the alumina is involved. Another method of preventing the photoproduced hole/electron pairs from photocatalyzing a reaction is to use redox couples as recombination centers. Such an approach was described in the preceding subsection. The semiconductor ZnO has been examined[58,59] as the photocatalyst, and the redox couple $Fe(CN)_6^{-4}$/ $Fe(CN)_6^{-3}$ (see Table 6.2) was used to recombine the hole/electron pairs before they could cause a photocatalytic reaction. These experiments will be described further in Section 9.2.

Cases of passivation of surfaces by natural causes are common. As indicated in Chapter 4, a clean surface of a covalent solid is highly reactive and shows a coverage of surface states approaching a monolayer. However, these reactive surfaces are seldom a problem. They are always passivated by natural occurrences, usually by oxidation. The resulting oxides are further passivated with respect to acidity or basicity, by hydration of the surface layer. Thus when an active surface is required, as, for example, when an active catalyst is required, extensive pretreatment is needed in general to remove these passivating layers.

Unwanted passivation of a surface, encountered often in catalytic processes, occurs with the adsorption of "poisons" on the catalyst surface. Common poisons are carbon and sulfur on catalysts used in the petrochemical industry. Sulfur, for example, will originate from decomposition of an impurity in the feed stream and will deposit on the active sites of the catalyst surface. Since the poisons themselves are nonvolatile or form a nonvolatile adsorbate, they present an expensive problem in catalyst regeneration. The analogy between catalyst poisons and corrosion inhibitors has been pointed out in the early literature.[60]

6.4.2.2. Surface Activation by Additives.

Two classes of surface activation by additives will be discussed. In the first, intentionally deposited additives are introduced onto a semiconductor surface to control the surface properties with the results analyzed in terms of the theories of Section 6.4.1. This class is still in the basic research stage. The second includes more empirical studies, where additives have been found to provide favorable activation of a surface. In this case the interpretation is not so simple, but the technological value to date is much greater.

The work of Bonnelle and Beaufils and their co-workers belongs to the first group. In one study using electron spin resonance (ESR) to monitor the Fermi energy they showed[45] that the Fermi energy of ZnO could be pinned by an iron cyanide redox additive during hydrogen oxidation independent of the hydrogen/oxygen pressure ratio. In another investigation[61,62] the influence of the additives chromium and iron in controlling the electrical properties of the semiconductor (ZnO) was tested again by ESR. Then these same additives were used to control the coadsorption of gases in a catalytic reaction. In the catalytic studies[63] they studied the coadsorption of CO and O_2. For this simple system satisfactory agreement was found between predictions and observations.

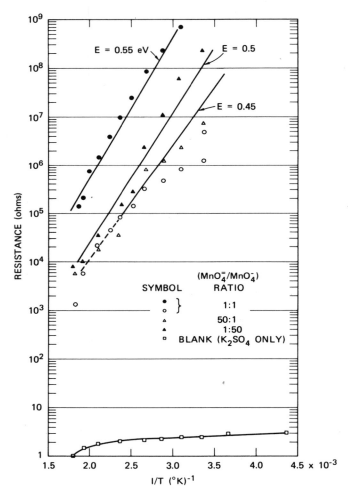

Figure 6.2. A plot of resistance versus temperature for a pressed pellet of ZnO with and without the manganate/permanganate redox couple deposited on the crystallite surfaces. The slope of the curve reflects the energy level of the surface state relative to the ZnO conduction band.

Another series of measurements, again on ZnO, was made[5] using the powder conductance measurement of surface state energy levels (Section 3.1.8), with a series of additives impregnated into the powder (Section 6.2.1). Figure 6.2 is included to illustrate the pinning of the Fermi energy of ZnO with the additive couple MnO_4^-/MnO_4^{2-} deposited on the surface. The resistance is much higher than the resistance of the blank, indicating electrons are withdrawn from the n-type semiconductor by the additive. The slope of the Arrhenius plot indicates [with Equation (3.14)] that the Fermi energy is pinned about 0.5 eV below the conduction band edge of the ZnO. The ratio of the two additives that is added makes no consistent difference, whereas by Equation (3.14) the measured resistance should be proportional to the ratio. It is concluded that uncontrolled chemical processes have occurred during the pre-treatment to define the ratio of reducing to oxidizing agent. Such lack of control over chemical factors illustrates clearly the complexity of the method and the need for a combined chemical and physical analysis.

Table 6.2 summarizes results of such measurements of Fermi energy pinning by use of several nonvolatile additives on the semiconductor ZnO. The value given as characteristic of the manganate complex is higher than the values from Figure 6.2 because another measurement[5] indicated a slightly higher value, and Table 6.2 gives the average of the two measurements.

In addition to the experimentally observed values, Table 6.2 shows the theoretical values obtained as described in the preceding section from electrochemical measurements and theories based on the redox potential. The agreement between the "theoretical" values from electrochemistry and the measured values is actually much better than it should be for these semiquantitative arguments and measurements. The major disagreement is in the observed value for chromium, where the disagreement is expected from the arguments of Chapter 5. The heavily complexed iridium, iron,

Table 6.2. Pinning of E_F of ZnO by Surface Additives[a]

	$IrCl_6^{-3}/IrCl_6^{-2}$	$Fe(CN)_6^{-4}/Fe(CN)_6^{-3}$	MnO_4^{-2}/MnO_4^{-1}	Cr^{+2}/Cr^{+3}
Theory	−0.8	−0.1	−0.4	+0.2
Experiment	−0.6	−0.1	−0.6	−0.8

[a] The Fermi energy of ZnO is pinned relative to E_c at the energy shown.

and manganese should not interact strongly with the ZnO surface, so the surface state energies are not modified by bond formation as the molecules adsorb. However, the uncomplexed chromium is expected to show a strong bonding interaction causing an energy level shift. The lower value at the solid/gas interface is attributed to this shift (see also Section 8.4).

Tests were reported[64] attempting to control the catalytic properties of ZnO with these additives, specifically in the CO oxidation reaction. Increased catalytic activity was found with the additives present, and the activity could be related to the Fermi energy position. However, within experimental error the increased activity could be accounted for by reaction on the surface of the additive crystallites, rather than reaction on the ZnO surface. The influence of the Fermi energy on the catalytic activity was indicated, despite the unknowns, for minor changes in composition that should not have appreciably affected the density of active sites caused correlated changes of Fermi energy and activity. However, the more recent results of Descamps et al.[63] are more instructive in demonstrating the control of a catalytic reaction by redox couple additives.

In the rest of this section we will be discussing activation by surface additives for cases in which the effectiveness of the additive was discovered empirically, with attempts to explain the effectiveness following. First additives for control of electrical properties, then additives for control of chemical reactivity, will be discussed.

The use of additives or activation in controlling the electrical properties of surfaces has been primarily for work function control and for increasing the photosensitivity of semiconductors.

The activation of solids by a monolayer of cesium, where a low work function or photoelectric threshold is required, is common, and clearly related to the low ionization potential (high negative redox potential) of cesium. An interesting example of the use of a cesium surface additive is its use in the generation of beams containing a high flux of negatively charged deuterium. A beam of D^+ is aimed at a surface, and with Cs adsorbed on the surface, the work function is lowered sufficiently for the reaction

$$D^+ + 2e \rightarrow D^-$$

to occur with reasonable efficiency. Beams of amperes per square centimeter, as can be generated by such a structure,[65] are of interest in the development of controlled fusion devices.

The surface treatment of photocells such as CdS is very critical to their operation, but the detailed function of the hole traps provided is not clear.

For example, the donor surface states associated with surface sulfide ions can be expected to be two-equivalent surface states (Section 5.5.3) and show the strong hole-trapping tendencies associated with such states. Then the surface treatments required to activate the photocell may be needed to minimize oxygen interference with hole trapping.

Additive dyes are used in standard photography or in electrophotography to increase the photosensitivity of the silver halide emulsion in standard photography or of the ZnO photoconductive layer in electrophotography. In both cases the action of the dye is to absorb light in the visible region, at the characteristic absorption band of the dye, and to contribute an electron to the conduction band of the solid. In the case of silver halide photography the electron contributed by the dye serves to reduce a silver ion to the metal, initiating the photographic process. In the case of electrophotography the electron contributed by the dye serves to move through the ZnO conduction band and decrease the electrostatic charge at the ZnO surface, providing a latent image. Dye additives are discussed further in Section 9.5.

The use of additives in increasing the chemical activity of a solid surface has been primarily of interest in catalytic work, with application both as promoters and for acidity control. The word "promoter" in catalysis can imply an additive for purposes other than control of the chemical properties of the substrate. For example, it may refer to an additive that provides more mechanically durable catalysts with therefore a longer active life. We are more concerned with activation of a surface than stabilization.

As an example of a promoter whose action appears to be the control of the properties of a semiconductor surface, we can consider the action of methyl bromide on the catalyst cuprous oxide, used for the oxidation of propylene to acrolein. The propylene oxidation reaction will be discussed further in Section 10.6.2; here it is of interest to refer to the studies of promoted copper oxide by Holbrook and Wise.[66] These authors found that methyl bromide stabilized the electrical properties of the cuprous oxide and the oxidation state of the cuprous ion against variation with oxygen pressure. Thus the promoter action could be attributed to the introduction of a surface state that pinned the Fermi energy. This viewpoint was supported by later work[67] in direct measurements of the Fermi energy of Cu_2O crystallites by the powder conductance measurement. In the latter work it was found indeed that the Fermi energy became pinned and independent of oxygen pressure when methyl bromide was present on the surface. From a practical point of view, with such stabilization much higher oxygen pressures can be used. With a stable catalyst and higher oxygen pressure the reaction rate is increased without altering the catalytic behavior (selectivity in particular) of the cuprous oxide.

An excellent example of acidity control in catalysis is the use of a fluoride additive on alumina. The general principles are discussed in the preceding theoretical section. The fluoride ion induces stronger acid sites associated with the aluminum cations. Webb[68] was the first to report that the strength rather than the number of acid sites increased with the addition of fluoride ions (by HF impregnation). Gerberich, Lutinski, and Hall[69] showed that the acidity and the catalytic activity for a cracking and for an isomerization reaction showed a maximum with about 1/4 monolayer of fluoride (where a complete monolayer is in this case considered to be with the fluoride ion replacing both the OH⁻ groups and any lattice oxygen ions exposed on the surface). Results of specific activity measurement with a full monolayer of F⁻ were identical to results with the solid AlF_3. In the above the acidity was measured by the retention of the gaseous base ammonia at 500°C. Thus the additive F⁻ does increase the acidity of the alumina as expected by the simple induction picture in the preceding section. However, this picture may be incomplete. Gerberich et al. suggest that a simple increase in acidity cannot explain their increase in isomerization rate, and believe Peri's dual acid/base sites (see Sections 4.3.3 and 10.3) may be the required active sites. However, the fundamental principles of how the acidity of sites is increased are still best described by the induction model.

Strong basic sites on a surface are caused by alkali metal additives. The generation of such centers, their action in forming carbanions (an olefin with a negative charge), and the resulting catalytic activity is reviewed by Pines.[70]

Another more general example of acidity control by additives is the use of mixed oxides. This technologically important method was discussed in detail in Sections 4.2.3 and 4.3.3, where it was shown that a foreign ion on the surface would be poorly coordinated and sites of extreme acidity may result. The use of the oxide of the foreign cation as a surface additive rather than mixing the oxides results in a somewhat different configuration, but the theories of such systems (Section 4.2.3) are not refined to the point of distinguishing the two cases. Examples of enhanced acidity with such additives are alumina on titania[51] and boron and other species on silica.[52,71,72]

6.4.2.3. Additives on Supports.

One can cite innumerable cases of importance where the additive is the important phase and the support has no role to play. Such cases in catalysis will be described in Chapter 10.

These are of less interest, however, in the present context. We shall restrict the discussion here to a few examples where the additive is the important phase, but the support does play a useful role.

In electrochemistry, one has the example of carbon supported platinum electrodes (in fuel cells). In solid state devices one has the example of epitaxial layers, where a thin single crystal phase is grown on a single crystal substrate. These are cases where the support provides electrical contact to the additive. With epitaxial layers the substrate in many cases provides an active electrical function in addition to its role as an electrical contact.

In catalytic chemistry one can cite bifunctional catalysts, where the additive catalyzes one reaction, the support another, or one can cite the novel and interesting phenomenon of spillover.[73]

To observe spillover a metal such as palladium is supported on an oxide. Hydrogen adsorbs on the palladium, becomes dissociated, and diffuses (spills over) onto the oxide surface. In catalytic reactions, the hydrogen so produced is active, and the support becomes an active catalyst by the spillover mechanism. Bianchi *et al.*[74] demonstrate the effect by showing that the catalytic activity associated with the hydrogen atoms is sustained after the metal is withdrawn. An application of spillover is in the hydrogen reduction of oxides—a small dispersion of noble metal additive supported on the oxide permits the reduction of the oxide to the base metal at much lower temperatures than would be possible with no additive.[75] Spillover of oxygen can also occur. A Pt additive has been used to dissociate O_2 and provide reactive O atoms which react with the support ZnS. Batley *et al.*[76] review such spillover cases where the surface additive activates a gaseous species for corrosion of the substrate. Sermon and Bond[77] review spillover in general.

6.5. The Real Surface

The term "the real surface" was coined in surface physics to identify the surface of a reactive solid that has been exposed to the atmosphere or to other reactive ambient media. Thus, for example, a "real" germanium surface is one with an oxide layer several tens of angstroms thick (and with water vapor usually adsorbed onto or absorbed into the oxide surface). Most materials can and will develop an oxide or an equivalent passivating surface layer if exposed to solutions or to the atmosphere. The passivation (oxidation) with metals and elemental semiconductors is obvious, but passivation will also occur on compounds. Cuprous oxide will form a thin layer of cupric oxide when exposed to air at room temperature. Sulfides will doubtless form sulfates or oxides at their surface. And so on. In general with real surfaces the exact thickness or composition of the passivating surface layer is not well defined.

Most of the phenomena observed on the real surface have been or will be described elsewhere in the text. For example, a real silicon surface has a layer of oxide on it, and most of its surface interactions with foreign species can be described as interactions that would normally occur on a silica surface. Such interactions have been and shall continue to be discussed in detail. However, there are differences both in chemical and electrical surface effects that arise on the "real surface" case because the surface layer is so thin.

The effect on the surface chemistry of a natural passivating layer on the solid has not been well studied, because in practice to speak of chemical reactions on a "real surface" is to mix concepts. Chemical reactions on the "real" surface usually remove, alter, or increase the passivating layer. For example, to activate a catalyst extensive pretreatments are used which can usually be associated with removal of oxides (or in some cases oxidation of the material) or otherwise homogenizing the catalyst. There is, however, one case where activity studies have been made on a configuration similar to that of the "real surface." Schwab[78] has attempted to examine the influence of a substrate of different composition under a thin layer of the catalyst, a configuration approaching that of the "real" surface. However the results to date are somewhat complex.

The net effect of a passivating layer on the electrical properties of a solid can be associated with surface states in three regions: surface states at the surface layer/semiconductor interface, traps and impurities in the passivating layer, and energy levels due to foreign species at the solid/gas interface. Such effects will be the main topic in this section.

The electrical behavior of the real surface has been examined in great detail. For its first decade (the 1950s) surface physics research was almost exclusively directed to this system, for this was the era of interest in germanium, and the behavior of the real surface of germanium was of great importance in transistor technology.

The electrical properties of the real surface (with a semiconductor substrate) have been analyzed in terms of two types of surface states, "fast states" and "slow states." The terms are particularly useful in the case of germanium, where after a disturbance the time constants to restore steady state in the electrical properties are easily separated into widely different values. The fast states reach electronic equilibrium in times the order of 10^{-6} sec, the slow states, on the other hand, require seconds to minutes. The fast states are attributed to surface states at the interface between the germanium and the germania arising probably from imperfections due to lattice mismatch. Because the electrons need not pass through the oxide, electron transfer between the conduction or valence bands and these states is very rapid. The density of such states with no impurities present at the interface is the order of $10^{15}/m^2$, a very low density. These states with their short time constants are primarily responsible for surface recombination of excess carrier pairs in the semiconductor. Slow states, on the other hand, are associated with foreign adsorbates (oxygen, water, etc.) at the germania/gas interface, and sometimes are also associated with traps within the oxide layer.[79,80] Electronic equilibrium between the germanium and such states is slow because the electrons have to go through the oxide layer. For example they may have to be promoted to the conduction band of the germania to transfer between the germanium and the surface. The rate of such transfer can be quite slow.

Despite the slow reaction, when steady state conditions are reached, the slow states control the surface barrier at the surface of the germanium because of the large density of states available. One of the first powerful experimental techniques of surface physics was the use of adsorbing gases to vary the surface barrier and measure the variation of the various electrical properties (Section 3.1) as a function of V_s. Such measurements and interpretation, particularly as applied to CdS, have been recently reviewed by Mark.[81,82]

Research is continuing on the real surface,[48,83,84] but still with the primary goal to study the electrical properties. There is some suggestion that the technique can be inverted and used to study gas/solid reactions involving ionic surface states and acid/base sites on insulating solids, using electrical measurements on the semiconducting substrate as a very sensitive probe of the effects of the reactions.[85] Work of this nature has been reported by Gatos and his group, who have provided a series of studies[86-88] of surface states on the "real surface" of compound semiconductors. They used surface photovoltage spectroscopy, field effect, and work function (Section 3.1) measurements of surface state energies. The interpretation of the measured surface states in terms of their chemical origin is still, however, not in the detail necessary to provide information about chemical processes at the real surface.

7 | Adsorption

In this chapter models of adsorption processes are reviewed. Both adsorption from the gas phase and from the liquid phase onto a solid are discussed, but adsorption from the gas phase is emphasized. Much of the material in earlier chapters is obviously pertinent background information for the topic of adsorption processes. In Section 5.2.4 the various degrees of interaction of gases at a solid surface, physical adsorption (physisorption), chemisorption, and finally the development of a new phase were described. In the remainder of Chapter 5, the forms of bonding of an adsorbate to the surface were reviewed. The present chapter will build on these concepts, moving on to discussions of amount adsorbed, the kinetics of adsorption/desorption, and electron transfer during adsorption (ionosorption).

7.1. Adsorption Isotherms and Isobars

7.1.1. Physical Adsorption

Physical adsorption is the adsorption of species with the least possible interaction. All species, even helium, show a van der Waals dipole/dipole interaction, which is the interaction generally assumed for physical adsorption theories. The adsorbing species polarizes slightly, the sorbent surface shows corresponding polarization, and interaction between these induced dipoles provides the heat of adsorption. Such mutual polarization was shown in the Lennard-Jones model[1] to lead to an energy decrease of the system varying as r^{-6}, where r is the atom/atom separation. More detailed models of van der Waals interaction have been developed,[116] but are beyond the scope of this book. The heat of adsorption is usually the order of 5 kcal/mole (0.2 eV/molecule). Actually any interaction with energy of this order of magnitude is often conveniently termed a "physical adsorption interaction." Physisorbed molecules are never dissociated.

A simple expression for the amount adsorbed on a solid as a function of pressure is obtained with the approximation that the rate of adsorption is given by

$$d\Gamma/dt = k_2(\Gamma_t - \Gamma)p - k_1\Gamma \tag{7.1}$$

where Γ is the concentration of adsorbate per unit area, Γ_t is the total concentration of sites available, p is the gas pressure, k_1 is the rate constant for desorption, k_2 is the rate constant for adsorption. The first term describes the rate of adsorption, proportional to the pressure and to the number of unoccupied sites. The second term represents the rate of desorption, proportional to the surface coverage Γ. The analysis implicitly assumes a uniform surface—that all sites are identical. As will be discussed in more detail below, such an assumption is better for physisorption than for chemisorption. In practice, because of surface heterogeneity or other reasons, the rate constants k_1 and k_2 are usually dependent on the fractional surface coverage θ:

$$\theta = \Gamma/\Gamma_t \tag{7.2}$$

particularly in the case of chemisorption.

For physisorption, the assumption that the k's are independent of θ is often acceptable as long as less than a monolayer is adsorbed. Solution of Equation (7.1) for steady state yields

$$\Gamma = \Gamma_t p/(p + b^{-1}) \tag{7.3}$$

where $b = k_2/k_1$. This is the Langmuir isotherm.

If the pressure is very low, the isotherm reduces to "Henry's law":

$$\Gamma = \Gamma_t bp \tag{7.4}$$

with the amount adsorbed simply proportional to the pressure.

In many of the discussions below we will implicitly assume Henry's law of adsorption, particularly for physisorption and at times for chemisorption, for the sake of simplicity. It restricts the cases to low adsorption, but realistic adsorption laws are not available, especially for chemisorption, so there is little gain in using a more complex expression. However, one must resist the temptation to expect the simple expressions thus obtained to be quantitatively valid at any but the lowest coverage.

The Brunauer–Emmett–Teller (BET)[2] adsorption isotherm is of particular importance in physisorption. In the derivation of this relation, a nonvarying value of k_1 (the rate constant for desorption) is assumed for the gas molecules adsorbed on the clean surface, but a different value is assumed for any gas molecules adsorbed on top of the first layer of adsorbate. By thus admitting that multilayer physisorption could occur at reasonably low pressure, Brunauer et al.[2] were able to develop a satisfactory isotherm for physisorption at coverages up to and exceeding a monolayer. This isotherm is of great importance in surface studies, for analysis of physisorption at low temperature using this isotherm to describe adsorption (of say nitrogen or krypton) yields the surface area of the adsorbent. Despite its importance in surface studies, we will not develop the expressions here, for we are more concerned in this book about cases of stronger interaction. The BET theory is well documented in most books on physisorption and adsorption.

Often even for "simple" physisorption the adsorption isotherms are not well represented by the Langmuir isotherms [Equation (7.2)] or even the BET isotherm. A general classification of physisorption isotherms has been developed, identifying five common shapes of isotherms as illustrated in Figure 7.1. Brunauer et al.[3] suggested this classification. Type I physisorp-

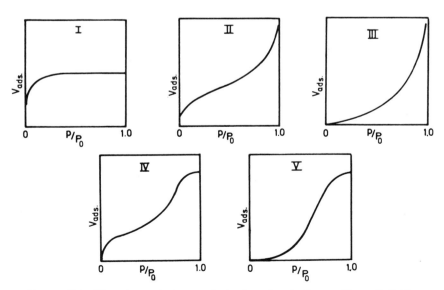

Figure 7.1. Five observed shapes of physisorption isotherms (from Ref. 3).

tion follows the shape of the Langmuir isotherm, Equation (7.5), with k_1 constant. An adsorption isotherm following the Langmuir relation must not only have k_1 constant, but be restricted to a single monolayer of gas. Such an adsorption isotherm may, for example, represent adsorption on small pores of a solid, where the amount of adsorption is strictly limited. Type II physisorption is that expected for the BET approximation, where multilayer adsorption occurs, so that at some pressure p_0, the "adsorption" goes to infinity as the gas begins to liquefy. Type III physisorption is observed if the interaction between the first layer and the substrate is much weaker than the interaction between the second layer and the first. Type IV is similar to type I, but the shape suggests there are two limitations on the amount adsorbed. Type V when observed suggests that a model applies combining that of type III (low sorbate/sorbent interaction energy) and type I (limited adsorption). Types III and V adsorption are not common.

A given adsorbate can usually be either physisorbed or chemisorbed, and the difference is detectable by means of the energy changes as the sorbate switches from one to the other. In the present book we are more interested in strong interaction, as for example in ionosorption or in strong covalent bonding, than in physisorption. The primary interest in physisorption arises because the first step toward chemisorbing a species may be its physisorption, and the amount chemisorbed may be determined by the equilibrium amount physisorbed. Thus it is of interest to connect physisorption and chemisorption. Lennard-Jones[1] provided a convenient visual aid in recognizing the connection as shown in Figure 7.2. Here the energy of a system is plotted against a coordinate related closely to the distance of the adsorbate from the surface. Curve (a) shows the energy of the adsorbate molecule as a function of distance, showing a weak minimum at a substantial distance from the surface that corresponds to the heat of physisorption. If the adsorbate is dissociated (at infinity) into two atoms, when it adsorbs it will show a stronger heat of adsorption Q characteristic of chemisorption bonding. Thus for chemisorption the energy/distance curve is represented by curve (b), where the energy at infinite distance is the dissociation energy. The model shows an activation energy for chemisorption E_A, the energy necessary to dissociate the molecule when it is physisorbed on the surface. It shows an activation energy for desorption $Q + E_A$. These activation energies will be discussed further below. The model shows nicely (if naively) the relation between physisorbed species as the "unoccupied surface state" or the "weakly bonded adsorbate"; one is almost tempted to relate the number of physisorbed to chemisorbed species by the Boltzmann relation with Q in the exponent. Indeed in the rest of this section we will for heuristic

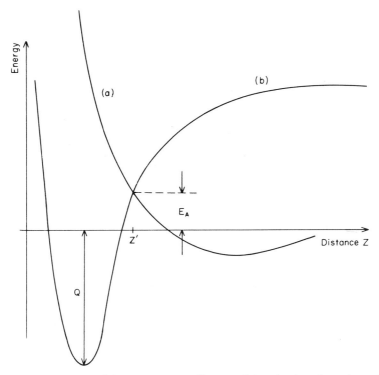

Figure 7.2. The energy of the system versus distance of the adsorbate from the surface. (a) Adsorption as a molecule, (b) adsorption as two atoms.

purposes consider the physisorbed species as the "unoccupied state" when discussing ionosorption. However it must be recognized that direct chemisorption of a bombarding gas-phase molecule cannot be dismissed so easily.

In this section we have presented a minimal description of physisorption. For more complete discussions, the reader is referred to the many treatises and articles on the subject, for example, the works of de Boer,[4] Hauffe and Morrison,[5] Neustadter and Bacigalupe,[6] Rees and Olivier,[7] Adamson,[8] or Aveyard and Haydon.[9]

7.1.2. Heat and Activation Energy of Adsorption, Irreversible Chemisorption

7.1.2.1. The Activated State.
The activation energy of adsorption, E_A, illustrated in Figure 7.2, represents the amount of energy that

must be supplied to a molecule to permit chemisorption. The energy can be supplied by a high temperature or by a nonequilibrium process such as illumination (Chapter 9) or the activation of the gas-phase molecule.[10] In general there are many reaction paths by which the same result, chemisorption, can be attained. For example, if the activation energy is required for molecule dissociation as in Figure 7.2, one path might be the dissociation of the molecule in the gas phase; the other might be the dissociation of the molecule when physisorbed. As indicated in Figure 7.2, the activation energy is much less in the latter reaction path. With each path there is an associated activation energy E_A and there is a "rate limiting step," the step in the adsorption process during which the system must obtain the energy E_A. The kinetically most active reaction path is of course generally the one with the lowest activation energy E_A.

The "activated state" of the system is that configuration when E_A electron volts has been fed into the molecule, the system is in its highest energy configuration, and the next step is exothermic. Thus, for example, in the Lennard-Jones model of Figure 7.2, the activated state is when E_A eV has been fed into the adsorbed molecule and it has reached the configuration when it can dissociate into two atoms with no further energy added. Once the dissociation has occurred, the subsequent chemisorption is exothermic all the way.

A great portion of surface chemistry is based on attempts to provide reaction paths with ever lower activation energies E_A of the activated state (or, when a passive surface is desired, to eliminate such reaction paths). There are innumerable ways in which the lowest-energy activated state on a surface can be altered. In essence all the information presented in this book, on heterogeneity, polar effects, bonding, additives, and surface sites, is in principle applicable knowledge. Unfortunately in general the solid surface is too complex and it is seldom that activated states can be identified.

Two examples where a lowering of E_A has been accomplished and the mechanism is clear are the introduction of surface steps and the use of a polar medium. As mentioned in Section 3.2.7 LEED can be used to measure quantitatively the presence of surface steps, so it has been possible to relate surface steps to the rate of gas adsorption. Both adsorption of oxygen[11] and of hydrocarbons[12] have been found to occur much faster in the presence of surface steps. The favorable reaction path becomes (a) adsorption at the surface step as an active site, and (b) diffusion of the adsorbate across the surface. The other example of activation energy lowering is the case of the addition of a polar medium. This was discussed in Section 2.3.3, where the mathematical formalism was developed showing how the presence of a polar medium increased the rate of electron transfer processes. In this case the favorable reaction path is (a) orientation of dipoles to shift the surface state energy level, then (b) electron transfer with no energy change. There will be more discussion of this latter mechanism in the next chapter.

In most cases, the form of the active intermediate cannot be identified. Worse, in many cases the route of the adsorption process that is kinetically most rapid (has the lowest activation energy) cannot be identified. For example, consider the adsorption of oxygen to form O^-:

$$e + O_2 = O_2^- \qquad e + O_2 = O_2^- \qquad e + O_2 = O_2^-$$
$$O_2^- = O + O^- \qquad e + O_2^- = O_2^{2-} \qquad O_2^- + O_2 = O_4^-$$
$$e + O = O^- \qquad O_2^{2-} = 2O^- \qquad O_4^- + e = O_3^- + O^-$$
$$O_3^- + e = O^- + O_2^-$$

provide a few example routes. Other more complex possibilities involving particular sites and complex formation at these sites can easily be suggested. It is a challenge in adsorption and catalysis to suggest the most likely reaction routes that will provide the lowest E_A and, more important, to develop methods of providing or of passivating the active centers that provide these low E_A values.

7.1.2.2. The Coverage Dependence of the Heat and Activation Energy of Adsorption.

In general practice it is found experimentally that for chemisorption the heat of adsorption Q decreases and the magnitude of the activation energy E_A increases with coverage θ. The variation of these parameters with θ will be shown critical in determining the general characteristics of adsorption. Two possible reasons for the Q and E_A variation based on specific models will be presented in Sections 7.2 and 7.3, but these reasons are not all-encompassing. Thus a general discussion is of value.

The variation of the heat of adsorption Q with θ can arise from many sources. The most obvious is heterogeneity of the surface. There will be some adsorption sites of exceptionally high energy. In general these will be expected to be occupied first, and the lower-energy sites will be occupied later. This alone can cause the large changes in Q with θ. Another source of the variation is adsorbate/adsorbate repulsion. As the surface becomes covered ($\theta \rightarrow 1$) the energy of adsorption will be less if further adsorbate species are repelled. A third source of the variation is the double-layer formation discussed in Chapter 2. If the adsorbate acquires a partial charge or is ionosorbed, the energy of adsorption will again be lowered as the double layer develops. This latter case will be discussed for semiconductors in detail, as it can be described analytically and is of great interest in the electronic/chemical surface processes. On metals also the influence of a developing double layer can be very important. Sargood, Jowett, and Hopkins[13] have shown direct analytic relation between the double layer formed and the electronegativity of the adsorbate.

The reason for the variation of E_A with θ is more difficult to justify for the general case. Referring to Figure 7.2, it is reasonable that by this diagram if Q decreases, E_A will increase, for the chemisorption branch of the curve shifts up while the physisorption branch is unaffected. However, to a great extent, Figure 7.2 is only phenomenological, and at best it only applies to the case where the activation energy arises due to dissociation of a molecule. However, it does provide a rational viewpoint that illustrates that E_A *could* depend on Q (and thus on θ) in the observed manner. For a satisfying demonstration that E_A will increase with θ with a particular mechanism, the model of that mechanism must be examined in each case. We will illustrate for the case of ionosorption (Section 7.2) that such an activation energy dependence exists. Thus we will show the rationale when electronic effects dominate, and the example of Figure 7.2 shows the rationale when molecular dissociation effects dominate. These two examples to some extent justify the generalization that E_A increases as θ increases.

If there is more than one form of the adsorbed species, the values and variation of Q and E_A can appear much more complicated. For example, if oxygen chemisorbs as ionosorbed O_2^- with negligible local surface reaction, but also chemisorbs as O^- with a strong local bond to the surface, then the apparent Q and E_A, as determined only from gas-phase measurements, will not follow a simple law. The apparent values will depend on the exact experimental procedure of adsorption (because, as will be shown, the ratio of concentration of the two species will depend on the exact procedure followed).

Experimentally, TPD (*t*emperature *p*rogrammed *d*esorption) measurements are of help to analyze whether several distinct surface complexes are formed or whether one is observing simply a variation of Q and E_A with coverage. The desorption process, as is indicated in Figure 7.2 and will be discussed in the next subsection, proceeds at a rate governed by the activation energy of desorption, E_D, given by reference to Figure 7.2 by

$$E_D = E_A + Q \tag{7.5}$$

Thus E_D is insensitive to variations in E_A and Q, for as Q decreases, E_A

increases. Thus the desorption rate is much less sensitive than the adsorption rate to coverage. With this in mind it can be seen that a TPD curve that shows well-resolved peaks, corresponding to discrete values of E_D, is suggestive of distinct sorbate/sorbent complexes. If the TPD curve shows a single peak, then a single complex may be the best model. If no peak at all is present, a large spectrum of sites associated with heterogeneity may be suspected.

In the next section we will discuss in more detail the rates of adsorption and desorption in order to suggest a "typical" adsorption isobar. We will assume in that discussion for simplicity that there is only one type of surface complex.

7.1.2.3. The Adsorption Isobar.

The rate of adsorption of a species X will be given by an expression of the form of Equation (7.6):

$$d\Gamma_x/dt = Ap_x(1 - \theta)\Gamma_t \exp(-E_A/kT) \qquad (7.6)$$

where for physisorption E_A is generally zero. Here Γ_x is the surface coverage of adsorbate X, Γ_t is the total site density available, $1 - \theta$ is the fraction of the surface available for adsorption, p_x is the pressure of X in the gas phase, and A is a constant that gathers the pre-exponential factors from rate theory and the kinetic theory of gases. Comparing Equation (7.6) to (7.1) we note that k_2 contains the factor $\exp(-E_A/kT)$.

The rate of desorption can be represented by the expression

$$d\Gamma_x/dt = - B\Gamma_x \exp(-E_D/kT) \qquad (7.7)$$

where again the pre-exponential constants are gathered into a constant B. We note that k_1 of Equation (7.1) contains the factor $\exp(-E_D/kT)$. Thus in Equations (7.3) and (7.4)

$$b = k_2/k_1 = C \exp(Q/kT) \qquad (7.8a)$$

with C a constant. A more detailed analysis in terms of rate theory and kinetic theory[8] leads to

$$b = N\sigma\tau_0(2\pi RTM)^{-1/2} \exp(Q/kT) \qquad (7.8b)$$

where τ_0 is the molecular vibration time (about 10^{-13} sec), M is the molecular weight, N is Avogadro's number, and σ is the cross-sectional area of the adsorbate molecule.

Chemisorption is characterized by heats of adsorption Q ranging from zero to hundreds of kcal/mole, and almost always shows an activation energy E_A. As discussed above, the value of Q decreases with increasing coverage, the value of E_A increases with increasing coverage. Often E_A is zero for a clean metal surface.[14,15] This behavior can be compared to that of physisorption, characterized by a low value of Q (the order of 3–6 kcal/mol or 0.1–0.2 eV/molecule) and a negligible value of E_A.

Because the values of Q and E_A and their variation with θ have characteristic behavior in both physisorption and chemisorption one can describe a typical adsorption isobar (the volume adsorbed as a function of temperature with the pressure constant). Figure 7.3 sketches the shape expected. The low-temperature portion, curve (a), represents physisorption. As there is no activation energy, physisorption by Equation (7.1) is possible at the lowest temperature. From Equation (7.8), the parameter b in Equation (7.4) decreases rapidly with increasing temperature, so the volume physi-

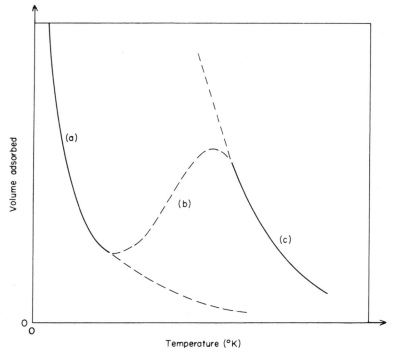

Figure 7.3. The adsorption isobar: adsorption at constant pressure. (a) Physisorption, (b) nonequilibrium chemisorption where the volume adsorbed depends on history, and (c) chemisorption.

sorbed [curve (a)] decreases with increasing temperature T. A similar expression (but with a higher value of Q of course) describes the *equilibrium* chemisorption branch [curve (c)] of Figure 7.3, where again the amount adsorbed decreases with increasing temperature. In the intermediate temperature region [curve (b)] we are in the region of irreversible chemisorption. In this temperature range the variation of the activation energy for adsorption E_A leads to limited coverage and the activation energy for desorption E_D is too high for desorption to be observed.

If adsorption is measured in the temperature range of irreversible chemisorption, equilibrium physisorption is rapidly reached in accordance with the lowest dotted line. Equilibrium chemisorption, which should be reached in accordance with the highest dotted line, ceases near the center line [curve (b)]. Qualitatively the cessation of adsorption is due to the variation in E_A and can be explained as follows: when a clean surface is initially exposed in this temperature regime to an adsorbing gas, the value of E_A may be very low (even zero), so the rate of adsorption by Equation (7.6) is very high. As the surface becomes covered, E_A increases, as described above, so the rate of adsorption decreases rapidly. The adsorption "pinches itself off."[16] At constant temperature, adsorption will continue and E_A will increase with θ until by Equation (7.6) the adsorption rate is negligible. If for example 10^{-3} monolayers/week is the limit of the scientist's patience in a particular experiment, then adsorption will occur until E_A is high enough to reduce $d\Gamma_x/dt$ to 10^{-3} monolayers/week. It is observed that the amount adsorbed increases with temperature in this region. This is expected by the following argument. If T is very low, a very low coverage will suffice to reach the pinch-off value of E_A. If the temperature is higher, E_A will be able to attain a higher value before the adsorption rate is so pinched off, and Γ_x will be higher when the experiment is terminated.

A numerical calculation of such a pinch-off process, where the surface double layer provides the activation energy limiting the rate of electron transfer for adsorption, was discussed in Chapter 2. The discussion was introduced shortly after Equation (2.47) to illustrate charge transfer across a double layer.

The reason the region (b) of Figure 7.3 is described as the "irreversible adsorption" region is that over most of this region, a simple evacuation will not remove the adsorbate. This can be shown by the following argument: The rate of desorption by Equation (7.7) is by definition less than the rate of adsorption [Equation (7.6)] until the coverage reaches the equilibrium value. As the rate of desorption is substantially less than the rate of adsorption, and on curve (b) of Figure 7.3 the rate of adsorption is by definition negligible, it is clear that the rate of desorption is also negligible. In fact the desorption rate is negligible at any coverage short of equilibrium in the region where equilibrium "cannot" be reached. To desorb the species, the temperature must be raised to the region of curve (c).

Thus if adsorption with the coverage shown as curve (b) is obtained in the intermediate region, lowering the temperature or pressure will neither increase nor decrease the amount chemisorbed.

As a final topic in this subsection, it is of interest to describe more quantitatively the rate of adsorption in this irreversible region, region (b) of Figure 7.3. The rate is of course given by Equation (7.5), but in Equation (7.6) the value of E_A is increasing with coverage, and so E_A is a variable, and such an implicit equation is not generally useful. Empirically it is found that for many cases of adsorption in this irreversible region the rate can be expressed by the "Elovich equation":

$$d\Gamma/dt = a \exp(-b\Gamma) \tag{7.9}$$

that was introduced in Chapter 2 as Equation (2.47) representing the rate of electron transfer to the surface under irreversible conditions. From the arguments in Chapter 2, if such electron transfer to an adsorbate is the rate limiting step in adsorption, then within certain limitations the Elovich equation should be the rate expression. This will be discussed further in Section 7.2, "Ionosorption." However, the Elovich equation is applied with success to many adsorption experiments,[17,18] where one is almost certain that heterogeneity must play the dominant role rather than the double layer in determining the E_A versus θ dependence. The explanation of the generality of Equation (7.9) is that most curves obeying Equation (7.6) with E_A increasing as any reasonable function of θ can be described fairly well by the Elovich equation. The two adjustable parameters a and b, together with the nonlinear exponential form, lead to great latitude in curve fitting. In other words there is probably no great theoretical significance in general to a fit to the Elovich equation. If one postulates a model to describe E_A versus a function of θ one must be cautious about attaching significance to the measured values of a and b unless one can demonstrate by another measurement that all other possible causes of the E_A variation with θ, other than the model proposed, do not participate.

7.1.3. The Adsorbate Superstructure

The forces involved in adsorption arise from both adsorbate/adsorbate interaction and adsorbate/solid interaction. If the temperature is high enough to permit the necessary atom movement, then such forces will determine the position of both the sorbate species and the outermost layer of substrate atoms, relative to the positions expected from a direct extension of the substrate crystal lattice. Such ordering need not require strong interaction; even krypton adsorption can cause reordering on alkali halide surfaces.[19]

The observation of adsorbate superstructure (reconstruction) or the observation of island formation (where, as discussed below, adsorption occurs in patches of adsorbate separated by patches of adsorbate-free substrate) depends on the adsorbate atoms being able to diffuse. Island formation requires major movement in response to adsorbate/adsorbate interactions, unless adsorption occurs only at the perimeter of the islands. On the other hand the generation (by sorbate/sorbate repulsion) of a super-structure only requires diffusion over a few lattice spacings.

Island formation is expected in cases where the adsorbate/adsorbate interaction is strong and attractive but the adsorbate/adsorbent interaction is weak. Because the adsorbate/adsorbent interaction is weak, the adsorbate molecules can diffuse over the surface, and because the adsorbate/adsorbate interaction is strong, the adsorbate will form islands of adsorbate. Clearly an island structure will lead to different adsorption energies and isotherms (or isobars) from those expected for random adsorption. In addition, such island formation may strongly affect the chemical and electronic activity of the resulting surface; for example, now the active sites may be at the perimeter of such islands.

A review of the theory of island formation is presented by Oudar[20] and by Schrieffer and Soven.[21] McCarty and Madix[117] have concluded that formic acid adsorbs as islands on graphite because of strong sorbate/sorbate interaction in the form of H-bonding. A clear case of island formation due to sorbate/sorbate attraction is described by Price and Venables,[22] who studied xenon adsorbed on the basal plane of graphite. They concluded that normally island formation initiates at particular active sites, perhaps with a nonvolatile contaminant as the nucleation point. Indeed, multilayer island formation is observed associated with such contamination. Many layers of adsorbate are adsorbed on an island while bare spots remain on the carbon substrate. In the cited article, the authors took great care to prevent contamination of the cleaved graphite surface, and found indeed that the first layer was then completed before the second layer was initiated. They note that the contaminant problem, where contaminants initiate multilayer island growth, seems to be common in many systems, for example, Cd on W, with the contaminant oxygen,[23] and Si on Si, where C is probably the contaminant.[24]

Ordering of adatoms due to repulsive interaction seems very common. Kaburagi and Kanamori[25] review the quantum theory of such repulsion and how, allowing sufficient atom movement, finite range of interaction can lead to various patterns of superstructure of the adsorbate.

Surface superstructure is reviewed by May[26] and by Adams[27] and in a nonmathematical presentation by Estrup.[28] Adams analyzed the effect of ordering by repulsive interaction on LEED and on TPD measurements. With even slightly mobile adsorbates, there will be no nearest neighbors up to 50% coverage (observable by LEED), and at this coverage there will often be a (2×2) superstructure that keeps the sorbate atoms separated as much as possible. Beyond 50% coverage the adsorption energy will suddenly decrease from H_0 to $H_0 - ZV_r$ (observable by TPD), where Z is the number of nearest neighbors and V_r is the repulsive interaction energy. The heat of adsorption is often observed to change abruptly with coverage due to such ordering and reordering of the adsorbate superstructure.[29-31] Figure 7.4 shows the results of a study by Tracy and Palmberg[29] of the isosteric heat of adsorption, as determined using the Clausius–Clapyron equation to analyze the dependence of θ on T and p. The sharp decrease in the heat of adsorption at $\theta \sim \frac{1}{2}$ is shown clearly. The value of V_r obtained, where V_r is the repulsive energy per pair in the above paragraph, is 1.8 kcal/mole.

In some cases the repulsive sorbate/sorbate interaction can dominate over the sorbate/solid interaction to the point that the adsorbate develops a structure out of registry with the substrate. An example[32] of this is CO on Ni(100). This shows a (2×2) structure to 50% coverage, but at greater coverage ($>60\%$) a hexagonal superstructure is formed which lacks good registry with the substrate.

With the common use of LEED in recent years, many other examples of adsorbate superstructure are known. Some examples of adsorption leading to superstructure are nitrogen on molybdenum;[33] ethylene on platinum;[34] oxygen on nickel,[35] ruthenium,[36] molybdenum,[37] and tungsten;[38] and chalcogens in general on nickel, as reviewed by van Hove and Tong[39] in their theoretical analysis. On the other hand, the adsorption of CO or NO on platinum destroys the reconstructed (1×2) structure of the platinum clean surface, restoring the (1×1) structure.[40] Silicon adsorbed on molybdenum leads to a (1×1) structure, when a monolayer is adsorbed, but

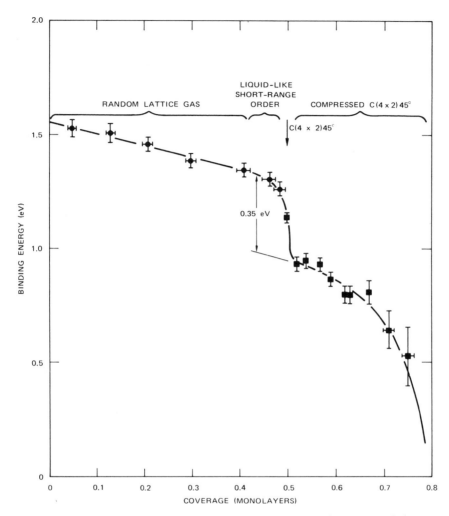

Figure 7.4. Binding energy of CO on Pd(100) as a function of coverage. Circles represent data determined from a work function method, while the squares are from LEED determined isosteres.

Ignatiev *et al.*[41] conclude that the adsorbed layer does not register above the Mo atoms, but the Si atoms are in four-coordinating positions in the hollows formed by surface Mo atoms. Oxygen on silver[42] is either disordered or is a (1×1) structure on the (110) or (100) surfaces, but shows a (4×4) structure on the (111) surface.

7.2. Ionosorption on Semiconductors

7.2.1. The Surface State Representation of Adsorbed Species

In general the ionosorption of a gas on a solid is a rather complex process involving chemical steps and involving electron transfer steps. For a theoretical description of ionosorption we start with the rigid band model where the electron transfer process is considered as the basic process, and then introduce the modifications required because adsorption and desorption are involved. Such modifications lead to three differences from the simple picture as viewed in Section 5.5 or as viewed in Chapter 6. First, because we are dealing with a volatile species, the total surface state density varies with temperature and pressure of the adsorbate gas. Second, the identification of the "unoccupied state" that should appear in the electron transfer equations is not clear. Third, the adsorption isotherm for the neutral species, be it physisorbed or chemisorbed, can become extremely complex, as outlined in the preceding section. We will assume for the neutral species that Henry's law of adsorption is suitable, but it must be kept in mind that more complex laws will apply as the surface becomes covered.

The complexity arising because ionosorption depends on double-layer effects combined with adsorption/desorption isotherms and isobars leads to very difficult mathematics. Fortunately, as will be described, the double layer can become so dominant that the amount ionosorbed becomes approximately independent of temperature and gas pressure, and the use of of this approximation simplifies the picture considerably.

7.2.1.1. Equilibrium Ionosorption. For the equilibrium ionosorption of an acceptor species [Region (c) of Figure 7.3] we may have

$$X_n = X_{ns} \tag{7.10}$$

$$X_{ns} = nX_s \tag{7.11}$$

$$X_s + S_1 = X_s \cdot S_1 \tag{7.12}$$

$$X_s \cdot S_1 + e = X_s \cdot S_1^- \tag{7.13}$$

$$X_s \cdot S_1^- + S_2 = X_s \cdot S_2^- + S_1 \tag{7.14}$$

where the subscript s signifies an adsorbed species. Here X_n is a molecule consisting of n atoms of X, and S_1 and S_2 are two surface sites: S_1 is the site energetically most favorable for the bonding of X_s; S_2 is most favorable for the bonding of X_s^-. The free energy per atom of chemisorption [Equations

(7.10)–(7.12)] is ΔG_c. $\Delta G''$ as before (Section 5.5.1) is the free energy change of Equation (7.14) as the bonding shifts from site S_1 to S_2. $(E_{cs} - E_{t1})$ is the change in electron energy during the electron capture.

Reactions 7.13 and 7.14 are equivalent to Equations (5.9) and (5.10) of Section 5.5.1 and have already been analyzed. The result of that analysis was expressed in terms of the density of states. Here the density of states is a variable, and the result must be expressed as a function of gas pressure. Thus Equations (7.10) and (7.12) must be taken into account to describe the surface coverage of the neutral adsorbate as a function either of temperature or of gas pressure.

With Henry's law for adsorption [Equations (7.8a) and (7.4)], with the mass action law for Equations (7.11), (7.12), and (7.14), and with Fermi function [Equation (2.32)] applied to the electron transfer relation (7.13), it is straightforward to determine that in the region where Henry's law applies, the density of unoccupied states $[X_s \cdot S_1]$ is

$$\ln[X_s \cdot S_1] = (1/n)(\ln p_X + \ln b\Gamma_t) + \Delta G_c/kT \qquad (7.15)$$

where p_X is the pressure of X_n in the gas phase. This expression can be introduced into Equation (5.16) to yield

$$kT \ln[X_s \cdot S_2^-] = E_F - E_m + \Delta G_c + (kT/n)(\ln p_X + \ln b\Gamma_t) \qquad (7.16)$$

where again [Equation (5.17)] we have let

$$E_m = \tfrac{1}{2}(E_{t1} + E_{t2}) + \tfrac{1}{2}(\Delta G'' - \Delta G')$$

Due to double-layer effects, the term $E_F - E_m$ is a function of $[X_s \cdot S_2^-]$. Thus Equation (7.16) is unsatisfactory. The relation between surface charge and Fermi energy must be introduced to obtain an expression for the surface charge (coverage) as an explicit function of temperature and pressure. From Figure 2.2 it is seen by inspection that the position of the Fermi energy at the surface is given by

$$E_F = -\mu + E_{cs} - eV_s \qquad (7.17)$$

In the most common case, that of a depletion layer, the relation between V_s and the surface charge is given from Equation 2.13 by

$$V_s = [e/2\varkappa\varepsilon_0(N_D - N_A)] \cdot ([X_s \cdot S_2^-] + N_0)^2 \qquad (7.18)$$

Here we have assumed that the total surface charge is $[X_s \cdot S_2^-] + N_0$, where N_0 is the net charge associated with species other than the volatile species X. Then with no ionosorption (of the species X) there is an initial surface barrier V_{s0}. We will define the change in V_s due to the ionosorption of species X as ΔV_{si}, and the corresponding change in Fermi energy ΔE_{Fi} is given by Equation (7.17). If ΔV_{si} is much less than kT/e, then from Equation (7.17) the corresponding Fermi energy shift ΔE_{Fi} is also much less than kT, and in Equation (7.16), E_F is a constant.

In the case of very low pressure, where $\Delta V_{si} \ll kT/e$, Equation (7.16) reduces to

$$[X_s \cdot S_2^-] = A p_X^{1/n} \tag{7.19}$$

where the constant A collects all the constants in Equation (7.16).

At moderate pressures, however, the coverage becomes independent of pressure. The charge introduced by the ionosorbing species will suffice to make $e\Delta V_{si} (= \Delta E_{Fi})$ equal to or greater that kT. Then to a first approximation the logarithmic terms of Equation (7.16) can be neglected. Then from Equations (7.16)–(7.18), neglecting N_A, we have the approximation

$$[X_s \cdot S_2^-] = -N_0 + (2\varkappa\varepsilon_0 N_D/e^2)^{1/2}(E_{cs} - \mu - E_m + \Delta G_c)^{1/2} \tag{7.20}$$

showing a surface coverage independent of pressure.

Equation (7.20) represents a saturation in the amount of chemisorbed species for the case of a depletion layer on the semiconductor. It is the Weisz[43] limitation (Section 2.1.2) again, due to double-layer formation. The limitation was discussed briefly following Equations (2.13) and (2.14), where it was suggested that for reasonable values of the double-layer voltage the surface charge would be severely limited. In Equation (7.20) we have introduced the feature that if there are other surface states that have a charge N_0, the surface charge possible on the surface state of interest is changed by $(-N_0)$. Also we have introduced several terms involving the energy of chemical transformations involved in ionosorption. However, since these terms are all under the square root sign, the net effect is not greatly different: we can still [as discussed following Equation (2.13)] only ionosorb the order of 10^{-3} monolayers of charged species before saturation if a depletion layer dominates the surface.

The limitation of Equation (7.20) applies only to ionosorption. This does not mean that a species such as oxygen can never chemisorb to more than 10^{-3} monolayers. Ionosorption leading to an accumulation or an inversion layer can result in a higher Γ. Also, as mentioned in Chapter 1, when there is local bonding without electron transfer the adsorption density can be much higher, for with local bonding the partial charge on the adsorbate is usually lower than in the ionosorption case and relocation of surface atoms may lead to negligible double layer. Even without relocation the charge separation in the double layer is only the order of a few angstroms rather than the 100 to 5000 Å involved in ionosorption.

7.2.1.2. Irreversible Ionosorption.

In region (b) of Figure 7.3, the general phenomenon of irreversible chemisorption is shown. In the general case only a qualitative understanding is possible, as discussed in Section 7.1.2.3, but in the case of ionosorption, if we assume the pinch-off is due to double-layer effects, an analytical description of irreversible adsorption is possible.[16] In fact, the analysis is a straightforward application of the Shockley–Read model of electron capture at trapping centers,[44] as given by Equation 2.29. It should be noted that similar expressions have been developed in the literature, in most detail by Volkenshtein and Peshev.[45]

We will discuss the ionosorption rate at constant temperature and pressure using the example of an n-type semiconductor and an oxidizing agent such as oxygen that becomes negatively charged. The density of unoccupied states, $N_t - n_t$ in Equation (2.29), is assumed to be the density of physisorbed species or of chemisorbed neutral molecules. The density of unoccupied states can be assumed constant. It is determined by the equilibrium of the sorption/desorption expressions such as Equations (7.10)–(7.12). This assumes that the rate limiting step in irreversible adsorption is the electron transfer, not a chemical process such as dissociation of the gas molecule. As described in Section 7.1.2.3 the rate of desorption [the second term in Equation (2.29)] is negligible in the irreversible region.

With $N_t - n_t$ constant the only variable in Equation (2.29) is n_s, which by Equation (2.38) depends on V_s. Thus with Equation (2.8)

$$d\Gamma/dt = K_n(N_D - N_A)\Gamma_u \exp(-eV_s/kT) \qquad (7.21)$$

where Γ_u is the density of unoccupied states. Of course the surface barrier V_s depends on the surface charge. As Melnick[46] first pointed out, if there is initially a substantial surface charge due, say, to intrinsic surface states, then the change ΔV_s due to adsorption of the oxidizing agent may be small compared to V_s. As described in Equations (2.43)–(2.47), the Elovich equation, Equation (7.9), follows. Thus in cases where ΔV_s is proportional to $\Delta\Gamma$, and the rate limiting step is exponentially dependent on ΔV_s, the Elovich equation should apply to irreversible ionosorption. As discussed following Equation (7.9), however, the reverse is not true: Compliance with the Elovich equation cannot be considered proof that the electron transfer model dominates the variation of E_A with θ.

A possible complication in the process of ionosorption which becomes particularly apparent in slow irreversible ionosorption is the field aided diffusion of impurity ions in the high electric field of the space charge region.[47,48] For example, in early work on the adsorption of oxygen on ZnO, it was found that the adsorption of oxygen depends upon the high-temperature pretreatment in a way that suggests that at the high temperature donor impurities (oxygen vacancies or interstitial zinc) move to or from the surface. In recent work by Seitz and Sokolz[49] these concepts have been expanded and shown to apply to other impurities (Li and Na) in ZnO. The influence of impurity movement on ionosorption has been discussed in detail by Hauffe.[50] He emphasizes the possibility that such impurity movement will lead to such reactions as

$$e + Zn_i^+ + \tfrac{1}{2}O_2 \rightarrow ZnO$$

(where Zn_i is the donor in ZnO, an interstitial zinc atom) and will be confused with adsorption).

7.2.1.3. The Theory of Oxygen Ionosorption.

In the experimental discussions of Section 7.2, the adsorption of oxygen will be emphasized. It is therefore of value to use this same case for a quantitative analysis (insofar as the theory can be quantitative) of ionosorption. For simplicity we will neglect Franck–Condon effects; inclusion of the Franck–Condon complications would make the mathematics unwieldy, so the discussion below will be restricted to equilibrium rather than kinetic effects, and the surface states will be designated to be of the form of E_m in Equation (5.17) or Equation (7.16). As equilibrium adsorption alone will be considered, the reaction route chosen is unimportant, for the occupancy of surface states and the amount adsorbed is determined by thermodynamics rather than by a particular route.

We can assume the simplest reaction route:

$$e + O_2 = O_2^- \tag{7.22}$$

$$O_2 = 2O \tag{7.23}$$

$$e + O = O^- \tag{7.24}$$

All reactants in the above equations are considered adsorbed. Thus the surface concentration of O_2 in Equation (7.22) may, for example, be given by Henry's law, Equation (7.4). The appearance of atomic oxygen in Equations (7.23) and (7.24) is not to be interpreted as a suggestion that any measurable amount of atomic oxygen will be observed on a typical ionic semiconductor. As discussed above, we are simply taking advantage of the fact that at equilibrium the reaction route chosen is unimportant. What we are interested in is the concentrations of O_2^- and O^-. These species, as will be discussed in the next section, are known to exist as ionosorbed species on ionic solids.

Application of the mass action law to Equation (7.23) and of Fermi statistics to Equations (7.22)–(7.24) yields

$$[O_2^-]/[O_2] = \exp[-(E_{O_2} - E_F)/kT] \tag{7.25}$$

$$[O]^2 = [O_2] \exp(-\Delta G/kT) \tag{7.26}$$

$$[O^-]/[O] = \exp[-(E_O - E_F)/kT] \tag{7.27}$$

where E_{O_2} and E_O are the values of E_m for O_2^-/O_2 and O^-/O, respectively, and ΔG is the free-energy change in Equation (7.23). A modest manipulation of Equations (7.25)–(7.27) yields

$$[O_2^-]/[O^-] = [O_2]^{1/2} \exp[-(E_{O_2} - E_O - \tfrac{1}{2}\Delta G)/kT] \tag{7.28}$$

which gives a useful expression for the ratio of the two species at equilibrium.

As discussed in Section 7.2.1.1, we can distinguish two regions of adsorption as a function of oxygen pressure.

The first region, that of extremely low oxygen pressure, is the region where so little is adsorbed that the change in surface barrier is much less than kT/e. From Equation (7.18) for a depletion layer this region is defined by the condition that

$$(N_t^- + N_0)N_t^- < \varkappa\varepsilon_0(N_D - N_A)kT/e^2 \tag{7.29}$$

where

$$N_t^- = [O^-] + [O_2^-] \tag{7.30}$$

Under this condition E_F is independent of the amount adsorbed, in other words $E_F = E_{F0}$, the position of the Fermi energy with zero adsorption. When E_F is not a variable, Equations (7.25)–(7.27) can be applied directly to obtain the amount adsorbed of each form of oxygen.

The second and more commonly encountered region is a saturation region, occurring at moderately high pressure. Here the surface charge N_t^- is essentially independent of temperature and pressure, and the magnitude of N_t^- is given by Equation (7.20). At moderate to high pressure the value of N_t^- can thus be considered known, and Equations (7.28) and (7.30) can be solved simultaneously to give the amount of each form of oxygen. For example, the concentration of O^- is

$$[O^-] = N_t^-/\{1 + [O_2]^{1/2} \exp[-(E_{O_2} - E_O - \Delta G)/kT]\} \qquad (7.31)$$

and at low oxygen pressure O^- will be the dominant species on the surface.

From Equation (7.28) the ratio of O_2^- to O^- concentration increases with the oxygen pressure. This ratio is of particular interest because it is measurable. The variation of the ratio with temperature depends on the relative energy levels E_O and E_{O_2} and, of course, on the energy of dissociation of the oxygen molecule ΔG. Obviously we expect the electron affinity of the oxygen atom to be much higher than that of the oxygen molecule, so $E_{O_2} - E_O$ in Equation (7.28) will be large and positive. However, the free energy of dissociation of the oxygen molecule ΔG is also large and positive, so the magnitude or even the sign of the exponent in Equation (7.28) or (7.31) is not obvious.

In the experimental section to follow it will be found that in general, at intermediate pressure, the concentration of O_2^- dominates at low temperature and O^- at high temperature. Whether the low-temperature behavior is equilibrium is not clear in most cases, but the high-temperature behavior is often measured in the reversible adsorption region. The experimental results available will be compared to Equations (7.28) and (7.31). A quantitative study of the coverage of O^- and O_2^- interpreted with such a model or an improved version thereof is still however not available.

7.2.2. Observations of Ionosorption

As an example to illustrate the major features of ionosorption, the adsorption of oxygen will be emphasized. Oxygen adsorption on n-type and p-type ionic semiconductors will be discussed, to illustrate, respectively, ionosorption with a depletion layer and with an accumulation layer at the surface. In general we will not discuss adsorption on covalent semiconductors such as germanium and silicon in this section, for clean surfaces of these materials do not ionosorb, but form a local bond. Adsorption with local bonding will be discussed in the next section. However, some of the semiconductors in the transition region between ionic and covalent solids have interesting ionosorption characteristics, and these will be mentioned.

7.2.2.1. The Oxidation States of Ionosorbed Oxygen. Electron spin resonance (ESR) has been used to demonstrate the presence of

O_2^- on many solids. A characteristic triplet of resonance peaks appears, in the region $g = 2.001$–2.08. With minor modifications this triplet appears on a long series of ionic solids when treated so that electrons are available to induce ionosorption [Equation (7.22)]. Lunsford,[51] in his review of ESR data for adsorbed oxygen, cites 25 or more solids where the O_2^- triplet has been observed associated with ionosorbed oxygen. The list includes both suitably excited insulators and n-type semiconductors, but includes none of the common p-type semiconductors.

Solid confirmation that the triplet is indeed due to the O_2^- species is obtained by the use of ^{17}O in the adsorbate. Tench and Holroyd[52] studied the O_2/MgO system. The presence of $^{16}O_2$, $^{16}O^{17}O$, and $^{17}O_2$ mixtures leads to a large array of lines due to hyperfine splitting effects. The lines observed have been analyzed in detail[53,54] and are clearly associated with O_2^-.

The reactivity of adsorbed O_2^- with other coadsorbing species has been followed by monitoring the variation of the intensity of the O_2^- triplet. For example, upon addition of NO at room temperature, the O_2^- signal disappears. However, the addition of hydrogen or ethylene does not lead to changes.

Another oxygen species that has been observed by ESR in a few cases is O_3^-. For example, an analysis of the hyperfine splitting with ^{17}O present has demonstrated[53,54] the existence of O_3^- on MgO. The MgO case is a rather special case of ionosorption. As with most insulators the conduction and valence bands cannot be expected to exchange electrons with most adsorbates, because the conduction band is too high and the valence band too low in the band diagram (Figure 1.1). In these experiments the electrons are excited onto energy levels on the surface by photons. The O_3^- species is obtained on MgO by a particular series of treatments: first, electrons are trapped at the surface using radiation, second, O^- is formed by interacting N_2O with the trapped electrons, and third, a low vapor pressure of oxygen is introduced so the oxygen molecule can become associated with the O^- species. Then O_3^- species are observed. Another case where O_3^- has been identified[55] is on V_2O_5, where as discussed below, the presence of O_2^- apparently can be suppressed by heat treatment.

The other important ionosorbed oxygen species is O^-. The O^- species itself is difficult to observe with ESR. As described in the preceding paragraph, the interaction of N_2O with MgO that has trapped electrons on the surface leads to an O^- spectrum. In this case the line appears as a doublet at $g = 2.042$ and $g = 2.0013$. Studies of hyperfine splitting using ^{17}O have confirmed[51] the identification.

Observation of equilibrium O^- on V_2O_5 (the species ionosorbed on

MgO is clearly not at equilibrium), has been made by the workers in Kazanski's group.[56] They observed O^- on carefully prepared V_2O_5 (an n-type semiconductor). The sample is oxygen treated at 300°C, evacuated to remove the excess oxygen [as per Equations (7.28) and (7.31), with the exponent negative, lowering the oxygen pressure should tend to remove O_2^-] and then cooled rapidly to 77°K. The g value for the O^- signal is 2.026.

As with the O_2^- species above, the reactivity of O^- with other adsorbing gases can be studied by the stability of the line. It is found that the O^- species is very reactive, the intensity of the ESR line changing with H_2, CH_4, C_2H_6, or CO adsorption. Naccache[57] has also studied the activity of O^-, in his study of O^- on MgO, and found that CO reacts to form CO_2^-, and C_2H_4 reacts to form $C_2H_4O^-$. Yoshida et al.[58] compare the activity of O_2^- and O^- for C_3H_6 oxidation on V_2O_5. They find O^- reacts rapidly, but the product does not desorb (at room temperature), whereas reaction at O_2^- sites leads to an easily desorbable product. In general, as expected, the O^- ion is much more reactive than the O_2^- ion.[51]

7.2.2.2. The Ionosorption of Oxygen on n-Type Semiconductors.

Oxygen sorption has been studied extensively on zinc oxide. We will therefore use adsorption on this n-type semiconductor as the first example to illustrate the various phenomena in ionosorption. The general behavior of oxygen ionosorption in removing conduction electrons and lowering the conductance of ZnO has been known for many years.[16]

The rate expression for the conductance change depends on the source of electrons for the oxygen reduction. Arijs et al.[59] have explored the use of donor surface states at the ZnO surface [essentially N_0 in Equation (7.20)] upon which to adsorb the oxygen. The surface donors were produced by photolytic decomposition of the ZnO in vacuum (see Section 9.3.2.1). The electrons due to the donors are observed by conductance, and the rate of their removal by oxygen admission is monitored. It was found that the rate is linear with oxygen pressure, which suggests simple formation of O_2^- at room temperature, and, as is reasonable considering the source of electrons is so near the surface, which suggests no barrier effects.

Adsorption rates for oxygen on ZnO with no intentionally added surface donors has also been examined in the irreversible region of adsorption by Melnick,[46] and by Enikeev and his co-workers,[60] and by Barry and Stone.[61] They observe an Elovich dependence and attribute it to surface barrier effects (Section 7.2.1.2). The demonstration that the Elovich behavior arises due to surface barrier effects rather than heterogeneity is only conclusive, however, in the work of Enikeev,[62] who related changes in work function to the rate of oxygen sorption on ZnO, and in the similar work of Komuro on TiO_2 described below.

There is substantial evidence that either O_2^- and O^- can be formed on ZnO. Sancier[63] found ESR evidence for the presence of two types of oxygen when adsorbed at room temperature. He observed the O_2^- triplet upon oxygen adsorption, and observed the simultaneous diminution of a line at $g = 1.96$ that is associated with conduction band electrons. Adsorption of CO following oxygen adsorption caused the 1.96 line to increase rapidly, but the O_2^- line was removed very slowly indeed. Thus it was concluded that a fraction of the captured electrons are on species other than O_2^- (presumably O^-) under the condition of the experiment. Others have also obtained ESR evidence of the simultaneous presence of the two species.[64–66]

Cunningham et al.[67] find that adsorption of N_2O at room temperature is followed by the desorption of the N_2 and a decrease in the conductance of the ZnO. They assumed that adsorbed O^- is formed. The assumption has been confirmed by Lunsford's group.[54]

The Weisz limit [Equation (7.20)] on the ionosorption of O^- was investigated[67] as a function of doping using ZnO doped with the donor In, undoped, and doped with the acceptor Li. However, the system is complex because μ, the energy difference between the conduction band edge and the Fermi energy, changes as well as $N_D - N_A$. It was found that with indium (high N_D and low μ), the concentration of O^- was high; with no doping (low N_D and intermediate μ), the density of O^- was intermediate; with the lithium dopant (indeterminate N_D and high μ), the density of O^- was low. Thus qualitative agreement with theory was claimed, but there is a great difficulty in assigning an effective donor density in the case of lithium doping. If the electrons trapped by the lithium are at an energy level sufficiently high, they must be considered as available donor levels.

Chon and Pajares[68] have elucidated the relative concentrations of the two forms of oxygen, O_2^- and O^-, in the equilibrium adsorption region. They used Hall effect measurements on a pressed pellet of zinc oxide to monitor the number of electrons removed for each oxygen molecule adsorbed. The Hall constant R_H for a single-crystal material is related to the electron density n_b by

$$R_H = 1/n_b e \tag{7.32}$$

when only electrons are current carriers and the Hall mobility equals the drift mobility of electrons. If it is assumed[69,70] that on a compressed powder pellet Equation (7.32) is valid so the Hall coefficient reflects the number of electrons per unit volume of the pellet, then the measurement can be used to determine the total number of conduction band electrons in a sample. Chon and Pajares[68] measured n as given by Equation (7.32) (with a small correction for lattice scattering) and measured volumetrically the number of molecules of oxygen taken up as n changed. They conclude that below 175°C, one electron is removed from the conduction band per oxygen molecule (that is, the oxygen is adsorbed as O_2^- below 175°C), but above 225°C two electrons are removed per oxygen molecule (that is, O^- is formed). The pressure used was in the micron range. The observation of O^- at high temperature, O_2^- at low temperature, thus suggests the exponent in Equation (7.28) is negative.

Tanaka and Blyholder[71] obtain two TPD peaks for oxygen desorption, one at about 185°C and one about 285°C. This agrees with the Chon and Pajares result, the peak at 185°C then being associated with desorption of O_2^-, and the peak at 285°C presumably being the desorption of O^-. Tanaka and Miyahara[72] find negligible isotope exchange in the case of oxygen associated with the low-temperature peak, confirming again this oxygen is adsorbed in the molecular form.

The rate of desorption of the two forms of oxygen as a function of temperature have also been studied by measurements of electron injection.[73] In these measurements the oxygen is adsorbed in high surface concentration by depositing potassium ferrocyanide to initiate the reaction:

$$Fe^{+2} + O_2 \rightarrow Fe^{+3} + O_2^- \tag{7.33}$$

In other words, with the ferrocyanide ions present, the concentration of O_2^- could be made high without building up a negative charge, and thus the Weisz limitation was avoided. Following the adsorption of oxygen, the surface is electrostatically charged to cause the O_2^- ions to inject electrons. An Arrhenius plot of the injection rate permits the estimation of the surface state energy. This was found to be 0.9 eV for O_2^-. If, by using Chon's pretreatment, the oxygen is formed as O^-, the rate of injection is characterized by an Arrhenius slope of 0.4 eV. Obviously, however, in the O^- case the electron injection is not a simple electron transfer, but must involve chemical combination of the oxygen atoms. Thus the value 0.9 eV may represent the surface state energy of O_2^-/O_2, but the value 0.4 eV must be described as the energy to reach the activated state for desorption.

The above observations show qualitative consistency with the behavior expected for oxygen ionosorption on an n-type semiconductor. In the pressure range of microns to Torr, the temperature at which the ratio $[O_2^-]/[O^-]$ given by Equation (7.28) goes through unity is apparently about 200°C. Below 200°C at equilibrium the surface should be covered by

O_2^-. Electron injection studies show an energy level of 0.9 eV below the ZnO conduction band for O_2^-, and this value is in excellent agreement, using Equation (3.18), with the observation of Tanaka and Blyholder[71] that O_2^- shows a TPD peak at 185°C.

In summary, the adsorption of oxygen on ZnO provides an excellent example to demonstrate qualitatively the expected features for ionosorption when a depletion layer is present in the semiconductor. It is unfortunate that some of the above measurements have not been put together in a quantitative study of oxygen adsorption with which to compare the simplest theories and determine their quantitative applicability. Certainly all the techniques are developed, and all that is required is a systematic study.

Titanium dioxide is another n-type semiconductor with properties similar to those of ZnO in many ways. Here again O_2^- is observed by ESR at low temperature. Schvets and Kazanski[74] observe slight differences in the g value depending on the concentration, and conclude these differences are due to differences in coordination of the titanium adsorption sites to which the O_2^- becomes associated.

Komuro[75] studied the rate of oxygen adsorption on TiO_2 in the irreversible region, and has shown (a) Elovich behavior and (b) donor movement and its influence. The procedure was (a) reduce the rutile slightly and then (b) lower the temperature and observe adsorption. The donors were provided by the reduction process (either oxygen ion vacancies or interstitial titanium ions could provide donors). If the sample was held a long time at high temperature, so that the donors could diffuse into the sample, the adsorption followed an Elovich equation, as described in Section 7.2.1.2. This behavior presumably was associated with adsorption at a normal depletion layer. If, on the other hand, the sample was first oxidized and then reduced for a very short time so that the donors could not diffuse into the sample, a local interaction between a Ti-rich surface and the adsorbing oxygen was expected. A parabolic rate law of adsorption was observed. The authors suggest that the parabolic dependence on time arises due to surface diffusion limitations.

The n-type semiconductor V_2O_5 is quite different from ZnO and TiO_2. In this case, as discussed in Section 7.2.2.1, the ESR signal due to O^- is easily observed, but, according to Schvets and Kazanski,[74] by a temperature pressure cycle described earlier, the presence of O_2^- can be avoided. Thus, for example, Shelimov, Naccache, and Che[55] observe on V_2O_5 the formation of O_3^- from $O^- + O_2$.

Using electrochemical techniques (Chapter 8), Gomes[76] has shown that the conduction band of V_2O_5 is much lower than the conduction band of ZnO, relative to the free electron energy. He finds that unlike the cases of ZnO or TiO_2, many adsorbed hydrocarbons will inject electrons at room temperature into the V_2O_5 conduction band. On the other hand, it is found that there is great difficulty in reoxidizing the solid.[77] At room temperature oxygen cannot extract electrons; to obtain a reasonable rate of oxygen reduction on V_2O_5, the temperature must be several hundred degrees. The low conduction band and the difficulty in reducing oxygen doubtless are closely associated with the ease of eliminating the O_2^- species on V_2O_5.

As a final remark on the adsorption of oxygen it is desirable to suggest caution in interpreting all adsorption on n-type semiconductors as ionosorption. On many n-type semiconductors, such as III–V compounds or chalcogenides, oxygen can adsorb by local bonding as discussed in Section 5.2.4 rather than by ionosorption. Thus, for example, on CdS, the formal oxidation state of sulfur can increase to $+4$ or $+6$, with groups resembling SO_3^{2-} or SO_4^{2-} forming upon adsorption. For example when Katz and Haas[78] observed 0.2 monolayers of oxygen uptake on CdSe, they were able to conclude with assurance this was not ionosorption.

7.2.2.3. The Ionosorption of Oxygen on Other Solids.

Due to the scarcity of p-type semiconductors with simple behavior, we have much less quantitative data on the oxygen adsorption characteristics of p-type than on n-type semiconductors. By the simple space charge models of Chapter 2, the adsorption of oxygen should cause an accumulation layer on p-type material. The question arises whether the formation of a simple accumulation layer with oxygen adsorption has been observed without being accompanied by chemical complications associated with strong local interaction. The ionosorption of oxygen on p-type semiconductors such as NiO and Cu_2O results in an increased conductance. However, it is difficult to be convinced in many cases that the observation is even simply chemisorption, rather than absorption of oxygen with concurrent formation of cation defects. In fact on NiO, Hauffe[50] concludes that most of the oxygen uptake is absorption rather than adsorption. In this case he concludes that motion of impurities rather than vacancies permits the absorption, but the result is the same.

Robert and his co-workers[79] have studied XPS signals from adsorbed oxygen on Cu_2O and find three different peaks, one at 530.4, one at 531.4, and one at 532.2 eV. By varying the conditions of oxygen admission, they conclude that the first is associated with lattice oxygen, the second with hydroxide oxygen, and the third with chemisorbed oxygen. This suggests that chemisorbed oxygen can be present; indeed, to show an XPS peak it must be present in large quantities as expected with an accumulation layer at the surface. However, the suggestion that the adsorption of oxygen is associated with an accumulation layer is not at all demonstrated. The oxygen could be covalently bonded, associated with ionic surface states (of Cu).

Adsorption of oxygen on ionic surface states on p-type materials may be common. Chromic oxide is another p-type semiconductor, and in this case also the adsorption of oxygen increases the conductivity.[80] However, in this case there is direct evidence[81,82] that the adsorption is associated primarily with the capture of (or perhaps a better phrase is "sharing of") electrons from ionic surface states that are localized on incompletely coordinated surface Cr^{+3} ions. The adsorption is *not* primarily associated with capture of electrons from the valence band. Davydov, Shchekochikhin, and Keier[83,84] have studied ir absorption and find a single peak associated with room temperature adsorption of oxygen. Using isotopic oxygen to induce interpretable structure in the ir absorption peaks, they were able to show that oxygen is adsorbed as a molecule at temperatures below 200°C, so the single peak at low temperature is associated with O_2^-. After a 250°C heat treatment a complex spectrum with several peaks is obtained, but then after heating beyond 370°C the pattern simplifies again to another single peak. This latter peak is presumably associated with dissociated oxygen.

Apparently on chromia[82] the surface Cr^{+3} act as bonding sites, giving, for example, at low temperature

$$Cr^{+3} + O_2 = (Cr - O_2)^{+3} \qquad (7.34)$$

with the chromium becoming formally in the $+4$ valence state, the oxygen in the O_2^- form. The quantity adsorbed, even at low temperature, approaches a monolayer. The conductivity is strongly affected by oxygen adsorption, but the effect arises[80] because with no adsorbed oxygen, the surface Cr^{+3} ions act as donor ionic surface states:

$$Cr^{+3} + h^+ = Cr^{+4} \qquad (7.35)$$

removing holes from the valence band to form a depletion layer. Adsorption of oxygen on the donor surface states simply restores the holes that were removed by the donor action, which results in flat bands so the conductance increases. An accumulation layer formation may occur by hole injection if the level associated with ionosorbed O_2^- is deep enough, but this cannot be determined from the data available.

Finally it is of interest to report adsorption on GaAs, the material we have cited in several instances as being in the transition region between an ionic semiconductor and a covalent semiconductor. According to Haneman's group,[85,86] at 77°K oxygen chemisorbs irreversibly on GaAs. By the resulting ESR signal it adsorbs as O_2^-. The presence of the adsorbed oxygen gives rise to a LEED superstructure, which indicates a high concentration. Again the low-temperature results may arise primarily due to a donor/acceptor complex between the donor arsenic and the acceptor oxygen molecules at the surface, analogous to the chromia case above, rather than being "real" ionosorption. This is suggested because a high coverage (compared to the Weisz limit for accumulation layers) is required to produce a strong LEED pattern. At room temperature the oxygen converts, forming a covalent bond with the surface As atoms.[87] Dorn, Lüth, and Russell[88] have reported an extensive study of the adsorption of oxygen on the GaAs(110) surface. They find the sticking coefficient varies with an Elovich behavior, and that one oxygen atom per surface arsenic atom becomes adsorbed.

In the above discussion we have reviewed the ionosorption of oxygen on various classes of semiconductor solids. For highly ionic n-type materials the features of ionosorption can be reconciled to the simple rigid band theory. However, with p-type materials and with more covalent n-type materials, local bonding interactions complicate any attempt to interpret the observations by a straightforward rigid band model.

7.3. Adsorption with Local Bonding

In the preceding section we discussed ionosorption on semiconducting ionic solids that involves either electron exchange with a band or electron exchange with a donor or acceptor ionic surface state. The latter could well have been included in the present section, but for convenience in presentation we arbitrarily chose to include cases of adsorption of oxygen as O_2^- with discussions of ionosorption. In the present section the discussion will center on adsorbate bonding where local bonding apparently dominates and the effect of the bands is a small perturbation. Thus we will discuss bonding to acid or base sites on ionic solids and covalent bonding to covalent or metallic solids.

7.3.1. Adsorption on Ionic Solids

7.3.1.1. Adsorption on Acid or Basic Sites. In the discussion of chemisorption, the interest has centered on the origin of the activa-

tion energy and its variation with coverage. In the case of chemisorption on acid or basic sites we have the unusual case of chemisorption that shows negligible activation energy. The reason for this general behavior doubtless stems from three factors: (a) the adsorbate normally chemisorbs as a molecule, so there is no direct Lennard-Jones (Figure 7.2) activation energy of dissociation; (b) there is no buildup of an electric field upon adsorption as was associated with the activation energy of ionosorption; in fact in the acid/base case it is more likely that any electric field is diminished; and (c) there is little interaction between the adsorbate molecules, as will be shown important for adsorption on metals. Each adsorbate molecule interacts only with its own active site, and for very strong solid acids or bases the active sites are not adjacent.

Because there is no activation energy, there is in principle no irreversible region in the adsorption isobar. In practice, of course, the rate of return to equilibrium is limited at low temperature by rate constants containing the factor $\exp(-E_D/kT)$ as in Equation (7.7). However, it is generally assumed—and no apparent complications arise—that when a sample is exposed to an adsorbing base (acid) at room temperature, for example, in preparing to make a TPD measurement, all the adsorption sites rapidly equilibrate.

The other aspect of chemisorption that has been emphasized is the heat of adsorption, or, better, the activation energy of desorption. Here again the characteristics for the present case differ. Instead of a reasonably monoenergetic activation energy of desorption [see the discussion of Equation (7.5)], for acid and base sites there is generally an extremely broad spectrum of heats of desorption (acid or basic strengths). The breadth may simply be associated with the fact that studies are usually made on fine powders, and these are expected to have greater heterogeneity in surface properties simply because so many crystal faces, steps, and corners are exposed. The heterogeneity may also arise because naturally heterogeneous surfaces (such as alumina or mixed oxides, see Figure 4.7) have more active acid/base sites and therefore are more often investigated.

The distribution of desorption energies of acid (or base) sites is often measured by TPD curves. The concept is simply to occupy all the active sites with the gaseous acid or base, then measure the amount desorbed as a function of temperature as the temperature is slowly increased. The volume of the adsorbed gaseous acid or base desorbed in each temperature interval reflects the relative amount of sites of each basic strength (or acid strength). Figure 7.5 from Webb[89] shows an example of such a measurement. The figure shows a series of TPD curves for ammonia desorption from flu-

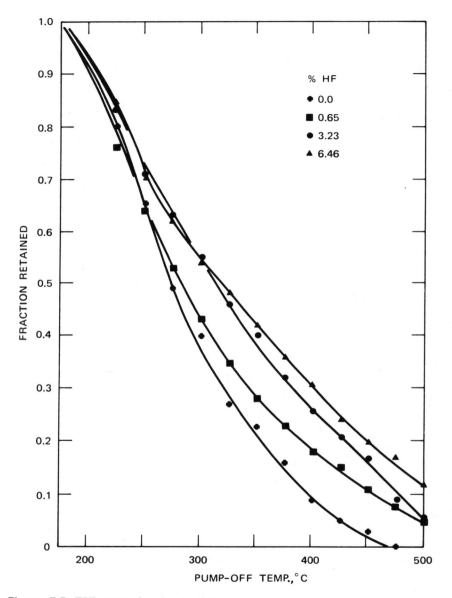

Figure 7.5. TPD curve for the desorption of ammonia from fluoridated alumina. The presence of the HF increases the acidity and, as indicated, increases the temperature needed to desorb the ammonia.

oridated alumina. The TPD was measured by Webb to show the increasing acid strength due to fluoride additives (Section 6.4.2.2), but it is shown in Figure 7.5 because the curves illustrate the typical TPD curves for desorption from acid sites. The curve shapes are clearly far from the single peak expected for a simple TPD curve following the theory of Equation (3.18) when a single type of site is present. The gradual decline indicates the broad heterogeneity as discussed above.

Their importance in catalysis has been the incentive for most of the interest in acid and basic sites up to the present time, so much of our knowledge about these important surface sites has been gleaned from the catalytic literature. The reason for the strong catalytic interest arises because of the ability of acid and basic sites to adsorb and dissociate hydrocarbon molecules, leaving the fragments in an exceptionally active form. This adsorption to form "carbonium ions" on acid sites or "carbanions" on basic sites is discussed in Section 10.3.

7.3.1.2. Adsorption from Solution.

In this section we will discuss the adsorption of ions (protons, hydroxide and metal ions) on ionic solids from aqueous solution. The interest in adsorption at the solid/liquid interface arises in part because such adsorption will affect electron and hole transfer processes at the semiconductor/electrolyte interface, and studies of these processes (Chapter 8) have been of exceptional value in understanding electron exchange reactions in general. But even more important, we are interested in adsorption out of liquids because often active surface additives (Section 6.2.1) are deposited onto a solid from the liquid phase. Thus to utilize or to prevent surface impurity effects, knowledge of the processes of adsorption from solution is desired.

When an ionic solid, in particular an oxide, is immersed in aqueous solution, protons and hydroxide ions are adsorbed on the surface, in a manner similar to the adsorption shown in Figure 4.4. When water is adsorbed from the gas phase, the number of OH^- and H^+ ions adsorbed must be equal. In aqueous suspension, however, there is no longer such a restriction, and indeed the surface of an ionic solid is seldom neutral. If one introduces a dry but hydroxylated (such as in Figure 4.4) powdered oxide into neutral water (which initially of course has 10^{-7} M each of protons and hydroxide ions present), there will generally be unequal adsorption (or desorption) of protons and hydroxyl groups. If the surface has a net acid behavior it will give up protons (or adsorb hydroxide ions), according to a reaction

$$H_a^+ + OH_{soln}^- = (H_2O)_a \qquad (7.36)$$

(where the subscript a indicates an adsorbed species) and the reaction will (i) cause the water to become acid, of pH less than 7, and (ii) induce a net negative surface charge on the oxide, which must be compensated by some positive ion in solution. In order to restore the solid to electrical neutrality, a soluble acid must be introduced into the suspension to decrease the concentration of OH^-_{soln} and drive Equation (7.36) to the left. If too much soluble acid is added to the suspension, the surface charge on the solid can be made positive, because Equation (7.36) is driven too far to the left.

If, on the other hand, the solid is a solid base, it will adsorb protons when suspended in neutral water, leaving excess of OH^- ions in the solution. Then the solution becomes basic, with the solid itself becoming positively charged. If the solution is made more basic by the addition of a soluble base, the surface can be made electrically neutral. A more quantitative discussion of the surface charge due to H^+ or OH^- adsorption is presented in Section 8.2.1.

Each ionic solid will have a pH value termed the IEP, the "isoelectronic point," at which there is no net surface charge in aqueous solution, i.e., the surface is electrically neutral. As discussed in detail by Cornell et al.,[90] if there is no adsorption of ions other than H^+ and OH^-, IEP = pzc, the "point of zero charge," where the coverage of H^+ equals the coverage of OH^-. Under these conditions, the IEP can be shown to be closely related to its acid/base behavior at the solid/gas interface, as discussed in the preceding two paragraphs.

The IEP has been shown to vary in a systematic manner with the properties of the cation for a series of oxides. Parks[91] examined a series of 18 oxides (hydroxides) and showed that the IEP varies according to

$$IEP = 18.6 - 11.5(Z/R) \tag{7.37}$$

where Z is the oxidation state of the cation and R is the separation distance in angstroms of an adsorbed proton from the cation (the diameter of the oxygen ion plus the radius of the cation). Now the expression for the IEP [Equation (7.37)] as a function of cation radius and charge and the expression for acidity as a function of cation radius and charge [Equation (4.20)] are not identical, but both are empirical and come from different fields. The fact that both show higher acidity with higher Z and show the same increase in acidity as a function of cation radius is expected because the two phenomena are related. The important thing is the similarity in concepts: in both cases we are interested in the attraction of an adsorbing proton or hydroxyl ion to the surface of an oxide and how this varies with the properties of the cation.

In summary, then, the condition of an ionic solid in aqueous solution can be represented by Figure 4.4, but with the occupation of the surface sites by OH groups dependent on the pH of the solution. If the solution has a pH more basic than the IEP of the particular oxide, H^+ ions will be replaced by adsorbed water [Equation (7.37)]. If the solution has a pH less than the IEP of the oxide, OH^- groups will be replaced by adsorbed water.

If other ions (than H^+ or OH^-) are present in the water, a process termed "ion exchange" can occur. The protons at the surface of the oxide an be replaced by other cations rather than water molecules, and the hydroxide ions at the surface can be replaced by other anions rather than water molecules. Such ion exchange processes are of great technological importance, particularly for ionic solids of the zeolite class, which have structure with a high internal surface area. A typical ion exchange process is described in Section 4.3.3.1, with reference to Equation (4.22). Other ion exchange processes are discussed in Section 6.2.1. As discussed for these cases, the concept is reasonably straightforward and will not be described further.

The adsorption from aqueous solution of hydrolyzable metal ions has been reviewed by James and Healy[92] in three articles. These are ions that can form complexes in solution of the form $\{M(OH^-)_n\}^{m+}$, a class of particular interest in the present discussions. Co^{2+} was studied primarily by James and Healy, although they also examined Ca^{2+} and Fe^{3+} ions. Such ions, in particular, the transition metals, are important surface additives of the class discussed in Chapter 6, and so we are particularly interested in their adsorption characteristics. James and Healy find that the adsorption of such ions cannot be described by the simple ion exchange laws, for other parameters, particularly pH, dominate. As indicated, the hydrolyzable metal ions of this class strongly attract (are hydrolyzed by) OH^- ions, and the hydrolysis depends on pH.

James and Healy conclude for such ions (a) adsorption and hydrolysis are closely related and (b) the adsorbed species is separated from the solid surface, upon adsorption, by at least one layer of H_2O molecules.

A model was developed for adsorption that agrees with experimental observation. The model is based on three contributing forces tending to attract or repel the adsorbate to or from the surface. The two attractive forces arise, respectively, from Coulombic and chemical interactions. The third is a repulsive force which arises because polarization energy (the dielectric part of the solvation energy as discussed in Section 2.3) is lost as the adsorbate approaches the surface. The first contributing force, the attractive Coulombic force, appears whenever the pH of the solution is more basic than the IEP, for in this case, as discussed earlier, the surface of the solid is charged negative. The Coulombic force is proportional to the charge m of the complex $\{M(OH^-)_n\}^{m+}$. The second contribution, the attractive chemical force, is ill defined, and the possibility of such a force is included only for completeness. The third force, the repulsive force, arises because near the solid and in the solid the dielectric constant is usually lower than the dielectric constant in the bulk solution, and from Equations (2.48) and (2.49) the energy of the system will increase (as ΔE_p decreases) if the effective $\varkappa_s$ decreases. Such an increase in the energy of the system as the ion moves toward the solid translates into a repulsive force. Now the reason that the dielectric constant is lower near the surface is two-fold. First, when the pH of the system is not near the IEP, there is an electric field near the surface which constrains the movement of the water dipoles and this lowers the effective dielectric constant. Second, when the ion is very near the surface, the dielectric constant of the solid becomes more important, and in general this is lower than the dielectric constant of the water.

According to Equations (2.48) and (2.49), it will be observed that the repulsive force will vary as the square of the charge on the solvated species, $\{M(OH^-)_n\}^{m+}$, and so varies as m^2.

In general the net effect of these attractive and repulsive forces leads to adsorption with the following characteristics. If the ion in solution is weakly hydrolyzed, in other words, if n is low so m is high, the repulsive energy term, associated with solvation, will prevent adsorption, because this energy varies as the square of m, where the attractive force varies linearly with m. The attractive force will only begin to dominate when the pH is raised to the point where perhaps m is the order of unity in Equation (7.38):

$$M^{+Z} + nOH^- = \{M(OH)_n\}^{m+} \tag{7.38}$$

that is, adsorption occurs when the ion is almost completely hydrolyzed, ready to precipitate as the hydroxide. It turns out that in general adsorption depends more on the pH at which hydrolysis of the ion occurs than on the IEP of the solid. The dominance of the hydrolysis of the ion will often mean adsorption will only occur in a small pH interval between complete solution of the ion on the one hand and precipitation as the metal hydroxide on the other hand.

The work of James and Healy seems particularly instructive with respect to the observations of dispersion techniques described in Section 6.1. As was discussed there, one of the best dispersion techniques was to adsorb the ion of interest, in particular platinum, by pH control. This observation is consistent with James and Healy's results. A simple impregnation approach will probably never during the drying process attain the right pH for a reaction by Equation (7.38) to occur, so never will attain the right conditions for adsorption. For highest dispersion of the additive, adsorption is needed, so deposition by a careful pH adjustment is preferable to simple precipitation of the additive during the evaporation of the solvent.

7.3.2. Adsorption on Platinum

The literature on the adsorption on metals is too extensive to attempt a general review. Therefore in this discussion the adsorption of various gases on platinum will be emphasized. This system provides a satisfactory example showing many of the characteristics dominating adsorption on metals. Using this example we will discuss heterogeneity of the surface or impurities on the surface affecting the activation energy of adsorption, variation of properties such as reconstruction, and finally discuss the heat of adsorption as a function of adsorption site and coverage.

Other cases of adsorption on metals and covalent semiconductors are discussed in Chapter 5, although in those discussions emphasis was placed on the bond formed during the adsorption, rather than the kinetic parameters.

The sticking coefficient of adsorbing gases is defined as the fraction of the impinging gas molecules that become adsorbed and is related to the activation energy of adsorption. The sticking coefficient s of oxygen on platinum is of particular interest as it illustrates the importance of heterogeneity. Values of s have been measured between 7×10^{-7} by Weinberg et al.[93] and 0.2 by Wood et al.[94] Other authors have obtained values between, but more often than not, the higher values are observed. The discrepancy has evoked substantial discussion. Joebstl[95] suggests that an apparent low sticking coefficient may be measured because, at low pressure, a background pressure of CO may suffice to remove adsorbed oxygen, forming CO_2. When the rate of adsorption is monitored by the volume desorbed upon heating (flash desorption), the rate would seem low because the oxygen is consumed by CO as fast as it is adsorbed. Others suggest surface heterogeneity is required to dissociate and adsorb oxygen, and the low values are observed on particularly homogeneous surfaces. Weinberg et al.,[93] from theoretical and experimental considerations, suggest that carbon contamination may provide the active sites that adsorb the oxygen. Other workers consider that the dominant surface flaw that permits oxygen adsorption is the surface step.

The surface step model is the one suggested by Somorjai's group,[96–98] who concluded that not only oxygen but H_2, D_2, OH, NH, and CH groups adsorb more rapidly on platinum with steps present. The adsorption of oxygen and nitrous oxide is reviewed by Alnot et al.,[99] who also favor the surface step as the active site for adsorption. Oxygen adsorption on silicon[11] and on GaAs[100] has also been shown to be more rapid in the presence of steps. Ducros and Merrill[118] are able to explain rapid O_2 adsorption on the Pt(110) surface by describing it as a series of steps and (111) terraces. In general it can be concluded that steps provide active sites for the adsorption of oxygen and other diatomic molecules on platinum.

There is evidence also that surface steps affect hydrocarbon adsorption on platinum. Somorjai's group[12] has studied adsorption on stepped surfaces and find that the hydrocarbons form ordered surface structures that depend both on the terrace width of the steps and the geometry of the hydrocarbon.

There is evidence that carbon is not the imperfection that induces oxygen adsorption, but it can be a promoter or poison in other reactions. With respect to oxygen adsorption, Völter, Procop, and Bernet[101] find on a carefully cleaned platinum surface that the value of s is still high. So factors other than impurities are indicated as the active sites. However, carbon seems to be important in NO_2 decomposition and in hydrocarbon adsorption. West and Somorjai[102] studied the dissociation of nitrogen dioxide on platinum, using the MBRS, beaming NO_2 molecules at the surface with low (thermal) energy, and observing the angular distribution of the emitted species. On a clean surface, NO_2 is accommodated, and the emerging beam is undissociated and has a cosine distribution. Such a cosine distribution indicates the adsorbate resides on the surface long enough that the species has no memory of its origin (Section 3.2.9). However, if carbon contamination is present, rapid dissociation of the NO_2 occurs. The dissociation must be rapid, for the emerging NO is not cosine in distribution, but is close to specular. Weinberg, Deans, and Merrill[103] in studies of propylene (C_3H_6) adsorption also find a strong influence of active sites associated with carbonaceous residue. They find the first layer of their adsorbate adsorbs irreversibly with sticking coefficient unity, ordering in a (2×2) array, and dissociating. When the carbonaceous residue forms a complete monolayer, a more passive surface is observed on which the next layer of propylene is adsorbed, this time nondissociatively and reversibly. McCarty and Madix[104] also observe a full monolayer of carbon acts as a poison rather than activating the adsorption. In this case the adsorption of CO, H_2O, CO_2, and H_2 on Ni was under investigation.

Reconstruction is observed on the Pt(110) surface by Lambert and Comrie.[105] They find that the clean Pt surface has a (1×2) structure, but is restored to the (1×1) form upon adsorption of either CO or NO. Other metals show changes in reconstruction with coverage even more clearly than platinum. For example, Sickafus and Bonzel,[106] show an abrupt change in LEED pattern [from random to (4×2)] together with an abrupt drop in the heat of adsorption when the Pd surface reaches half-coverage with bridge bonded CO. Such behavior is discussed in more detail in Section 7.1.3.

The heat of adsorption of hydrogen on platinum decreases with coverage. Norton and Richards[107] found the heat of adsorption on a clean surface is 21 kcal/mole. The heat decreases with coverage according to a relation interpretable either on the basis of a uniform surface model, where one assumes a repulsive potential energy between neighboring hydrogen atoms, or a heterogeneous model. For adsorption at 77°K a repulsive potential energy of 2.8 kcal/mole would fit Roberts' model[108] of sorbate/sorbate repulsion. Actually, Norton and Richards[107] point out several indications that Roberts' model dominates. Procop and Völter[109,110] observe 16 kcal/mole heat of adsorption (with zero activation energy) on a clean surface. The heat decreases with increasing θ to 5 kcal/mole, which value they attribute to physisorption. Kubokawa et al.[111] observe two heats of adsorption for hydrogen on platinum, one at 22 and one at 9.5 kcal/mole. Thus the initial heat of adsorption, about 20 kcal/mole, and a decrease with coverage, seems common to all observations, but there are differences in details that need to be resolved.

Part of the discrepancy in the above measurements may be due to the presence of several forms of the adsorbate on the Pt surface. Dixon, Bartl, and Gryder[112] used supported platinum so that infrared measurement of adsorbed hydrogen could be made (Section 3.3.3). They found that the appearance of ir active adsorbed hydrogen at 77°K was dependent on the pretreatment. For example, cooling the Pt in helium from 300°C led to a passive surface with no adsorption, but annealing the sample at room temperature in hydrogen reactivated the surface. They found two ir bands, suggesting two hydrogen species. Clewley et al.[113] also found at least two forms. They studied the low-temperature form of adsorbed hydrogen, adsorbing the low-temperature form over a saturated coverage of the high-temperature form.

Multiplicity of adsorbed species is also observed with oxygen on platinum. For example Alnot and his co-workers describe measurements of TPD of oxygen from platinum ribbons. As others have found,[109,110,114] they find several desorption peaks. In Alnot's studies the oxygen was adsorbed by decomposing N_2O on the surface. Two desorption peaks, one at 580°K and one at 625°K showed a desorption rate first order in oxygen coverage, so were assumed to be in the form of adsorbed molecular oxygen. A third desorption peak at 740°K is second order, so probably arises from oxygen adsorbed in the form of atoms. Norton[115] in an XPS study observed a TPD peak at 300°K and 600°K, and found two separate XPS peaks corresponding to these strongly and weakly bound oxygen species.

In summary it appears that, as expected, adsorption on covalent and metallic surfaces can be very dependent on imperfections and deposited impurities. The overall behavior described suggests that the adsorption generally occurs on a few sites of very low activation energy and, with the surface otherwise fairly uniform in its bonding, the adsorbate can move across the surface to free the active site for the next impinging gas molecule. This, in principle, should make the mathematics of adsorption much easier, except that the nature and number of the active sites is very hard to generalize for a useful theory. The heats of adsorption may be in some cases more amenable to simple models than are the activation energies, at least according to the one observation of Norton and Richards.[107]

8 | The Solid/Liquid Interface

8.1. Introduction

Studies at the solid/liquid interface have proved particularly valuable in improving our knowledge of the chemical physics of surfaces, particularly when the rigid band model is appropriate. In such studies it is found possible to (a) determine the relative positions of band edges for various semiconductors, (b) determine the surface state energy levels of well-behaved surface species relative to these band edges, and (c) relate these "surface state energy levels" with the chemical properties (particularly the redox potential) of the foreign species.

With knowledge of these energy relationships and with a few particularly powerful experimental techniques limited to use at the solid/liquid interface, it has been possible to analyze relatively complex chemical processes that occur upon electron exchange at the surface.

The determination of energy levels as described above in general depends on special properties observed only on semiconductor electrodes, and so most of the discussion in this chapter will relate to semiconductor electrochemistry. Metal electrode behavior as it can be described with the energy level representation will be outlined, but a review of the wealth of experimental data and experimental techniques that has been developed through the years using metal electrodes will not be made. These data have not in general provided the detailed picture of electronic energy levels that the recent semiconductor results have done.

An ion in solution, particularly in aqueous solution, becomes "solvated," which means that the solvent medium responds to the electrostatic and ligand field forces associated with the ion and forms strongly bound shells around the ion. There are three such solvation processes that are of particular interest. First there is normally an "inner coordination sphere" of very strongly bound water molecules or other polarizable groups (OH^-, NH_3, etc.) surrounding an ion. Second, there is an "outer sphere" of polarized water molecules, attracted by a polarization energy based on Equation (2.48), also around the ion. Thirdly, of interest here, there are similar hydrolysis and water layers at the surface of the ionic solid electrode.

One result of this solvation activity is that seldom are ions adsorbed at the semiconductor surface. This provides a substantial simplification in interpretation of electron exchange reactions, eliminating an important unknown of most systems, the adsorption energy. The reason for the lack of adsorption is that the energetic spheres described above, both on the solid and around the ion, must be broken for adsorption to occur, and this energy must come from the solid/ion bond. Thus only the strongest interactions can lead to adsorption. An analysis of one of the more important systems, the adsorption of hydrolyzable metal ions, was presented in Section 7.3.1.2, showing that adsorption only occurs under very special conditions.

In the discussions to follow we will assume in general not only that there is no specific adsorption, but also that there is no change of the inner coordination sphere during the electron transfer process. Where this latter assumption is possible, a second substantial simplification in models results. The second simplification develops because with outer sphere effects as the only form of chemical transformation, the chemical reorganization energy is very similar for all ions in solution. It is the reorganization of the polar medium (water) discussed in Sections 2.3.3 and 5.5.2, where the important parameter is the rearrangement (reorganization) energy λ. In our theoretical analysis below we will assume that outer sphere reorganization is the only chemical process, and choose systems where inner sphere changes should be minimal. It should be recognized, however, that this assumption is never completely valid even for the simplest systems (iron cyanide, for example). Some allowance will have to be made for the fact that a varying fraction of the chemical reorganization energy for each system will not be quantitatively described by the simple theory.

With the assumption that only outer sphere effects contribute to the reorganization energy, the interpretation of the results becomes particularly rewarding. With all the species having the same reorganization energy, to a fair approximation, the effect of other properties of the ions (such as the electron affinity or redox potential) on the electron exchange behavior can be easily distinguished and evaluated.

In earlier chapters we have reviewed many concepts which provide direct background for the discussions to follow. In Section 2.3.3 the kinetics of charge transfer to an ion in aqueous medium was described. The energy levels associated with ions in a polar medium were shown to fluctuate with time over what could be a broad energy span (many tenths of an eV) and such fluctuations were shown to control the kinetics of electron transfer between the solid and ions at its surface. In Section 5.5.2 the polarization of the medium after electron transfer was shown to cause a splitting of the energy levels associated with the occupied and unoccupied states (reducing and oxidizing agent) of the ions (redox couple) in solution. The reducing agent of the redox couple was shown to have a lower energy level than the oxidizing agent of the redox couple as illustrated in Figure 2.5. In these earlier analyses, as in the present, changes in the inner solvation sphere were not considered.

Finally, in the preceding chapter (Section 7.3.1) the form of an ionic solid in solution was described, with emphasis on the double layers present on the surface being due to the adsorption of protons or hydroxide ions from water, and to some extent being due to the adsorption of other ions. In the present chapter, the development will be continued. Section 8.2 will be devoted to developing the concepts from Sections 2.3.3, 5.5.2, and 7.3.1 into a semiquantitative model of electron or hole transfer between a solid and a redox couple immersed in an aqueous solution. In Section 8.3 the model will be compared to experimental observations, and in Section 8.4 a short discussion will be presented of the relation between measured energy levels in solution and the surface state energies of the same ions adsorbed at the gas/solid interface.

8.2. Theory

8.2.1. Double Layers and Potentials in Electrochemical Measurements

In addition to the double layers present in the solid, as analyzed in Chapter 2, there are double layers present on the solution side of the solid/liquid interface, whether the solids be metal, semiconductor or insulator. These double layers are the Helmholtz and the Gouy double layers.

We will dismiss the Gouy layer with a few sentences. The Gouy layer is a diffuse space charge region extending out from the electrode, in which the Poisson equation in one dimension [Equation (2.1)] applies. The density of charge however varies with distance from the surface as the ions in solution are, of course, mobile. The density of charge to be inserted in Poisson's equation is a function of potential, given by the Boltzmann factor, $C_i \exp(-Ze\phi/kT)$ (with Z the ionic charge, ϕ the potential, with $\phi = 0$ at infinity, and C_i is the concentration of the dominant ion i at infinite distance from the solid). Application of this Gouy–Chapman model to a solution of ions shows that the thickness of the double layer becomes negligible and indistinguishable from the Helmholtz double layer for moderate ion concentrations. Since most of the measurements to be discussed have nonreactive ions such as K^+ and Cl^- present in concentration sufficient that the Gouy double layer is of no concern, it will be neglected in the following discussions.

The Helmholtz double layer, on the other hand, plays a major role in the electrode processes. It is a dominant factor in all the electrode reactions to be discussed. This double layer arises between the charged solid and the ions of opposite sign attracted to the neighborhood of the solid to neutralize its charge. We will restrict our discussion of the double layer to the case where ions of the host solid do not dissolve and reprecipitate, in other words, are not exchanged with the solution. In a few cases below corrosion of the solids due to electron exchange reactions will be discussed, but not in cases where these processes influence the double layer.

With a metal electrode in solution with no net current passing through it, where a redox couple is dissolved in the solution, the Helmholtz double-layer potential is defined by the energy levels of the ions. Consider Figure 2.5, in particular Figures 2.5a and 2.5b. The diagram shows the energy levels approximately at equilibrium. If when the metal is first dipped into the solution the system is not at equilibrium, for example if the Fermi energy E_F is below E_{red}, then electrons from the reducing agent will transfer to the empty levels above the Fermi energy. This transfer will cause a double layer to form (the Helmholtz double layer), and the sign of the double-layer voltage will be such that E_F moves up (relative to E_{ox} and E_{red}). Equilibrium will be reached when the double-layer voltage suffices to raise E_F to the point where the rate of electrons transferring from the occupied levels in the solid to the unoccupied E_{ox} levels equals the rate of electrons transferring into the solid. Thus the Helmholtz double layer is determined (at zero net electrode current) by the electron transfer process. Adsorption of ions if it occurs will be determined by the Helmholtz double layer, rather than the reverse.

At an insulator surface, on the other hand, there are no important redox processes. In this case the Helmholtz double layer arises from the process described in the preceding chapter (Section 7.3.1.2), namely, the adsorption/desorption of ions. In particular, the adsorption of protons and hydroxide ions will usually dominate in charging the surface, and the Helmholtz double layer will develop between this charge on the surface and counterions (K^+ or Cl^- in a solution of KCl, for example) in the solution near the surface. In this case the magnitude of the Helmholtz double layer varies with the pH, as the adsorption of protons or hydroxide ions varies with the pH of the solution. A simplified derivation will suffice to indicate the relationship for insulators between the Helmholtz double-layer potential V_H and the surface charge $[H_s^+]$, where the latter symbol represents the net density of adsorbed protons. In the reaction

$$H_3O^+ = H_s^+ + H_2O \qquad (8.1)$$

where H_3O^+ is a hydronium ion (essentially a proton) in solution, the enthalpy of the reaction, ΔH, varies linearly with the double-layer potential, for the proton must surmount this potential to become adsorbed. With this proportionality, equilibrium in Equation (8.1) leads to

$$\frac{[H_s^+]}{[H_3O^+]} = \exp\left(-\frac{\Delta G}{kT}\right) = A \exp\left(-\frac{eV_H}{kT}\right) \qquad (8.2)$$

where A is a constant. The double-layer potential V_H in turn is proportional to the charge adsorbed if we make the simplifying approximation that C_H, the capacity (per unit area) of the Helmholtz double layer is independent of V_H:

$$e[H_s^+] = C_H V_H \qquad (8.3)$$

Inspection of Equations (8.2) and (8.3) shows $[H_s^+]$ will vary slowly with $[H_3O^+]$, and to the approximation of this analysis we have

$$eV_H = B + kT \ln[H_3O^+] = B - 2.3kT \,(\text{pH}) \qquad (8.4)$$

and the Helmholtz potential decreases about 60 mV per pH unit.[1]

The semiconductor case turns out to be similar to the insulator case with respect to the origin of V_H. At the semiconductor/electrolyte interface there are two variable double-layer potentials, V_s and V_H, the surface barrier and the Helmholtz potential. As there are two processes that must come to equilibrium, electron exchange and adsorption/desorption of ions, both equilibria requirements can be accommodated. It is found that the Helmholtz double layer is determined by the adsorption/desorption of ions, and the depletion double layer adjusts to make the net rate of electron transfer equal to zero between the band and the redox couple in the electrolyte.

The reason the Helmholtz double layer is insensitive to electron transfer is that the thickness of the space charge layer in the semiconductor is the order of $100\times$ to $1000\times$ the thickness of the Helmholtz double layer. So if charge is transferred from the bulk to the surface, giving a voltage change, only a small fraction of the voltage change appears in V_H. Even when the voltage is externally applied, almost all the applied voltage appears across the space charge region of the semiconductor, not across the Helmholtz double layer.[2] Normally V_H is found experimentally to be a function only of pH (Section 8.3.3.1), varying approximately in accordance with Equation (8.4). However, if other ions than H^+ or OH^- are also adsorbed, such adsorption can[3] affect V_H.

The fact that on semiconductors V_H is usually independent of applied voltage and independent of nonadsorbed ions in solution will be shown to be an important simplifying factor in analyzing electron exchange at a semiconductor surface. The relative positions of the energy levels in solution and the energy levels in the solid are not affected by electron exchange, in contrast to the case of a metal electrode. Thus on a semiconductor the measured currents can be shown related more directly to the energy levels associated with the ions.

Figure 8.1 shows an energy level model[4] for a semiconductor in solution. If indicates the energy of an electron as a function of distance: in the semiconductor, in the solution, and in a reference electrode immersed

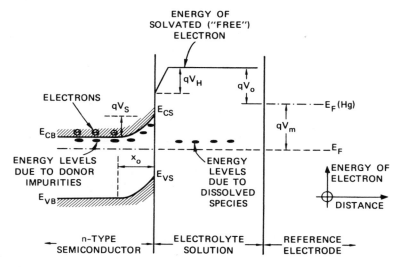

Figure 8.1. Allowed electronic energy levels in three adjacent phases: an n-type semiconductor, an electrolyte, and a reference electrode.

in the solution. A reference electrode is an electrode that reflects the potential in the solution independent of any ions present in solution. Thus the parameter in Figure 8.1 defined as V_0 is a constant by the nature of the reference electrode. In the bulk of the solution the potential is as indicated a constant, assuming that there is a reasonably high concentration of ions in the solution. Thus any electronic energy level could be used as a reference in the solution phase. For specificity, and to provide an easy relationship to the energy of a free electron usually used for reference in a vacuum, we have chosen the energy of a solvated electron as the reference energy in the solution. A solvated electron is a close analog to a free electron but is actually surrounded by the usual polarization shell of polar water and in practice this "free" electron, the solvated electron, becomes associated with a proton. At the surface of the semiconductor the potential drop across the Helmholtz double layer V_H is shown. The distance is not to scale—the Helmholtz double layer is only the order of an angstrom in thickness. On the other side of the semiconductor/solution interface the surface barrier double layer, normally the order of 1000 Å thick, is shown. The difference in Fermi energies between the solid (E_F) and the reference electrode [$E_F(\text{Hg})$], is measured as qV_m, where V_m is the voltage measured, q is the electronic charge. The diagram does not indicate the sign of V_m. Recalling that the band model gives the energy of an electron, where the potential is defined in terms of a positive charge, we have

$$V_m = -[E_F - E_F(\text{ref})]/q \qquad (8.5)$$

where we have generalized to a relation defining the measured voltage at an inert electrode relative to the Fermi energy of any reference electrode, $E_F(\text{ref})$.

The model of Figure 8.1 will be different if the electrode of interest is a metal rather than a semiconductor. The differences with a metal electrode are the lack of a space charge double layer in the solid and of course the dominance of the Fermi energy in separating the occupied and unoccupied levels in the solid (Figure 2.5a).

8.2.2. Charge Transfer between the Solid and Ions in Solution

In Section 2.3.3 it was shown that any energy levels in aqueous solutions will fluctuate with time over a wide span of energy, and in Section 5.5.2 it was shown that due to Franck–Condon splitting occupied levels will be shifted by an amount 2λ below the unoccupied energy levels. In Section

5.5.3 it was shown that if the redox couple is two-equivalent, two energy levels must be considered, each of which shows Franck–Condon splitting. A combination of Figure 8.1 showing the potentials through the system, and Figure 2.5, showing the Franck–Condon splitting and the fluctuations of the energy levels in solution, provides a general qualitative picture of the aqueous electrolyte system containing a one-equivalent species. ·

The current across the interface measures electron transfer between the solid and ions in solution. The particular current/voltage relationship depends on whether the applied voltage appears across the Helmholtz double layer or across the space charge region of the semiconductor. In the case of metals, there is no space charge region in the solid and any change in applied voltage appears across the Helmholtz double layer. In the case of a semiconductor we have shown earlier that the voltage change will usually appear across the space charge region within the semiconductor, assuming no adsorption or intrinsic surface state effects.

8.2.2.1. The Metal Electrode.   Marcus[5–7] has developed expressions for the current through a solid electrode using his classical models. Dogonadze and his co-workers have studied the system more recently with a quantum mechanical analysis.[8,9] We will present a simplified version of Marcus' analysis for metals, in order to compare the use of the fluctuating energy level model to describe a metal electrode with its use (next section) to describe a semiconductor electrode.

Marcus calculated by methods closely related to absolute rate theory that the injection current from a one-equivalent reducing agent into an inert metal electrode is given by

$$J_a = eZc_{red} \exp(-\Delta G^*/kT) \tag{8.6}$$

where Z is a constant, determined by analysis to be the thermal velocity of the ions, and ΔG^* is the change in free energy to reach the activated state, the state at which electron transfer to the solid can occur with no change in energy. The use of Equation (8.6) assumes a transmission coefficient of unity: if the energy level of an ion at the surface fluctuates to an energy high enough for electron transfer (in the case of metals to an energy E_F), then electron transfer will occur with a probability of unity. A more accurate form of Equation (8.6) contains a factor depending on the mean square distance from the surface at which electron injection occurs divided by the mean square deviation from a reaction hypersurface. Marcus[7] and Hale[10] suggest this factor should be the order of unity, so it has been omitted from Equation (8.6).

If we assume that there is no adsorption of reactants, no formation or breaking of chemical bonds, no changes in the first coordination sphere of the ion, and zero work is required for the ion of interest to reach the electrode, then the energy to reach the activated state is simply the energy necessary to shift the energy level of the reducing agent from E_{red} to the energy of the activated state, E^*. This energy is given by Equation (2.54):

$$\Delta G^*(E) = (E^* - E_{red})^2/4\lambda \tag{8.7}$$

where in the case of an inert metal electrode $E^* = E_F$.

Now, with a metal electrode, the applied voltage appears across the Helmholtz double layer. This means that the difference $E_F - E_{red}$ varies directly with the applied voltage. If we let $E_F = E_{F0}$ when the net electrode current is zero, then

$$E^* - E_{red} = E_F - E_{red} = E_{F0} - E_{red} - \eta e \tag{8.8}$$

where η is the "overvoltage," the voltage in excess of that required to maintain zero current. Equations (8.8) and (8.7) in (8.6) give an expression for the anodic current. An expression for the cathodic current can be derived in an entirely analogous manner, if we consider that the probability of electron transfer to an unoccupied level in solution is unity when the ion is at the metal surface and the unoccupied level fluctuates to an energy below E_F. Then the net current, the anodic minus the cathodic, is given by

$$J = eZ\{c_{red} \exp[-(E_{F0} - E_{red} - \eta e)^2/4\lambda kT]$$
$$- c_{ox} \exp[-(E_{ox} - E_{F0} + \eta e)^2/4\lambda kT]\} \tag{8.9}$$

where we have assumed $Z_{ox} = Z_{red} = Z$. At equilibrium $J = 0$, whence using the relation $E_{ox} - E_{red} = 2\lambda$ [from Equation (5.18)] we have

$$E_{F0} = \tfrac{1}{2}(E_{ox} + E_{red}) + kT \ln(c_{red}/c_{ox}) \tag{8.10}$$

which can be reformulated as

$$E_{F0} - E_{red} = \lambda + kT \ln(c_{red}/c_{ox}) \tag{8.11}$$

The introduction of Equation (8.11) into (8.9) gives

$$J = eZ\big(c_{red} \exp\{-[\lambda + kT \ln(c_{red}/c_{ox}) - \eta e]^2/4\lambda kT\}$$
$$- c_{ox} \exp\{-[\lambda - kT \ln(c_{red}/c_{ox}) + \eta e]^2/4\lambda kT\}\big) \tag{8.12}$$

and if $\lambda \gg e\eta$ and $\lambda \gg kT \ln(c_{red}/c_{ox})$, this simplifies to

$$J = eZ(c_{red}c_{ox})^{1/2} \exp(-\lambda/4kT)[\exp(\tfrac{1}{2}\eta e/kT) - \exp(-\tfrac{1}{2}\eta e/kT)] \qquad (8.13)$$

$$= J_0[\exp(\tfrac{1}{2}\eta e/kT) - \exp(-\tfrac{1}{2}\eta e/kT)] \qquad (8.14)$$

which is the "Tafel equation." The factor J_0 as defined by Equations (8.13) and (8.14) is (at low η) termed the "exchange current," the current passing both anodically and cathodically across the electrode surface at equilibrium. The first term of Equation (8.14), representing the anodic current, is usually found written with the exponent $\alpha\eta/kT$, where α is called the "transfer coefficient." As was shown by Marcus,[6,7] if the only contribution to the energy is the rearrangement energy λ associated with the dielectric constant of the medium, and if the approximation $\lambda \gg \eta e$ is valid, the transfer coefficient becomes $\tfrac{1}{2}$ as in Equation (8.14). Experimental values of the transfer coefficient α are commonly[11] close to $\tfrac{1}{2}$. The exponent in the cathodic term of the Tafel equation can be shown[12] to be $(\alpha - 1)\eta/kT$, leading of course to $(-\eta/2kT)$ when $\alpha = \tfrac{1}{2}$, as found in Equation (8.14) in accordance with the theoretical model.

8.2.2.2. The Semiconductor Electrode.

In the case of a semi-conductor electrode, the corresponding Tafel equation is derived most easily by starting with Equation (2.60). We will make the same assumptions as those made in deriving Equation (8.7). As described above, the important difference between this and the metal electrode is that in the semiconductor case any change in voltage appears across the semiconductor space charge region rather than across the Helmholtz double layer. Thus with changing voltage the variable is no longer $E_{cs} - E_{ox}$, but rather n_s, the density of current carriers at the semiconductor surface. In the development of the Tafel equation for semiconductors we will continue to consider an n-type semiconductor electrode in our analysis for specificity. In principle, the concepts are carried over directly to p-type materials, although in practice there seem often to be surface state or corrosion problems rendering the simple analysis less useful with p-type semiconductors. We will assume that the rate-limiting step is electron transfer rather than diffusion of ions or electrons to and from the interface. With this assumption the surface density of reducing agent $n_t = c_{red} d$ in Equation (2.60) is independent of voltage, i.e., $n_t = n_{t0}$, and the electron density at the surface in equilibrium with the bulk density in the conduction band, i.e., Equation (2.38) is valid. The use of Equation (2.38) means the variable n_s is simply related to the over-voltage.

With these assumptions the net current density, from Equation (2.60), becomes

$$J = \sigma \bar{c} e \, d(kT/\pi\lambda)^{1/2} c_{\text{ox}} \exp[-(E_{cs} - E_{\text{ox}})^2/4\lambda kT](n_s - n_{s0}) \qquad (8.15)$$

where n_{s0}, the surface density of electrons necessary to make the cathodic current equal to the anodic current at equilibrium, is a function of the ratio of $c_{\text{red}}/c_{\text{ox}}$. By the model ($V_H$ independent of applied voltage) the anodic current is independent of voltage, for the density of "occupied states" is independent of voltage and the energy level E_{red} is independent of voltage. With

$$E^* = E_{cs} \qquad (8.16)$$

the anodic current obtained from Equations (8.6) and (8.7) is

$$J_a = Zec_{\text{red}} \exp[-(E_{cs} - E_{\text{red}})^2/4\lambda kT] \qquad (8.17)$$

The value of n_{s0} can be calculated from Equations (8.15) and (8.17) and shown to be

$$n_{s0} = [Zc_{\text{red}}(\pi\lambda)^{1/2}/\bar{c} \, dc_{\text{ox}}(kT)^{1/2}] \exp[-(E_{cs} - E_{\text{redox}}^o)/kT] \qquad (8.18)$$

where we have used the relation that $E_{\text{ox}} - E_{\text{red}} = 2\lambda$, and have defined a new parameter, introduced by Gerischer,[13] which will be useful later:

$$E_{\text{redox}}^o = \tfrac{1}{2}(E_{\text{ox}} + E_{\text{red}}) \qquad (8.19)$$

With the current through the electrode given by Equation (8.15), with n_{s0} given by (8.18), and recalling that the only variable in (8.15) with voltage is n_s, we can write Equation (8.15) in terms of an exchange current J_0:

$$J = J_0(1 - n_s/n_{s0}) \qquad (8.20)$$

We have that n_s by Equation (2.38) is proportional to $\exp(-eV_s/kT)$ and we have shown the overvoltage η appears across the space charge region so that $V_s - V_{s0} = \eta$ (where V_{s0} is the surface barrier at equilibrium). Then Equation (8.20) becomes

$$J = J_0[1 - \exp(-e\eta/kT)] \qquad (8.21)$$

the Tafel relation for a semiconductor electrode.

The "transfer coefficient" α, the coefficient of η/kT in the anodic term, is thus zero for an n-type semiconductor electrode (unity for p-type). The transfer coefficient is $\frac{1}{2}$ for metal electrodes because the rate of electron transfer depends on the fluctuating energy levels—their position relative to the Fermi energy in the metal is determined by the overvoltage. In the semiconductor case, the overvoltage does not affect the position of the fluctuating energy levels relative to the band edges but only the density of electrons at the surface. Thus the origin of the Tafel relation is entirely different and the "transfer coefficient" is different. It is of interest to note that on a "semiconductor" electrode where there is a strong accumulation layer—where the surface is degenerate, almost metallic—then the transfer coefficient becomes $\frac{1}{2}$ again. In this case the applied overvoltage appears across the Helmholtz double layer. Tench and Yeager's[14] study of NiO with high anodic currents provides a fine example of this case.

In practice, equilibrium as described by Equation (8.21) seldom obtains at a semiconductor electrode. A behavior analogous to irreversible chemisorption often occurs: J_0 is often too small to be experimentally observed,[15,16,4] and in fact currents due to other, perhaps heterogeneous, processes, are larger than J_0. Experimentally it is found that cathodic current flows from an n-type semiconductor in accordance with the last term of Equation (8.21); the cathodic current increases exponentially with applied negative voltage. Anodic current is observed to flow when a sufficiently positive potential is applied to the electrode, but the anodic current is independent of the concentration of reducing agent, so is presumably associated with flaws in the electrode surface[17] rather than with J_0 of Equation (8.21) or J_a of Equation (8.17). Thus J_0 cannot be easily measured, and the interpretation of current/voltage characteristics of the semiconductor electrode in accordance with the Tafel equation [Equation (8.21)] must be made with caution.

Gerischer, in several publications, has developed the Tafel equation for semiconductors in much more detail than has been presented here. He has included, for example, hole exchange with the valence band,[13] the analysis when intrinsic surface states are present at the semiconductor surface, and electron transfer to or from excited ions in solution (dye molecules).[18,19]

8.2.3. Energy Levels of Surface Species Relative to Band Edges

In this section we want to explore more deeply the relations (a) of the energy parameters E_{ox}, E_{red}, and E_{redox}^o of ions in solution to the redox potential scale which assigns a standard oxidation potential E^o to these ions, and (b) of the energy of the band edges E_{cs} and E_{vs} to the same redox potential scale. The possibility of placing the energy levels of both the ions and the band edges on the same energy scale by independent measurement is perhaps the most valuable contribution of semiconductor electrochemistry to surface science, for it establishes (within the limits of the necessary assumptions) the relationship between the chemical properties of foreign species and the energy of the surface states that they provide. In addition, the analysis provides a simple measurement relating the energy bands of various semiconductors to each other. Each energy level in these studies is referred to the zero of the redox potential scale, the Fermi energy of the hydrogen reference electrode. The value of E^o, the standard oxidation potential (redox potential) of a redox couple such as Fe^{2+}/Fe^{3+} in solution is a measure of its tendency to give up or accept electrons. The value is determined by measuring the zero current potential of an inert metal electrode in a solution containing equal concentrations (or more accurately equal activities) of the two species, with the solution standardized to 1 M acid, and the potential measured relative to the hydrogen reference electrode. If the reducing agent has a tendency to give up electrons, the metal electrode will acquire a negative potential and E^o will show a high negative value. If the oxidizing agent of the couple is active in accepting electrons, it will extract electrons from the metal electrode and E^o will be positive. Thus the value of E^o, as tabulated, for example, in Latimer,[20] is a measure of the tendency of the redox couple to give up or accept electrons.

The value of E^o_{redox} for a couple [Equation (8.19)], can be related to E^o through Equations (8.10) and (8.19). From Equation (8.10) under standard conditions (with $c_{ox} = c_{red}$) the expression

$$E_F = \tfrac{1}{2}(E_{ox} + E_{red}) = E^o_{redox} \qquad (8.22)$$

is obtained, where the last equality stems from Equation (8.19). With E^o the potential of an inert metal electrode immersed in the solution, the relation between E^o and E_F is obtained from Equation (8.5), where the Fermi energy of the hydrogen reference electrode is defined as zero. Then substituting for E_F in Equation (8.22), gives

$$E^o_{redox} = -eE^o \qquad (8.23)$$

Equation (8.23) is only valid for the simplest systems, where such factors as multiequivalent reactants, complex formation, reactive electrodes, or inner sphere changes do not complicate the process. However, if we restrict our interest to simple outer sphere reorganization, Equation (8.23) provides a useful determination of E^o_{redox}.

A direct relation between energy levels in solution and E^o was first proposed by Beck and Gerischer.[21] It was later pointed out[16] that the relation could only apply to one-equivalent species because the measurement of one redox potential E^o cannot provide values for the two energy levels involved with a two-equivalent species. The parameters to describe the energy levels for two-equivalent species must be determined by additional measurements, as will be discussed in Sections 8.3.2 and 8.3.3.4.

Next we will relate the energy of the band edges of a semiconductor to the Fermi energy of the hydrogen electrode (the zero of the redox potential scale). The analysis is made with the help of Figure 8.2, which shows the important elements of Figure 8.1, but shows the semiconductor bands with a surface barrier (dotted lines) or with the bands flat (solid lines). Now measurement of the potential difference V_m between the semiconductor electrode and a reference electrode (for simplicity the hydrogen electrode) gives by Equation (8.5) the Fermi energy of the semiconductor relative to the Fermi energy of the reference electrode. If the reference electrode is the hydrogen electrode, $-eV_m$ is the Fermi energy in the semiconductor on the

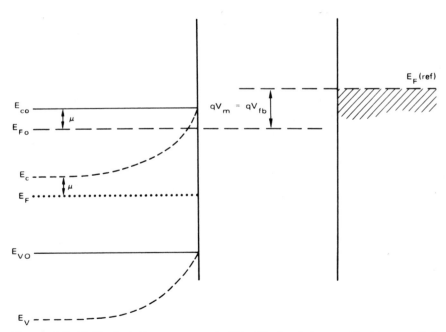

Figure 8.2. To illustrate the measurement of E_{cs}. The bands (at, say, E_c with no applied voltage) are shifted by an applied voltage V_m to the flat band condition where $E_c = E_{co}$. The value of $E_{cs} - E_F(\text{ref})$ is then obtained knowing the value of μ.

redox potential scale. In the case where the bands are flat, the potential is termed the "flat band potential" V_{fb}. It is seen from Figure 8.2 that knowing μ for the semiconductor (from the bulk properties) and V_{fb}, the energy on the redox potential scale of the band edge E_{cs} can be immediately obtained[22,23] as

$$E_{cs} = -qV_{fb} + \mu \tag{8.24}$$

The flat band potential V_{fb} is determined in practice by varying the applied voltage V_m, monitoring the capacity C of the space charge layer, and extrapolating with the help of Equation (3.11) to find the potential at $(1/C) = 0$ (flat bands).

The value of E_{cs} obtained must depend on V_H, for, as is clear from Figure 8.1, any variation in V_H shifts the value of E_{cs} relative to the reference energies in solution. Specifically, it is clear by inspection of Figure 8.1 that [using $E_F(\text{ref})$ instead of $E_F(\text{Hg})$ for a general reference electrode]:

$$E_{cs} - E_F(\text{ref}) = A - qV_H \tag{8.25}$$

with A a constant. On semiconductors, V_H depends on pH [Equation (8.4)] and on the presence of adsorbable ions. Particularly the pH must be specified when reporting a value for E_{cs}.

Much of our information regarding the relative position of surface energy levels for one-equivalent species and bands for various semiconductors has been determined by Gomes and his group, and will be reviewed in the next section.

8.3. Observations with Semiconductor Electrodes

8.3.1. Measurement Methods

In Chapter 3 most of the most valuable surface measurements for general use were discussed, including the space charge capacity measurement that is used both at the solid/gas and solid/liquid interface. In this subsection a few more measurements are introduced that are specific to the semiconductor/liquid interface, and were thus not described in Chapter 3. We will be primarily interested in measurements that describe electron transfer to or from the surface rather than measurements that describe the semiconductor space charge region. Space charge measurements have been described in Chapter 3, in the review by Pleskov[24] and in older books[25,26] and reviews.[4]

Electron transfer to or from the surface results in redox reactions at the interface and is measured directly by the current. For cathodic current to flow, for example, electrons (or holes) must pass through the semiconductor, and reduce an ion at the interface, changing its valence by at least -1. The ion must move into the solution, and it, or more likely another ion, must give up an electron to the metal counterelectrode. The external connection between the metal counterelectrode and the semiconductor completes the circuit. Figure 8.3 shows a schematic of a circuit for such a measurement, where the potential of the semiconductor electrode is measured relative to a reference electrode, and the current is passed through a counterelectrode made of platinum. To ensure that ions released at the counterelectrode are not allowed to contaminate the solution that contains the semiconductor electrode, an apparatus can include a salt bridge, a divider that blocks the passage of foreign ions, conducting by the motion of inoffensive ions such as potassium.

The "transistor" measurement determines whether an electron transfer reaction occurs via the conduction band or the valence band, that is, whether electrons or holes are primarily responsible for the current across the interface. A schematic of the circuit is shown in Figure 8.4. Germanium is indicated as the semiconductor, and actually it is the first[27] and only semiconductor that has been used to date with this configuration. The

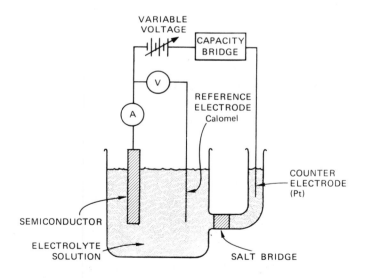

Figure 8.3. Schematic of an electrochemical cell. V represents a voltmeter and A an ammeter.

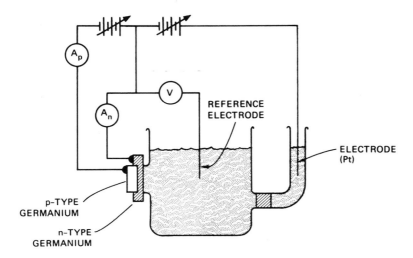

Figure 8.4. Schematic of the electrochemical cell adapted to use the "Transistor Method." The meter A_p measures hole flow to and from the electrolyte interface, the meter A_n measures electron flow.

required long diffusion length is available on either germanium or silicon, but the formation of silica makes work with silicon difficult. The transistor configuration is used to indicate the energy levels of ions in solution relative to the semiconductor bands. A high energy level such as E_{ox} shown in Figure 2.5 could only capture conduction band electrons. Valence electron capture (hole injection) would indicate on the other hand a low energy E_{ox}.

With the configuration of Figure 8.4, with an n-type material facing the electrolyte, there are both conduction band electrons and valence band electrons available to reduce an ion in solution. A p-n junction is provided near the surface, and the junction is reverse biased. If a hole is injected, and the hole diffusion length is long enough, the hole will reach the junction before recombining with an electron, and the hole will proceed into the p-type region. Thus any current associated with valence band processes will be measured as current through the contact to the p-type material. Any current associated with conduction band processes will be measured as current through n-type material. Thus the ammeters A_p and A_n measure directly the current associated with valence band and conduction band processes, respectively.

Another measurement technique that in principle provides quantitative values for the energy levels in solution relative to the semiconductor bands is the tunneling technique. In this measurement the semiconductor is doped

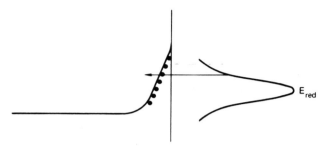

Figure 8.5. Tunneling from a reducing agent, through a thin surface barrier region, into the conduction band of a semiconductor. The thin surface region is obtained by a small bias on a heavily doped surface layer.

very heavily, so that the depletion layer is very narrow. The object is to make the electric field at the surface so high that tunneling can occur as indicated in the sketch of Figure 8.5. The tunneling current must dominate over any breakdown current, and so the high surface electric field must be obtained with very low applied voltage, perhaps 1 or 2 V, to avoid avalanche multiplication. To obtain such a thin surface barrier with a low voltage, it is clear from Equation (2.11) that a high N_D (for n-type material) is required.

As indicated in Figure 8.5, when such tunneling is obtained, current can be obtained from energy levels that otherwise would be inactive because they are too far below E_{cs}. By analysis of the current versus the voltage as was given by Memming and Möllers[28] values for E_{red} and λ can be obtained.

Another powerful tool in the studies of electrode behavior is the current doubling effect. It is difficult to discuss separately the measurement technique and the results obtained therefrom, so both technique and results will be reviewed together in Section 8.3.2 below.

Cardon and Gomes[29] have used noise measurements to study correlation effects in current doubling. The interpretation of noise measurement will not be discussed, although some of the results will be referred to. The reader is referred to the original article for measurement details and interpretation.

8.3.2. Radical Generation (Current Doubling)

In Section 5.5.3 the behavior of two-equivalent surface states when reacting with a (one-equivalent) hole or electron was discussed. The reaction was identified[30] at the semiconductor/electrolyte interface, and because of the effect on the electrode current, was called "current doubling."

The experiments can be summarized as follows. A single crystal n-type ZnO electrode with the (000$\bar{1}$) face exposed to the solution was anodically biased. As discussed with reference to Equation (8.21), the dark anodic current is usually negligible (10^{-10}–10^{-9} A/cm^2). Band gap illumination, however, produces holes and then anodic current is observed proportional to the hole generation rate. With no active species in solution, the reaction associated with this anodic current is the photolysis of the ZnO according to

$$h^+ + H^+ + ZnO \rightarrow Zn^{++}(soln) + \tfrac{1}{2}H_2O_2(soln) \qquad (8.26)$$

In the cited reference the H_2O_2 was observed in solution; others have observed the dissolved zinc.[31] Thus a reaction of the form of Equation (8.26) is strongly indicated in inert solutions. However, upon addition of the formate ion, $HCOO^-$, to the solution, the generation of H_2O_2 ceases, and the current increases by about 70%. The reaction was interpreted in terms of "current doubling" by the two-equivalent formate ion:

$$h^+ + HCOO^- \rightarrow HCOO \cdot \qquad (8.27)$$

$$HCOO \cdot \rightarrow e^- + H^+ + CO_2 \qquad (8.28)$$

where Equation (8.27) represents the generation of an unstable radical by hole capture, Equation (8.28) represents its dissociation by electron injection. With an apparatus as in Figure 8.2, each hole h^+ passing through the surface causes the injection of an electron by Equation (8.28) and with the two equivalent $HCOO^-$ present, the current "doubles" over its value when Equation (8.26) represents the reaction.

The measurement of current doubling has been made with many two-equivalent species and with many semiconductors, both n- and p-type. Table 8.1 summarizes some of the investigations.

Micka and Gerischer[40] have investigated current doubling as a function of the anion present in solution and found effects that they attribute to anion adsorption. They were able to attain exact current doubling, attaining this value in a solution containing only ClO_4^- as the anion. This ion does not easily form complexes so may have less tendency to exchange with surface OH^- groups (cf. Chapter 5). The details of why such an adsorption would block the current doubling process are not clear, but the observation may suggest the participation of acid or basic sites at the solid surface.

The observation of current doubling, as mentioned above, has provided a powerful tool to study hole or electron transfer steps and to study the reactions of radicals. Its use as a tool to measure the capture cross section

Table 8.1. Observations of Current Doubling

Semiconductor	Type	Ions in solution	References
ZnO	n	Arsenic, sulfide, formate, alcohols, other organic species	Morrison and Freund[30,32] Gomes, Freund, and Morrison[33,34] Freund and Gomes[35]
CdS	n	Alcohols, aldehydes	Gerischer[36,19]
TiO$_2$	n	Alcohols, aldehydes, acids	Benderskii et al.,[66] Dutoit et al.[67]
CdSe	n	H_2CO, $(CHO)_2$, but not with alcohols	Vanden Berghe, Gomes, and Cardon[38]
GaAs	p	Bromine	Gerischer[19]
GaP	p	$S_2O_8^{2-}$, H_2O_2	Memming[39]

for minority carriers will be discussed in Section 8.3.3.3, where hole reactions are described.

The reactions of radicals on the surface can be studied as follows. Such reactions are observed by monitoring the excess current due to electron injection [Equation (8.28)]. If a reactant is present in the electrolyte solution that interacts rapidly with the radical, then current doubling will not be observed. Gomes et al.[33] have examined the interaction of the ammoniated cupric ion with radicals using this technique, where the radicals were produced by oxidation of (hole capture by) various alcohols and formate. It was found that the interaction between the radical formed and $Cu(NH_3)_4^{2+}$ showed the same characteristics independent of the source. For this reason it was concluded that the radical was the same independent of the organic species, and specifically was a hydrogen atom, the only possible radical species that could be produced by oxidizing any of the two-equivalent species tested. This suggests, for example, that in the oxidation of formate [Equations (8.27) and (8.28)] the CO_2 would be better represented as separating off in Equation (8.27) rather than (8.28).

Cardon and Gomes[41] used pulsed current measurements and noise measurements to estimate the lifetime of the radicals on ZnO and found $\tau < 10^{-5}$ sec, so clearly the radical species are quite unstable.

The appearance of excess current due to injection by unstable intermediates is not restricted to carrier capture by free ions in solution. The effect was observed in the first modern study of semiconductor electrodes, that of Brattain and Garrett on germanium.[27] They noted even at that time that when holes reached the surface, electrons were injected into the conduction band. However, the process being much more complex (the germanium was oxidized from a valence state of 0 to $+4$), it was, and still is, not amenable to simple interpretation.[4]

8.3.3. Measurements of Energy Levels and Band Edges

In this section we examine the experimental determination (using the models of Section 8.2.3) of the relative position of the energy levels of ions in solution and the energies of semiconductor band edges for various semiconductors. The energies are given relative to the Fermi level of the hydrogen electrode. In addition, measurements of the reorganization (rearrangement) energy λ will be reviewed.

8.3.3.1. Measurement of V_H. A good starting point toward building a picture of the energy level model of the semiconductor surface in an electrolyte is the evidence available regarding the variation of the Helmholtz double-layer potential with the appropriate parameters. From Figure 8.1 it is clear that the variation of V_H with electrolyte composition can be easily measured by the variation of E_F with electrolyte compositions when the bands are flat.

From Equation (8.4) it was concluded that for insulators V_H should vary with pH, the proportionality being 60 mV per pH unit. This is also found to be the case for ionic semiconductors. Measurements by Lohmann[42] on ZnO, Boddy[3] on germanium, Laflere et al.[43] on GaAs, Möllers and Memming[44] on SnO_2, and Fujishima et al.[45] on TiO_2 all showed this behavior. Actually germanium is a little unusual because it is found that the double-layer voltage also depends on the adsorption of radicals, as discussed below (Section 8.3.4.3). Boddy found other ions can also affect the value of V_H when adsorbed, and in particular found iodide ions adsorbed on germanium caused shifts in V_H.

The lack of variation of V_H with applied potential and with redox species in solution permits the assignment, to be discussed below, of relative energy levels of ions in solution and of semiconductor energy bands. The energies E_{cs} of conduction band edges of a series of semiconductors can be compared on the redox potential scale, if one uses the same pH (or allows for variation in pH) even though the measurement is made with different ions in solution and different voltages.

8.3.3.2. Measurement of Energy Levels of Surface Species. The determination of the energy level of one-equivalent ions in solution, as represented by E^o_{redox} [$= \frac{1}{2}(E_{\text{ox}} + E_{\text{red}})$ by definition from Equation (8.19)] is not made with semiconductor electrodes, but follows from classical electrochemistry by the simple application of Equation (8.23) relating the standard redox potential of one-equivalent species to the parameter of interest. The standard redox potential applies at pH $= 0$, so the value of E^o_{redox} obtained is the value associated with the ion hydrolyzed accordingly. A correction must be made if the ions are hydrolyzed differently (different number of OH$^-$ ligands). Also corrections must be made [with Equation (5.13)] for other complications in chemical form: (a) if there is more than one dominant hydrolyzed form at pH $= 0$, so that for neither form does $c_{\text{ox}} = c_{\text{red}}$, which invalidates Equation (8.22) or (b) there are inner sphere changes upon reduction or oxidation. For simplicity in the results discussed below, redox couples are chosen such that such corrections can be assumed unnecessary.

8.3.3.3. Measurement of the Reorganization Energy.

With a measurement of E^o_{redox} as above and a measurement of the reorganization energy λ, the complete picture of the fluctuating surface states associated with species in solution (Figure 2.5) can be represented. Values of λ have been obtained by two semiconductor measurements, a tunneling measurement as described in the preceding subsection, and a measurement of the variation of cathodic current with the Helmholtz double-layer potential. It should be pointed out that λ can in principle be determined directly from the current/voltage characteristics. However, a much more detailed evaluation of the pre-exponential factor in Equation (8.13) must be made. Hale[10] has made such a calculation of λ for many species from exchange current data available on a metal electrode, and, for example, has concluded for ferricyanide that $\lambda \sim 1.2$ eV.

Vanden Berghe, Cardon, and Gomes[46] evaluated λ by observing the variation of the cathodic current on ZnO when V_H, the Helmholtz potential, was shifted with pH.

From Equations (8.25) and (8.4), and by inspection of Figure 8.1, the conduction band E_{cs} of ZnO shifts with V_H or pH relative to E^o_{redox} (and therefore relative to E_{ox}). Thus the cathodic term of Equation (8.13) shows a strong dependence on pH. In Equation (8.15) it is clear that the most important unknown that appears in the relationship, with E_{ox} known, is the reorganization energy λ. Thus in principle λ can be calculated by measurement of J (with n_s constant) as a function of pH. Analysis of this variation for the electron transfer to the ferricyanide ion led to a value of 0.75 eV for λ. This approach is valuable with the couple ferricyanide/ferrocyanide, for with this ion there should be negligible hydrolysis change with pH, so the same form of the ion should be present, independent of pH.

The tunneling measurement of λ depends on the analysis of the model illustrated in Figure 8.5. As the applied voltage is increased, the current can originate from energy levels at lower and lower energies. The density of levels increases sharply at lower energy, as observed in the sketch of Figure 8.5 or the analytical expression for $W(E)$, Equation (2.56). In an analytical expression for the tunneling current, $W(E)$ is therefore a dominant feature, and the parameter λ dominates $W(E)$. Thus analysis of the tunneling current versus voltage characteristics can provide a value for λ.

With this analysis Memming and Möllers[28] investigated the redox couples Fe^{2+}/Fe^{3+}, Ce^{3+}/Ce^{4+}, and $Fe(CN)_6^{4-}/Fe(CN)_6^{3-}$, using SnO_2 as the semiconductor, and determined $\lambda = 1.2$, 1.7, and 0.4 eV, respectively, for these ions. From Equation (2.49) λ should increase as the radius of the ions decreases, so the observed differences in λ are not inconsistent with expectations.

The value of λ for ferro/ferricyanide, 0.75 eV, obtained by Vanden Berghe *et al.* using the pH variation method, is somewhat higher than the value 0.4 eV obtained by Memming and Möllers by the tunneling technique. The discrepancy may originate simply because both methods of λ determination are rather indirect and the several sources of uncontrolled errors prevent higher accuracy by these techniques. For example, in the tunneling measurement the heavy doping may lead to variation in V_H with applied voltage that is not accounted for in the calculations.[47]

8.3.3.4. Measurement of the Energies of Semiconductor Band Edges.

The determination of the energy of the band edges for a series of semiconductors has been made by Gomes and his group as reviewed by Gomes and Cardon.[48] The method used as discussed in Section 8.2.3 was to measure the electrode potential when the bands of the various semiconductors were extrapolated to the flat band case.

Figure 8.6 shows the result of this investigation. The value of E_{redox}^o for iron cyanide as determined from E^o is indicated. The distribution function for the fluctuating energy levels is derived using $\lambda = 0.75$, as suggested in the preceding subsection. The parameter D is the density distribution, with $D_{ox} = c_{ox}W(E)$, and $D_{red} = c_{red}W(E)$. The energy of the band edges from measurement of V_{fb} is shown for ZnO, GaAs, GaP, CdS, and CdSe.

The dependence of the band edge on the pH of the solution as shown in Figure 8.6 for GaP is expected from Equations (8.25) and (8.4), which combined, yield

$$E_{cs} = \text{const} + 2.3kT \,(\text{pH}) \tag{8.29}$$

showing reasonable agreement with the measured shift in E_{cs} with pH.

Other illustrations of the measurement of band edges and energy levels of surface species have been provided by Memming and Möllers.[49,68] In reference 49 they report studies of a two-equivalent surface species. In this case the results not only show further information on semiconductor band edges, but provide a firmer concept of the energy level model for a

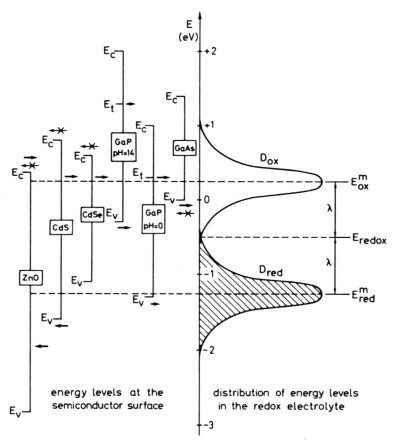

Figure 8.6. Energy level diagram indicating the energy positions of the conduction and valence bands for various semiconductors relative to each other and relative to the fluctuating energy levels of the ferrocyanide/ferricyanide redox couple in aqueous solution. The energies E are given with the hydrogen reference electrode as the zero of energy.

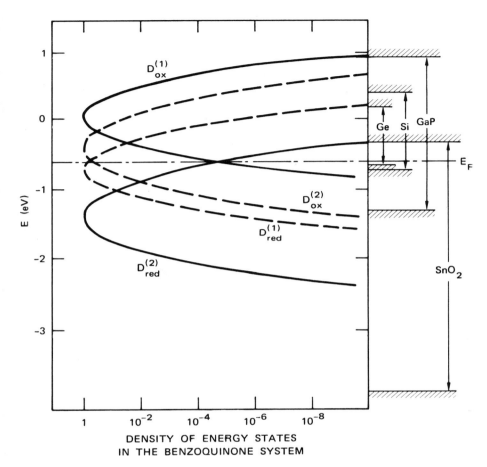

Figure 8.7. The fluctuating energy levels of the quinone/hydroquinone two-equivalent couple. The superscript [2] represents the hydroquinone/semiquinone couple, the superscript [1] represents the semiquinone/quinone couple. The energy values (relative to the hydrogen electrode) are obtained by comparison of the GaP case (pH = 0) with Figure 8.6.

two-equivalent species. Figure 8.7 is from Memming and Möllers' exhaustive study of the oxidation of the two-equivalent species hydroquinone (through semiquinone to quinone). Several semiconductors were examined, as indicated in the figure. The model for the energy levels associated with the hydroquinone and the semiquinone were determined by a detailed analysis of the current/voltage characteristics, the current doubling characteristics, the transistor method (with germanium), and tunneling experiments. In other words, about all the experimental techniques available were used to provide a self-consistent model for this complex system. We have matched the ordinate in Figure 8.7 to that of Figure 8.6 by equating the band edges of GaP.

In Figure 8.7 the "density of states" curve for the quinone system, in other words, the probability curves for the energy of the fluctuating energy levels, shows four branches. The need for four branches can be recognized by comparing Figure 5.7, and recognizing that each of the two levels shown must be split by Franck–Condon effects, separating the E_{ox} from the E_{red}. Thus two branches in Figure 8.7 represent the hydroquinone/semiquinone couple, showing the distributions $D_{red}^{(2)}$ and $D_{ox}^{(2)}$ for this species (compare Figure 8.6), and the other two branches represent the semiquinone/quinone couple showing the distributions $D_{red}^{(1)}$ and $D_{ox}^{(1)}$ for this species. It is noted that E_{redox}^o is higher for the intermediate than for the reduced form. The standard redox potential E^o of the two-equivalent couple is indicated by E_F. At equilibrium this parameter should adjust to be equal to the Fermi energy of the solid electrode, as discussed in Section 8.2.3.

The model of Figure 8.6 provides a convenient illustration of the basic band diagram of the semiconductor/electrolyte system, where to illustrate the levels of ions in solution, an ion with the least possible complication, the ferrocyanide/ferricyanide couple, is chosen. Figure 8.7 is also helpful but more complex. A glance at Figure 8.6 permits one to suggest, for example, which semiconductors can reduce ferricyanide with conduction band electrons (all of them) and which will more likely reduce ferricyanide by hole injection, the capture of valence band electrons (GaAs and GaP at pH 14). In real cases, however, a model such as that in Figure 8.6 can only provide a starting point in describing the possible reactions between carriers and ions at the solid surface. Reactions of the solid, such as dissolution, together with complicating reactions of the ions such as hydrolysis, complexing, dimerization, the generation of unstable oxidation states, make the picture much more complicated. However, even in the discussion of charge transfer with more complicated chemical transformations, a model as illustrated by Figure 8.6 can be used as the starting point.

8.3.4. Other Charge Transfer Measurements, Capture Cross Section

8.3.4.1. Electron Capture and Injection. Qualitatively the theories of Sections 8.2.2 and 2.3.3 indicate that for electron injection into an electrode, the energy level E_{red} must fluctuate enough in energy to be equal to or higher than the unoccupied levels in the solid, and for electron capture the energy level E_{ox} must fluctuate enough to be equal to or lower than the occupied energy levels in the electrode.

Vanden Berghe and Gomes[50] have summarized the experimentally observed requirements for electron injection from one-equivalent species in solution. They conclude empirically that if the conduction band energy of a semiconductor is less than 0.8 eV above E_{redox} $(= -E^o)$, electron injection is observed. In the future more accurate analysis will presumably show, as expected from models such as that in Figure 8.6, that such a rule depends on the value of λ for the species. However, the above rule applies for the species studied to date.

For example, Freund[51] has studied injection into the ZnO conduction band from Cr^{2+}, Eu^{2+}, V^{2+}, Co^{2+} (ethylene diamine), and Ti^{3+}, all of which have $E^o \leq 0.1$ V $(E^o_{redox} \geq -0.1$ eV) and all of which inject electrons. On the other hand he found Co^{2+}(hexamine), Fe^{2+}(ethylenediamine-tetracetic acid), Fe^{2+}(hexacyano), Fe^{2+}, and Mn^{2+}, all of which have $E^o \geq 0.1$, do not inject. Laflere et al.[43] studied electron injection into GaAs and found Eu^{2+} injects but V^{2+} or Ti^{3+} do not. Thus, as shown in Figure 8.6, E_{cs} is higher for GaAs than for ZnO. Gomes[52] studied injection into the V_2O_5 conduction band and found injection from I^-, Fe^{2+}(cyano), and Fe^{2+} in both acid and basic solutions, and from H_2O_2, Mn^{2+}, and SO_2^{2-} in acid solution and BH_4^- and $S_2O_4^{2-}$ in basic solution. Br^-, Co^{2+}, and Cl^- do not inject. Unless the semiconductor is atypical and does not follow the simple theory (there is evidence of electrode dissolution during the experiments), these results suggest that V_2O_5 has a very low conduction band energy.

Electron capture on ZnO by ferricyanide was studied by Freund and Morrison.[15] It was shown that Equation (8.30), derived from Equation (2.29), with $K_n = \bar{c}\sigma$, is obeyed:

$$J = e\bar{c}\sigma[X]n_s \tag{8.30}$$

that is, the cathodic current is proportional both to c_{ox} (assumed proportional to the density of available unoccupied states, $[X]$), and n_s, where n_s is obtained by capacity measurements [Equations (3.11) and (2.38)]. It was shown that electron capture is irreversible—no electron injection could be observed from ferrocyanide—as discussed in Section 8.2.2.

As described above, Vanden Berghe and Gomes have studied the pH variation of the reduction of ferricyanide on ZnO to evaluate the reorganization energy λ.

A series of one-equivalent oxidizing agents was studied[16] at the ZnO electrode to establish the relation between the effective capture cross section

σ for electrons (electron reactivity) and the redox potential E^o. The rate of reduction [cathodic current as per Equation (8.30)] was compared for the various species with n_s and c_{ox} held constant. In other words, the experimental value of $\sigma[X]$ was determined from a measurement of J with n_s known from capacity measurement. Now a comparison of Equation (8.30) with (8.15) shows theoretical value of $\sigma[X]$ is

$$\sigma[X] = d(\pi\lambda/kT)^{-1/2}c_{ox}\exp[-(E_{cs} - E_{ox})^2/4\lambda kT] \qquad (8.31)$$

Figure 8.8 shows the experimentally determined[16] [from Equation (8.30)] variation of $\sigma[X]$ as a function of the redox potential E^o for a series of ions. The value of c_{ox} was 10^{-2} M in all cases, and $\bar{c}$ as usual is assumed to equal 10^5 m/sec. Then experimental results illustrate how the current, as affected by the factor in Equation (8.31) will vary depending upon the chemical properties of the ion.

By the model [Equation (8.31)], if the ion has a value of E_{ox} equal to that of E_{cs}, the value of $\sigma[X]$ should be maximum. If the ion has a value of E_{ox} different from that of E_{cs}, the capture cross section should decrease exponentially with $(E_{cs} - E_{ox})^2$. The shape of the experimentally observed capture cross section curve is in reasonable agreement with the model, considering the implicit approximation that λ is the same for all species. The present interpretation of Figure 8.8 was not made in the cited reference[16] where an error in analysis led to neglect of the reorganization energy λ. However, the experimental data provide supporting evidence that the general concepts can be applied to a variety of redox species.

To suggest how the various experimental data are consistent we will analyze the results of Figure 8.8 to provide independent confirmation of the position of the conduction band edge of ZnO. By Equation (8.31) the maximum of the curve of Figure 8.8 should be centered around the point where

$$E_{cs} = E_{ox} = E^o_{redox} + \lambda \qquad (8.32)$$

where the second equality is derived from Equations (5.18) and (8.19). Now the experimental observations suggest a maximum at $E^o = 0.8$ V. With [Equation (8.23)] $E^o_{redox} = -E^o$, we find from Equation (8.32) that on the redox potential scale the conduction band of ZnO is at an energy

$$E_{cs} = -0.8 + \lambda \text{ (eV)}$$

If we assume λ is the order of 1 eV, this places E_{cs} for ZnO on the order of $+0.2$ eV on the redox potential scale, comparable to the value determined

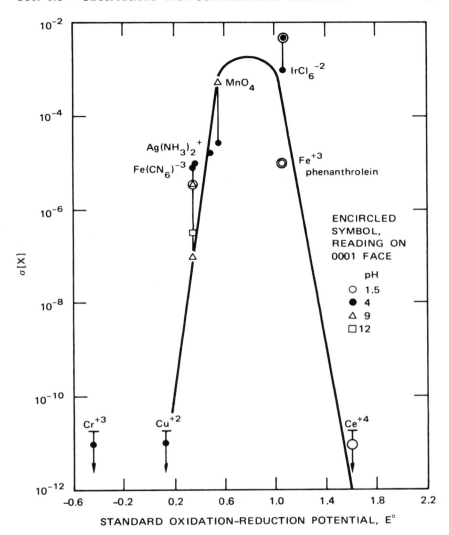

Figure 8.8. The variation in electron reactivity for reduction at the ZnO surface as a function of the standard redox potential (the energy level) for various one-equivalent oxidizing agents.

by Gomes and his co-workers by V_{fb} measurements and reported in Figure 8.6.

Further investigation showing the energy levels associated with ions in solution relative to the ZnO bands have been reported by Gerischer's group,[53] which made studies of tunneling across the surface barrier at the

ZnO surface. These workers conclude that the I^- level is close enough to E_{cs} of ZnO to become a source of electrons by tunneling as in Figure 8.5. The ion Fe^{2+} also will provide electrons for tunneling. The potential required is of the order of 0.5–1.5 V (versus the saturated calomel reference electrode). Bromide ions, on the other hand, do not have a high enough energy level to cause electron injection. In the cathodic direction there is evidence that electrons can tunnel out of the ZnO and reduce the $IrCl_6^{-2}$ ion over a limited potential span. This is not inconsistent with Figure 8.8 and the general theory. With $E_{F,redox}$ for $IrCl_6^{-2}/IrCl_6^{-3}$ at about 1.1 V, and with λ the order of 1 V, the energy level E_{ox} for this species would be about 0.1 V, just a few tenths of a volt below the ZnO conduction band, perfectly located for cathodic tunneling.

Another valuable measurement technique for electron capture is the comparison (on germanium) of electron capture rates from the conduction band versus electron capture rates from the valence band (hole injection). This comparison, which provides direct valuable information on the energy levels in solution (relative to E_{cs} of germanium) is discussed in the next section.

8.3.4.2. Hole Capture and Injection. One of the more powerful techniques to compare, for various species, the capture cross section of holes is the current doubling measurement, described in Section 8.3.2 above. In the original measurements[30] it was shown qualitatively that hole capture from ZnO by formate ions is highly favored over hole capture by lattice oxide ions (dissociation of the ZnO crystal). Gomes and his co-workers[33,34] utilized the current doubling technique to provide much more quantitative data for the hole reactivity (capture cross section) of various species in solution.

The analysis is based on the following arguments. Both a two-equivalent species R and a one-equivalent species S are introduced into the solution with the concentration ratio variable. If the two-equivalent species captures a photoproduced hole, there will be an electron current J_e [Equation (8.28)] calculable from

$$J_e = J - J_p \tag{8.33}$$

where J is the measured current and J_p is the current of photoproduced holes (measurable when no current doubling species is present). If J_e holes are captured by the two-equivalent species, then the remaining holes, $J_p - J_e$, will be captured by the one-equivalent species. However, the ratio of hole capture on the two species should be proportional to the ratio of their hole reactivities and to the ratio of their concentrations. Thus the ratio of the capture rates on the one-equivalent species S to that on the two-equivalent species R is given by

$$\text{Rate}_S/\text{Rate}_R = (J_p - J_e)/J_e = K_S[S]/K_R[R] \qquad (8.34)$$

with K_S and K_R the respective hole reactivities. This K_S/K_R can be evaluated by measurement of J_p and J_e as a function of the ratio $[S]/[R]$.

The simple model is found to give a straight-line relationship, and the hole reactivities (which are proportional to the hole capture cross sections σ if there is no specific adsorption), are given by Table 8.2.

These results suggest the possibility that the fluctuating energy level model does not describe the capture of holes. Consider, in particular, the series of three halides, chloride, bromide, and iodide. One expects, barring unforseen chemical interactions such as hydrolysis or adsorption on the ZnO surface, that the energy level E_{red} of chloride will be the lowest of the three, that of iodide the highest. As discussed above, Gerisher and his co-workers[53] have shown by tunneling experiments that E_{red} for I$^-$ is near the ZnO conduction band (tunneling injection of electrons is observed from I$^-$). With such an E_{red} level for I$^-$, the energy level picture of Figure 8.6 suggests that the order of hole reactivity should be highest for the chloride ion and lowest for the iodide ion, because the chloride energy level should be closest to the valence band. Thus from Table 8.2 the order of the hole reactivity of I$^-$, Br$^-$, and Cl$^-$ is reversed from that predicted by the simple theory. A possible reason for the failure of the simple model to describe the hole capture cross section is that hole capture is a two-step process. Holes may be first captured by ionic surface states associated with lattice

Table 8.2. Hole Reactivities

Substance	Reactivity, K
I$^-$	1 (by definition)
SO$_3^{-2}$	0.23
(CH$_3$)$_2$CHOH	0.18
C$_2$H$_5$OH	0.18
CH$_3$OH	0.08
(CHOHCOOH)$_2$	0.04
Br$^-$	0.01
Cl$^-$	0.01

oxygen ions (Section 4.2.4); then in a second step electron exchange may occur between the surface state and the ion in solution.

Freund in a series of experiments[54] measured hole reactivity by the current doubling technique for a series of one-equivalent species, where in many cases Equation (8.23) can be expected to apply. The results are shown in Figure 8.9. The same one-equivalent species were "titrated" against three different two-equivalent species to check the self-consistency of the current doubling method of measurement of hole capture cross section. As observed, the trend was common in all cases although the quantitative results were not identical. Again in these results it is seen that, contrary to the fluctuating energy level theory, the hole reactivity increases with decreasing E^o (increasing E^o_{redox}), which suggests that the rate controlling step is not the hole capture, but another process such as the electron exchange between atoms (between the ionic surface state O^- and the ion in solution).

Vanden Berghe and his co-workers,[38] again using current doubling, compared the ratio of hole reactivities of $Fe(CN)_6^{-4}$, I^-, and Br^- on ZnO and found $10^2:1:10^{-2}$ confirming the above trend. They compared these results on ZnO with results on CdSe, finding for the latter $10^3:1:0$. An examination of Figure 8.6 permits the CdSe case to be rationalized. The valence band of CdSe is high, very close to the value of E_{red} for ferrocyanide, and it is not unreasonable that a trapped (or even a valence band) hole would not have enough energy to oxidize Br^-.

Capture of the holes from TiO_2 by water molecules is a particularly interesting reaction; detailed discussion of this process will be reserved for Chapter 9.

Hole capture at the surface, when there is no reducing agent in solution with which the hole can react, often leads to corrosion because the holes are captured by lattice ions which then go into solution. Such corrosion was referred to in earlier discussions for the case of ZnO [Equation (8.26)]. The process seems common to many materials, both n and p type. Gerischer[19,55] discusses the process in some detail. He points out that a hole trapped at a surface ion weakens the bonding of the surface ion to the solid, and the energy to be gained by solvation can exceed the remaining energy in the bond. He suggests the corrosion reaction is most probable at a crystal step or a "kink" site. For most solids the species to be oxidized is or can be two-equivalent. The anion oxygen is sometimes an exception, for it can go into solution either in the -1 or 0 oxidation state. Gerischer points out that in this anodic dissolution process germanium is the only material examined (in the group Ge, Si, GaAs, CdSe, GaP, CdS, and ZnO) in which the two-equivalent reaction involves electron injection. In all other cases

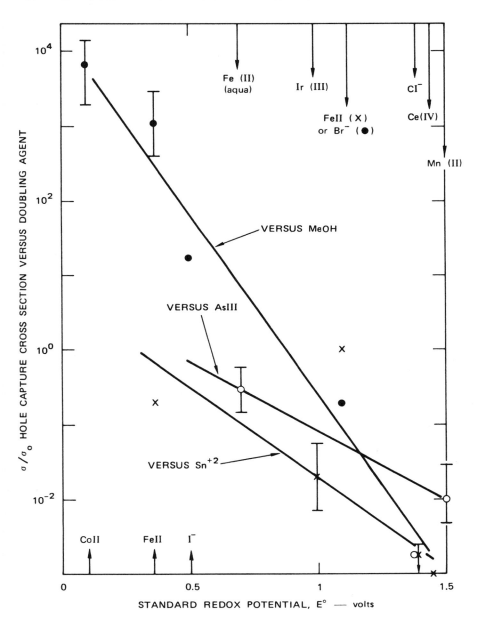

Figure 8.9. The reactivity toward holes from ZnO for various one-equivalent reducing agents as measured by the current doubling technique, titrating against a two-equivalent species. The curves for three titrants, methanol, arsenic, and tin, are shown. The higher hole reactivity of the stronger reducing agents is unexpected by the simple models.

dissolution depends entirely upon hole capture to break the bonds. In the case of CdSe and CdS, the product of the hole capture is insoluble (Se or S in the zero-valent state). The lack of solubility provides complications in the behavior of CdSe and CdS as electrodes when holes are involved in the electrode processes.

Hole injection (valence electron capture) has been studied extensively on germanium, particularly using the transistor technique. This technique (Section 8.3.1) has provided valuable information about the ratio of the electron capture rate from the valence band to the electron capture rate from the conduction band of germanium. Pleskov[56,57] as reported by Turner[26] and by Myamlin and Pleskov[25] report values of y, the ratio of hole injection to conduction band electron capture for various oxidizing agents. Gerischer[55] reports results obtained by his group where hole injection is detected by the resulting corrosion reaction. The latter technique, although not so quantitative as the transistor technique, indicates whether the dominant reduction mechanism for a given oxidizing agent is injection of holes or capture of conduction band electrons. All these results are summarized in Table 8.3.

It is observed that reduction of one-equivalent species follows the redox potential series as expected from Equation (8.23)—a low E^o-means a high value of E_{ox} and so conduction band electrons are necessary. On the other hand, two-equivalent species give results that do not correlate E_{ox} with

Table 8.3. Reduction of Species in Solution^a at the Germanium Electrode (pH $= 0$)

One-equivalent species	E^o	y	Multiequivalent species	E^o	y
V^{+3}	-0.35	cb^b	H^+	0	cb
$Fe(CN)_6^{-3}$	$+0.36$	0.66–0.80 (vb)^c			
MnO_4^-	$+0.56$	0.78–0.88 (vb)	I^-	$+0.54$	0.42 (both)
Fe^{+3} (aquo)	$+0.64$	vb	Quinone	$+0.7$	0.38 (both)
			$C_6H_4O_2$	$+0.9$	vb
Ce^{+4}	$+1.39$	vb	$Cr_2O_7^{-2}$	$+1.33$	0.03–0.08 (cb)
			HO_2^-	$+1.77$	

^{a} y = ratio of reduction by valence band to conduction band electrons.
^{b} cb = reduction dominantly by conduction band electrons (Gerischer, Ref. 55).
^{c} vb = reduction dominantly by valence band electrons (Gerischer, Ref. 55).

E^o and in some cases require contributions both from the valence and the conduction bands, as may be expected for a species with two energy levels. The case of H_2O_2 is complex,[4] involving corrosion of the germanium.

8.3.4.3. Surface States due to Insoluble Species.

As was pointed out by Dewald[2] and Brattain and Boddy[58] effects due to ionic surface states (Section 4.2.4) or due to surface states from insoluble impurities are seldom observed at the semiconductor/electrolyte interface. There are several reasons that could account for this lack of surface state effects, and equally good reasons for the appearance in the cases where they are observed. The problem may arise because the detection methods (in particular, surface state capacity measurements described in Section 3.1.6) are ineffective at the solid/liquid interface. In order that electron or hole trapping on surface states be observed by surface state capacity measurements, the trapped carrier must have a reasonable probability of being reinjected into the band. A capacity measurement is a measure of stored charge, and if the charge is not reinjected, it is not measured as stored charge. There are two reasons why reinjection may seldom be observed. First when the charge is trapped on a surface ion, the ion may dissolve into the solution. If the "surface state" thus moves into the solution, reinjection is unlikely. Second, hydrolysis and reorganization effects will make reinjection difficult, for the trapped charge becomes stabilized. Compare the couples that can capture electrons from ZnO with those that can inject electrons into ZnO as reported in Section 8.3.4.1. In order to capture electrons the energy level must have a value of $E^o_{\text{redox}} < -0.1\,\text{eV}$ ($E^o > 0.1\,\text{V}$), but in order to inject electrons the reduced species must have a value of $E^o_{\text{redox}} > -0.1\,\text{eV}$ ($E^o < 0.1\,\text{V}$). By this oversimplified argument no species can qualify as a potential surface state, because experimentally on ZnO there are no redox couples that have a high cross section both for electron capture and for electron injection.

A few cases of intrinsic surface state effects have however been observed. Tyagai and Kolbasov[59] and later Vanden Berghe, Cardon, and Gomes[60] have reported surface state effects on CdS and CdSe, the latter authors using tunneling measurements. For these materials, as discussed in the preceding subsection, an ionic surface state associated with the sulfur or selenium atom (zero valent) is not removed into solution, and remains available to be reduced (able to recapture electrons). Tyagai and Kolbasov thus observed that in this case the rate of electron transfer to ferric or ferricyanide ions in solution is no longer expressed by the cathodic term of Equation (8.20), which, with Equation (8.15), is linear in c_{ox}.

The cathodic current is given by the more complex relation:

$$J = Kc_{ox}^{0.7} c_{red}^{-0.2} \exp[(1 - \alpha)\eta/kT] \tag{8.35}$$

with $1 - \alpha$ about 0.7. The results were interpreted in terms of a surface state, the charge of which changed with applied voltage and the concentration of reducing agent. Because of surface state effects the transfer coefficient α is no longer zero. Gomes and his group[60] examined tunneling through the CdS surface barrier, analyzed the results using the theory of Memming as described in Section 8.3.1, and concluded that a surface state energy level was present at 0.12 eV below the conduction band energy.

Two types of surface states have been studied on germanium. One is the intrinsic surface state, the dangling bond; the other is an adsorbed impurity, copper. The dangling bond on the germanium surface was shown by Gerischer, Maurer, and Mindt[61] to be normally shared in a covalent bond to either hydrogen atoms or hydroxyl radicals. Passage of current, however, can remove these species, and in particular it was shown that the reactions

$$Ge-OH + e + H^+ = Ge\cdot + H_2O \tag{8.36}$$

$$Ge\cdot + H^+ = Ge-H + h^+ \tag{8.37}$$

occur on the surface upon application of current. Thus as a cathodic current is passed through a surface with OH· radicals adsorbed, reactions (8.36) and (8.37) will occur and the bonds to OH· will be replaced by bonds to H·. Conversely, an anodic current will reverse the process. In this series of reactions, the Ge· reacts with species from solution as indicated, injecting or capturing holes and electrons. Memming and Neumann[62] have shown that the Ge· dangling bond, when not associated with either the OH· or the H· radicals, acts as a normal surface state recombination center (see Section 9.2), recombining holes and electrons at a rate proportional to the concentration of Ge·. A review of the behavior of this system is found in Ref. (4).

Copper ions in solution apparently adsorb, and in particular become reduced to the metal, at a germanium electrode. Thus copper becomes an impurity surface state on the germanium surface. Boddy and Brattain[63] and Memming[64] have shown the influence of copper surface states on the capacitance/voltage curves for germanium. There is a substantial surface state capacity [Equation (3.8)] even with only the order of 10^{-8} M copper ions in solution. The interpretation of the exact chemical form of the copper surface states is still not clarified.[4]

Bernard and Handler[65] report results on GaAs suggesting recombination centers. At low cathodic currents they observe hole injection by monitoring luminescence from band/band recombination. At high cathodic currents, the luminescence ceases, and it is suggested that cessation occurs because of surface recombination at surface states.

8.4. Comparison of the Solid/Liquid with the Solid/Gas Interface

Because of the ability to measure easily electron transfer to and from the surface, and because of the lack of complications due to highly variable adsorption interactions, a much more self-consistent picture of surface states and electron transfer at the solid/aqueous solution interface has been developed than is available for electron transfer at the solid/gas interface. Thus the question arises whether energy level values determined for ions in solution can be related to the surface state energy of the same ions adsorbed on the solid surface with no liquid ambient.

If in the solid/gas case there is negligible local bonding of the adsorbed ion, it is possible that the energy levels at the solid/gas interface may have a very close relationship to the parameter E_{redox}^o (for the ion of $V_H = 0$). The Franck–Condon splitting of the levels by 2λ is large in aqueous solution, and at the gas/solid interface the splitting may be relatively negligible. This is why E_{redox}^o rather than E_{ox} or E_{red} provides a better reference. Thus, for example, with ZnO, in Table 6.2 the surface states at the gas/solid interface associated with $Fe(CN)_6^{-4}/Fe(CN)_6^{-3}$ and $IrCl_6^{-3}/IrCl_6^{-2}$ are reported.[4] The difference in surface state energy is about equal to the difference in the redox potential of the two species, and the energy levels at the gas/solid surface are close to the values (indicated as "theory" in Table 6.2) of E_{redox}^o for the species in solution. The agreement can be rationalized because with the complete ligand shell around these two ions, very little local interaction with the solid is possible with the species adsorbed at the gas/solid interface.

If such an argument is valid, then indeed the results of energy level measurements for one-equivalent redox couples at the solid/liquid interface can be related to energy levels for the same ions at the solid/gas interface by the following steps: The solid/liquid theory gives, as discussed in relation to Equation (8.23), the energy level for the species as $E_{\text{redox}}^o = eE^o$, assuming no local interaction. If there are interactions other than outer sphere changes in solution, these must be accounted for. Examples of such complications are (a) changes in inner sphere complexing or hydrolysis upon electron transfer, (b) several identifiable forms of the solvated ion, (c) processes such as adsorption or dimerization that occur before or after electron transfer, and of course (d) multiequivalent behavior such that E^o does not represent a single electron process.

After identifying the energy level of the species in solution by electrochemical means, the next step is to use this value to estimate the surface state energy of the species at the gas/solid interface, by considering the effect of local interaction. In such an endeavor we return to the discussion of Section 5.5, where the influence of bonding on the surface state energy levels was examined in detail. As discussed in that chapter, the interaction can at times become very complex, and the use of electrochemical "theory" to predict energy levels of foreign ions at the gas/solid interface may be practical only when the adsorption interaction is minimal.

Although a quantitative determination of the surface state energy levels stemming from observations at the solid/liquid interface may be difficult, a qualitative evaluation can be very helpful in an initial selection of possible species to use as surface states at the solid/gas interface.

9 | Photoeffects at Semiconductor Surfaces

9.1. General

Photons absorbed by a semiconductor can induce three types of electron transitions involving the bands of the solid: (i) electron transitions from a filled band to a conduction band, (ii) electron transitions to or from an adsorbate or an equivalent surface state, or (iii) electron transitions to or from energy levels in the forbidden gap region in the bulk material. The last case is generally of little interest in surface studies even though it is of great interest in fluorescence and other bulk properties. Absorption associated with surface states is of interest because it can provide information about the energy levels at the surface.

Absorption due to band-to-band transitions, where a hole and an electron are thus photoproduced, is of most interest, and most of the present chapter will deal with the effects of light thus absorbed. In particular, we will be interested in the effects when the photoproduced hole/electron pair is generated close enough to the surface that one or both of the current carriers can reach the surface to interact with surface states.

When holes and electrons are produced and move toward the surface, initially the capture of one of the two may predominate. However, eventually, independent of the associated chemical reactions, the steady state flow of photoproduced holes and electrons to the surface must be equal in order to preserve electrical neutrality. Any initial transients will decay. If, for example, holes initially flow to the surface at a greater rate than electrons, eventually the resulting charge buildup will tend to repel hole flow and attract electron flow until at steady state the rates of hole and electron flow are equal. The only exception to this is the electrochemical case, where an external circuit is introduced for current continuity and steady state can occur with a net flow of carriers of one type toward the surface. With this exception we can generalize that at steady state

$$U_p = U_n = U \tag{9.1}$$

Where U_p is the net current of holes to the surface, U_n is the net current of electrons to the surface.

There are two classes of reaction that can occur as the holes and electrons reach the surface. The first is simple recombination, the second is recombination with a chemical reaction: photocatalysis or photolysis. The case of simple recombination will result with a surface state A^n/A^{n+1} such that the surface group is A^n if the surface state is occupied, A^{n+1} if the surface state is unoccupied [see Equation (1.1) and Figure 1.1], and the reactions of the electron and hole are

$$e^- + A^{n+1} \rightarrow A^n \tag{9.2}$$

$$h^+ + A^n \rightarrow A^{n+1} \tag{9.3}$$

Here it is clear that after the hole and electron are both captured, there is no chemical change in the system. Such recombination is illustrated in Figure 9.1, where the arrows show the electron returning to an unoccupied state in the valence band—a hole.

The second class of reaction is photocatalysis, where the electron and hole are captured by different surface groups:

$$e^- + X^{n+1} \longrightarrow X^n \tag{9.4}$$

$$h^+ + M^m \longrightarrow M^{m+1} \tag{9.5}$$

or:

$$X^{n+1} + M^m \overset{h\nu}{\longrightarrow} M^{m+1} + X^n \tag{9.6}$$

where the $h\nu$ indicated with Equation (9.6) indicates a photoinduced reaction. Actually a photoinduced reaction may have many steps. A typical example might be

$$2h^+ + O_L{}^{2-} \rightarrow O \tag{9.7}$$

$$CO + O \rightarrow CO_2 \tag{9.8}$$

$$2e^- + \tfrac{1}{2}O_2 \rightarrow O_L{}^{2-} \tag{9.9}$$

where $O_L{}^{2-}$ is a lattice oxygen ion of a solid oxide. Thus the holes activate

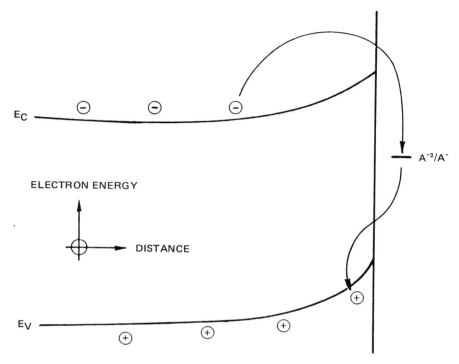

Figure 9.1. Electron/hole recombination, the return of a photoexcited electron to the empty state in the valence band via a surface state.

the lattice oxygen, which oxidizes CO, then the electrons reduce gaseous oxygen, restoring the lattice oxygen. The point is that in both Equations (9.2) and (9.3) and in Equations (9.7) through (9.9), holes and electrons have been "recombined"—the electron that was excited from the valence band to the conduction band has been restored to the valence band. However, in the series (9.2) and (9.3) the photon energy has been dissipated finally as heat, while in the series (9.4) and (9.5) or the series (9.7)–(9.9) the photon energy has been used to cause a chemical reaction to take place which could not otherwise have occurred because of an activation energy barrier. In some cases some of the photon energy is actually converted to chemical energy, although in most cases the photon energy is just used to overcome an activation energy.

In Section 9.2 we develop the equations for hole and electron capture and simple recombination, using essentially the Shockley–Read notation,[1,2] and then in Sections 9.3 and 9.4 discuss photochemical effects that can result from the capture of photoproduced holes and electrons.

9.2. Simple Hole/Electron Recombination

9.2.1. Theory

In Chapter 2, Section 2.3.1, the equations were developed describing the rate of electron flow and of hole flow to the surface. Equations (2.40) and (2.33) can be rewritten as

$$U_p = dp_t/dt = K_p[p_s^*n_t^* - p_1(N_t - n_t^*)] \qquad (9.10)$$

with

$$p_1 = N_v \exp[-(E_t - E_{vs})/kT] \qquad (9.11)$$

from Equation (2.42). From Equations (2.29) and (2.36) we have

$$U_n = dn_t/dt = K_n[n_s^*(N_t - n_t^*) - n_1n_t^*)] \qquad (9.12)$$

with

$$n_1 = N_c \exp[-(E_{cs} - E_t)/kT] \qquad (9.13)$$

where, in Equations (9.10)–(9.13) we have introduced asterisks as a reminder that the electron and hole densities under discussion are not necessarily equilibrium values. With Equation (9.1) the density of electrons in the surface states n_t^* can be eliminated from Equations (9.10)–(9.13) yielding for steady state:

$$U = \frac{K_n K_p N_t (n_s^* p_s^* - n_b p_b)}{K_n(n_s^* + n_1) + K_p(p_s^* + p_1)} \qquad (9.14)$$

We will assume the surface density of holes and electrons is at equilibrium with the bulk density, whence by analogy with Equation (2.38) we have

$$n_s^* = n_b^* \exp(-eV_s/kT) \qquad (9.15)$$

$$p_s^* = p_b^* \exp(eV_s/kT) \qquad (9.16)$$

Two final simplifying assumptions are helpful; the first is that the increase in the bulk electron density equals the increase in the bulk hole density (no bulk carrier trapping). Then we have

$$\delta n_b^* = n_b^* - n_b = \delta p_b^* = p_b^* - p_b \qquad (9.17)$$

The second is that the fractional increase of the carrier densities is small:

$$\delta n_b < n_b \qquad \delta p_b < p_b \qquad (9.18)$$

Then the surface recombination velocity s can be calculated from Equation (9.14):

$$s = U/\delta n_b = U/\delta p_b$$
$$= [(K_p K_n)N_t(n_b + p_b)/2n_i]\{K_n n_b \exp(-eV_s/kT) + K_p p_b \exp(eV_s/kT)$$
$$+ K_n n_b \exp[(E_t - E_F)_0/kT] + K_p p_b \exp[(E_F - E_t)_0/kT]\}^{-1} \qquad (9.19)$$

where the first equalities define s. Here $(E_F - E_t)_0$ is the difference between the surface state energy level and the Fermi energy with no band bending and n_i is the density of electrons for the semiconductor if intrinsic. The above equation appears complex, and indeed a more elegant form is available,[2] but Equation (9.19) permits easy visualization in terms of varying V_s and E_t.

It is clear from Equation (9.19) that there is a maximum s_{max} in the surface recombination velocity for an intermediate surface barrier. If V_s is large and negative, the first term in the denominator is large, so s is small. If V_s is large and positive, the second term in the denominator is large, so s is small. An intermediate V_s leads to a maximum in s. A similar observation holds for the relative position of the Fermi energy and the surface state through which recombination is occurring. If the Fermi energy (at zero band bending) is much above the surface state, the last term of the denominator is large, so s is small. If the Fermi energy is much below the surface state energy, the remaining term is large, so s is small. Allowing for significant variation because the pre-exponentials in the denominator are not equal and depend on whether the sample is p or n type, etc., we can say in general that for an effective hole/electron recombination center, we should look for: (a) a modest surface barrier V_s, and (b) a surface state in the neighborhood of the Fermi energy at flat bands, that is, a small $(E_F - E_t)_0$.

9.2.2. Experimental Results

A few illustrative experimental results will be presented to indicate the behavior expected with simple recombination centers. For more detailed description of experimental results the reader is referred to the earlier books on surface physics.[2,3] Most of the work on surface recombination has been done on surface states at the semiconductor/oxide interface, specifically at the germanium/germanium oxide interface. For example, Figure 9.2 shows results of Many and Gerlich[4] for surface recombination velocity s as a function of surface potential ($U_s = eV_s/kT$) on germanium.

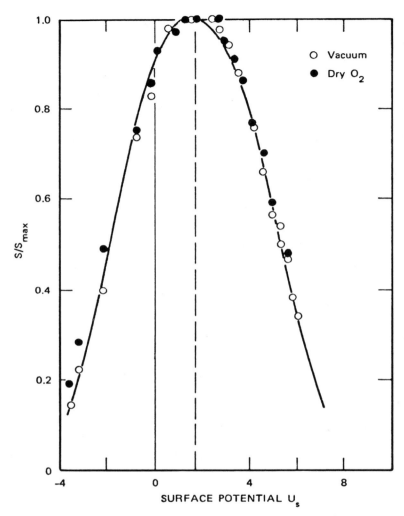

Figure 9.2. The surface recombination velocity s as a function of surface potential $U_s \, (= eV_s/kT)$, showing the maximum at intermediate surface barrier.

This curve provides an excellent illustration of the maximum that is found for s when the surface potential V_s is at a value such that both holes and electrons can move relatively freely to the surface. The results of Figure 9.2 were obtained with a surface treatment such that the exact chemical groups that form the surface states are unknown. As discussed in Section 6.5, the form of the electrically active groups at the Ge/GeO_2 interface is still not clarified in general.

In some cases, however, impurities have been identified as the groups that cause recombination centers. For example, copper on germanium causes a high density of recombination centers, as was discussed in Section 6.1. Frankl[5] shows results analogous to Figure 9.2 for copper, finding a value of $s_{max} = 6000$ cm/sec, with copper present, whereas if the surface is washed with KCN to remove the copper ions, the value of s_{max} is about 100 cm/sec. By analyzing the results [in accordance with Equation (9.19)] he concluded there are two surface states associated with copper, one at about 0.15 eV above the valence band, the other at 0.55 eV or so above the valence band. A third level is also possible.

From the discussion of Section 9.1, in particular, from Equations (9.2) and (9.3) as compared to (9.4) and (9.5) it is clear that there can be a competition for photoproduced minority carriers (holes in n-type material) between surface states where hole capture would lead to chemical change and surface states where hole capture would lead to simple recombination. With this concept in mind the intentional addition of recombination centers has been tested[6] as a means of stabilizing a surface, specifically to prevent photolysis. The action of iron cyanide as a surface state on ZnO has been discussed earlier, with the energy level reported in Table 6.2 and a discussion of the properties of the surface state in Section 8.4. This redox couple is clearly of the one-equivalent form desired for recombination according to Equations (9.2) and (9.3). As will be discussed, in Section 9.3.2.1 it is found that in vacuum after oxygen exposure uv irradiation on ZnO will either cause oxygen desorption or cause photolysis according to

$$2h^+ + O_L{}^{2-} \rightarrow \tfrac{1}{2}O_2 \tag{9.20}$$

$$2e^- + Zn_L{}^{2+} \rightarrow Zn_i{}^+ + e^- \tag{9.21}$$

Here the second equation describes the fate of the lattice zinc ion that remains when the oxygen leaves. The excess zinc often dissolves interstitially into the lattice and becomes a donor, contributing one electron to the conduction band of the ZnO. Because these photolysis reactions continually increase the density of conduction band electrons, the effect is easily observed with a simple conductance measurement. Collins and Thomas[7] studied the reaction on single crystal ZnO and showed that the quantum efficiency of photons (with energy sufficient to produce hole/electron pairs) approaches unity on undoped ZnO with a freshly oxidized surface, although the quantum efficiency decreases rapidly as the excess interstitial zinc builds up with continuing photolysis.

In Figure 9.3 is shown the quantum yield for the production of electrons

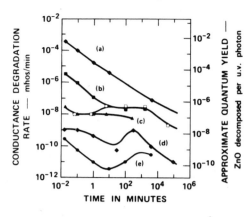

Figure 9.3. The use of simple recombination centers to prevent photolysis of ZnO. Curve (a) is the blank, with no iron cyanide recombination centers. Curves (b)–(e) show the lower quantum yield as the density of deposited recombination centers is increased.

by photodecomposition or photodesorption where the iron cyanide recombination center was introduced to compete with the photolysis reactions. Conductance measurements were made on a lithium doped single crystal sample.[6] The upper curve of Figure 9.3 shows photolysis (in vacuum) of the freshly oxidized, iron-cyanide-free surface. Curves (b)–(e) show the much lower quantum yield with increasing surface concentration of the recombination center iron cyanide on the surface. For the measurements of curve (c) 10^{19} molecules/m² of iron cyanide was deposited, for curve (d) 5×10^{19}/m², and for curve (e) 10^{20}/m². As discussed in Section 6.1, the additive cannot be present in a monomolecular layer, but rather in the form of crystallites, probably at least 100–500 Å in diameter. The curves clearly demonstrate a much lower quantum yield for the photolysis reaction [Equations (9.20) and (9.21)] and therefore demonstrate the effectiveness of surface additives applied as recombination centers in preventing photochemical activity. Similar results, using ESR to monitor the effectiveness of the recombination centers, have been reported.[8]

9.3. Photoadsorption and Photodesorption

9.3.1. Theory

The phenomenon of photoadsorption or photodesorption of ionosorbed species at the surface of a semiconductor is experimentally well established, as will be discussed in the next subsection. The reason for the effect is quite straightforward: band gap radiation changes the concentration of the electrons or holes available at the semiconductor surface and thus changes the rate processes affecting ionosorption. Unfortunately there are many features complicating any attempt at quantitative analysis. First, the dark ionosorption may not be at equilibrium. In fact, during photodesorption or photoadsorption experiments the adsorption is usually in the low-temperature irreversible region of chemisorption [region (b) of Figure 7.3], because it is in this region where the greatest effects are observed. Second, the density of both holes and electrons is increased. If, for example, we are considering photoadsorption or photodesorption of oxygen, the photoproduced electrons will induce oxygen adsorption:

$$e^- + O_2 \rightarrow O_2^- \tag{9.22}$$

but the photoproduced holes will induce oxygen desorption:

$$h^+ + O_2^- \rightarrow O_2 \tag{9.23}$$

so whether the net effect is photoadsorption or photodesorption is not predictable without a detailed knowledge of the rate constants and the system parameters for the dark case.

Photodesorption is not restricted to ionosorbed species. Another form of photodesorption to be expected is the desorption of adsorbates with high partial charge that are locally bonded to a semiconductor. Examples might be oxygen bonded to CdS forming a local covalent bond to the sulfur atom, or OH^- groups bonded to acid centers on TiO_2. There is evidence, as discussed below, that with the latter example photodesorption of OH radicals occurs. We will not explicitly analyze photodesorption depending on ionization of a local bond—the mechanism is expected to be quite analogous to the mechanism for photodesorption of ionosorbed species.

We will discuss three of the most important classes of the photoadsorption/photodesorption of ionosorbed species, each of which illustrates a particular approximation. The three examples are not intended to cover all possibilities and are given identifying letters simply to facilitate later reference to them when we discuss experimental observations. The first example, Class A, is the case in which the ionosorption is at equilibrium [region (c) of Figure 7.3] and the ionosorbed species provides the only surface state. The second example, Class B, is the case where the system is in the irreversible chemisorption region [region (b) of Figure 7.3], and the assumption is made that the ionosorbed species provides the only surface state. The third example, and usually the most realistic, Class C, is the case of a system where photosorption involves capture of one carrier, but where traps for the other carrier are present on the surface. The last case, as will be discussed, is only one step from photocatalysis.

Even in the first example, Class A, where we assume that the ionosorption in the dark is at equilibrium, we will present an oversimplified analysis[9] that only provides the sign of the effect—whether adsorption or desorption occurs. We are de-emphasizing detailed analyses such as those presented by Volkenshtein and his co-workers[10] because even in the simplest case there are too many unknown parameters to test the theory properly. Quantitative comparison of analysis with experiment must await measurements on much better defined surfaces than are available today.

We will assume for specificity an n-type semiconductor with a negatively charged adsorbate (say oxygen ions O_2^- on ZnO), assume there is only one surface state involved, and assume that the density of unoccupied surface states (physisorbed oxygen in the example) is controlled by rapid adsorption/desorption kinetics and is independent of the density of charged surface states.

In the analysis of Class A effects, the assumption of equilibrium in the dark means that Equations (9.10) and (9.12) (without asterisks) each are identically zero, whence from Equation (9.10):

$$p_s[X^-]_0 = p_1[X]_0 = p_1[X] \tag{9.24}$$

where we have designated the neutral adsorbate as X, and the ionosorbed species as X^- and used these symbols in square brackets to designate the density of unoccupied and occupied surface states, respectively. The subscript zero indicates the value of $[X]$ in the dark. The last equality in Equation (9.24) arises from the assumption that the density of unoccupied surface states depends only on the physisorption isotherm, so that $[X]_0 = [X]$. Similarly, from Equation (9.12) we have

$$n_s[X]_0 = n_1[X^-]_0 \tag{9.25}$$

The parameters upon illumination are changed by an amount

$$n_s^* = n_s + \Delta n_s \tag{9.26}$$

$$p_s^* = p_s + \Delta p_s \tag{9.27}$$

and

$$[X^-] = [X^-]_0 + \Delta[X^-] \tag{9.28}$$

where Δn_s and Δp_s are always positive, and the sign of $\Delta[X^-]$ is the parameter of interest.

As always at steady state the rate of hole capture is equal to the rate of electron capture, and so

$$U_n = U_p \tag{9.29}$$

and we can substitute the expressions from (9.24)–(9.28) into (9.29), where the expressions for U_n and U_p are given by (9.10) and (9.12). Thus, for example, with the use of Equations (9.24) and (9.25) to simplify the expression we obtain

$$K_n(\varDelta n_x[X] - n_1 \varDelta [X^-]) = K_p(\varDelta p_s[X^-]_0 + p_s \varDelta [X^-] + p_s \varDelta [X^-])$$

and with further manipulation we obtain:

$$\varDelta [X^-]/[X^-]_0 = (\varDelta n_s/n_s - \alpha \varDelta p_s/p_s)/(1 + \alpha + \alpha \varDelta p_s/p_s) \qquad (9.30)$$

where

$$\alpha = (K_p/K_n)(p_1/n_1)([X]/[X^-]_0) \qquad (9.31)$$

It is clear that the density of ionosorbed species X^- can either increase or decrease depending on the sign of the numerator. This was the object of the calculation: to show that in Class A effects either photoadsorption or photodesorption can occur depending on the values of the various parameters given in Equation (9.31). Unfortunately there are so many interdependent parameters, for which the relationship depends on the details of the model assumed, that it requires an extended model, such as used by Volkenshtein, to describe the system in sufficient detail to predict whether adsorption or desorption should occur upon irradiation.

The second example, Class B, of photoadsorption/photodesorption is the somewhat more realistic case in which the ionosorption in the dark is irreversible. This is more likely to be the regime in the systems studied (many of which are at room temperature).

Consider, for example, the adsorption of oxygen as O_2^- on an n-type semiconductor for the case where the system is in the irreversible region— the temperature is so low that electron injection from the O_2^- into the conduction band of the semiconductor can be neglected. When the sample is illuminated, O_2 will be assumed the only surface species capable of capturing electrons and O_2^- will be the only species capable of capturing holes. Then we can write for electron capture from Equation (9.12)

$$U_n = K_n(n_b + \varDelta n_b)[O_2] \exp(-eV_s/kT) \qquad (9.32)$$

For hole capture we make the assumption that there are few holes available in the dark, so $\varDelta p_b > p_b$. We will make the further assumption that all holes reaching the surface are captured, namely, the photogeneration of holes is the rate limiting step. Then the rate of hole capture equals the rate at which holes are generated within a diffusion length L_p of the surface (more accurately, with $V_s > 0$, within $L_p + x_0$, where x_0 is the thickness of the space charge region). Then we have

$$U_p = \gamma I \qquad (9.33)$$

where I is the light intensity, and γ is essentially a constant.

The assumption that holes reaching the surface react extremely rapidly, and hence that the rate of hole capture is independent of V_s (for V_s positive) is probably a realistic assumption for most wide band gap n-type semiconductors. In these, as discussed above [for example, Equation (9.23)] the holes are usually so energetic that if no other oxidizable species are available, they will attack the host crystal. The fault in the model is that it does not take into account other recombination centers (such as the anionic surface states of the host crystal) competing for the holes, or perhaps acting as hole traps, storing the holes until they are transferred to the O_2^-. (This latter case, of course, belongs to the third class of photoeffects, Class C, discussed below.)

It is easily shown that, with no hole storage and with an n-type material of reasonable donor density, the model leads to photodesorption. Specifically, if in Equation (9.32) $n_b \gtrsim \Delta n_b$, illumination can only cause photodesorption. By definition (see Sections 7.1.2.3 and 7.2.1.2) U_n is negligible in the dark in the irreversible region, so if the density of electrons is not appreciably increased, then the only way U_n can increase is if V_s decreases. But a decrease in V_s can only occur if the concentration of O_2^- is decreased. This implies photodesorption of oxygen. A description of the physical process may make the conclusion clearer. With n_b of reasonable magnitude in the dark, the only initial effect of the illumination is to increase U_p, the rate of photodesorption, by Equation (9.33). As the concentration of O_2^- decreases by hole capture and desorption, V_s will decrease. The value of V_s will decrease until at steady state U_n [from Equation (9.32)] equals U_p [from Equation (9.33)]. At steady state it follows that

$$V_s = (kT/e) \ln\{K_n n_b [O_2]/\gamma I\} \qquad (9.34)$$

and the greater the light intensity, the lower the value of V_s. The value of $[O_2^-]$ can be calculated from V_s with the Schottky relation, and clearly also decreases with increasing light intensity.

On the other hand, if $\Delta n_b > n_b$—particularly in the case where $[O_2^-]$ is low in the dark because n_b is so low, the increase in U_n can exceed the increase in U_p, and there can be a net photoadsorption. Thus it is possible for photoadsorption to occur in the irreversible region (without trapping effects), but it is expected only if the semiconductor is of low majority carrier density in the dark.

The Class C case, on the other hand, where we admit there are traps present on the surface, will lead to photoadsorption in most cases. The traps may be intrinsic surface states[11,12] or may be another adsorbed species. A very important case, which will be discussed further in the experimental section, is the case of oxygen photoadsorption where organic molecules are present on the surface and these species provide the "traps" which capture holes.

Consider the example where the hole activity (capture cross section) of the organic molecule is orders of magnitude higher than the capture cross section for hole capture by the O_2^-. Then during the time the light is shining on the sample, all the holes produced will in principle be captured by the organic molecules and all the electrons will be captured by the oxygen. If there is no chemical interaction between the oxygen and the positively charged organic molecules, the amount of adsorbed oxygen will increase indefinitely. Realistically, if there is no chemical interaction, eventually there will be enough O_2^- present to dominate the hole capture and the system will asymptote to a steady state situation. Even more realistically a reaction will occur between the ionosorbed oxygen and the ionized organic molecules and, after a moderate amount of excess oxygen is adsorbed we must consider the oxygen being "consumed in a photocatalytic reaction" rather than being photoadsorbed.

The analysis of Class C photoadsorption is extremely important, but clearly the general case is extremely complex, involving many chemical as well as electronic steps. The electronic steps themselves would be difficult to analyze. They involve nonlinear equations where the rates vary with the amount of each species ionosorbed both in the exponent (from V_s) and in the pre-exponential, and are not easily soluble. The importance of the general picture of how traps will affect photoadsorption/photodesorption processes cannot be overemphasized, however, and for specific cases it should be possible to describe the processes mathematically. However, to obtain a manageable formalism, simplification must be introduced by use of specific physical models such as the examples provided above for Classes A and B photoeffects.

9.3.2. Experimental Observations of Photoadsorption and Photodesorption

9.3.2.1. Photodesorption and Photodecomposition.

Photodesorption of ionosorbed species has been detected primarily by its influence on the electrical properties of the semiconductor. Baidyaroz et al.[13] provide an extensive list of references where photodesorption was detected by conductance changes. Probably the system most studied is the photo-desorption of oxygen from the semiconductor ZnO.

The early work of Miller's group[14-16] on this system led to many of the currently accepted models. The dominant measurement was conductance; a pressure increase was observed,[16] but the gas was not analyzed. The photodesorption was attributed[14] to hole capture that neutralized negatively charged ionosorbed oxygen. This model is essentially as described in the preceding section as Class B photodesorption.

Collins and Thomas[7] in studies on single-crystal ZnO concluded that the oxygen photodesorption from ZnO is accompanied by photolysis, the evolution of lattice oxygen from the material as described following Equations (9.20) and (9.21). Because single crystals were used permitting quantitative calculations, the authors could show that the density of excess electrons being provided to the conduction band (the density of holes captured to neutralize the oxygen ions) was far greater than that permitted by the Weisz limit on the maximum equilibrium ionosorption [Equation (7.20)]. From this and other work on ZnO,[6,17-19] it must be considered that unless shown otherwise, the photodesorption of oxygen will be a mixture of photodesorption and photolysis. In the series of measurements of Figure 9.3, for example, it was suggested[6] that the discontinuity in the curves for the quantum yield (versus time) is associated with the exhaustion of O_2^- from the surface and the initiation of photodecomposition of the lattice. It is to be noted that the suggestion implies that the hole capture cross section for O_2^- is much higher than that for O_L^{2-}, the lattice ion.

In recent studies of irradiated ZnO oxygen has been clearly detected in the gas-phase associated with photodesorption or photodecomposition from single crystals. Hirschwald and Thull[20] have shown that after exposure of ZnO to O_2 (720°C, 100 Torr O_2), irradiation in vacuum results in photo-desorption of oxygen atoms. No photodesorption is detected below 350°C. They also find thermal desorption at $T > 350°C$, and find photodesorption and thermal desorption are proportional. The suggestion is made that in order to obtain measurable quantities of oxygen from a single crystal, photodecomposition must occur as well as photodesorption. In order to obtain continuous photodecomposition, the zinc atoms that remain after the oxygen is desorbed must be disposed of. Hirschwald and Thull suggest that 350°C is a high enough temperature for volatization of the Zn atoms. It is also known that 350°C is a high enough temperature for the zinc to dissolve and diffuse into the crystal.[21]

Cunningham et al.[22] also observe oxygen photoremoved from ZnO.

They find no corresponding oxygen from TiO_2. Steinbach and Harborth[23] find that the time constant for this photodesorption process is reasonably fast and use the time constant difference to distinguish photodesorption from thermal desorption. They use illumination chopped at 5–20 Hz, and find O atom photodesorption; with O_2 and CO simultaneously present they find CO_2 photodesorption from ZnO.

Hauffe and Volz[24] have introduced a novel method of examining photodesorption. They examine the change in the electrophoresis associated with the semiconductor particles in solution. The electrophoresis effect is associated with the net charge at the surface of the particles. An applied field will cause particles suspended in an insulating liquid to move due to the net charge on the particles. Thus, for example, ZnO was studied suspended in toluene with isopropanol present. Illumination decreases the negative charge at the surface of the ZnO by photodesorption, photolysis, or a reaction between the holes and the isopropanol (the exact reaction is not easily determined), and the movement of the particles in the electric field is altered.

Illumination of CdS, as studied by Mark's group[13] caused CO_2 rather than oxygen photodesorption. This is a common observation[25] presumably associated with hole trapping by adsorbed organic molecules which then "photocatalytically" interact with the adsorbed oxygen, as discussed in Section 9.4. Photodesorption of oxygen from CdSe has been observed by Chung and Farnsworth.[26] For both CdS and CdSe the oxygen may well be bonded by a local covalent bond rather than (or as well as) ionosorbed.

9.3.2.2. Photoadsorption.

In Section 9.3.1 it was suggested that Class B photoadsorption on an n-type semiconductor in the irreversible adsorption region can occur only if the change in the electron density by the illumination is comparable to or greater than the electron density in the dark. Thus in the case of ZnO, Tanaka and Blyholder[27] observe photoadsorption for a fully oxidized sample with a low donor density. They find if the sample is heated to 400°C and cooled to room temperature in oxygen without subsequent illumination, a TPD curve shows negligible desorbable oxygen on the surface. However, if after the heat treatment the sample is illuminated at room temperature in oxygen, then both O^- and O_2^- are photoadsorbed, as determined by a subsequent TPD curve (the O_2^- desorbs at 185°C, the O^- at 290°C). Wong et al.[28] have found that irradiation of ZnO at 77°K in the presence of N_2O leads to photoinduced adsorption of O^-. In both these cases the fate of the photoproduced holes is not clear; Lunsford and his group conclude that the holes are trapped in the bulk. Alternately, the holes could be trapped at lattice oxygen ions at the surface, providing another source of desorbable O^-. In either case, the result may correspond more closely to Class C photoadsorption than to that of Class B.

Photoadsorption on other semiconductors has also been observed. Balestra and Gatos[29] observe photoadsorption of oxygen on a clean CdS surface. No adsorption occurs without illumination. By contact potential measurements they conclude that the photoadsorption may be reversible. The contact potential is restored to its initial value if the sample is pumped under illumination. The fate of the holes is not known in these photo-adsorption experiments. It is possible that the sulfide ions at the surface are being oxidized, for it is clear that stable oxidation states of the sulfur from -2 to $+6$ can be present. Thus the photoadsorption may involve hole trapping, (the formation of groups resembling sulfate groups) and thus again approach Class C. The effect of water (OH groups) in hole trapping will be discussed further below, but it is of interest that these authors find water increases the photoadsorption of oxygen (as measured by contact potential). Wentzel and Monteith[30] observe the photoadsorption of oxygen and[31] of SO_2 on CdS.

Nelson and Hale[32] find photoadsorption of oxygen to form O_2^- on MgO when irradiated. Again the low density of electrons in the dark makes photoadsorption reasonable.

Another particularly illustrative study on the photoadsorption and desorption of oxygen has been made by Petrera, Trifiro, and Benedek.[33] They studied the n-type semiconductors SnO_2 and TiO_2, observing photo-adsorption versus photodesorption while cycling the oxygen pressure. At low pressure, particularly as the pressure was being decreased, photo-desorption of adsorbed species was observed. At high pressure, photo-adsorption was observed. The method was particularly of interest—pressure cycling techniques allowed photoadsorption to be studied after a period of photodesorption and *vice versa*. Unfortunately only the pressure was measured, so no information about the presence of hole traps in the n-type semiconductors could be obtained. However, the switch from photoadsorp-tion to photodesorption is not inconsistant with the mechanism of Section 9.3.1 termed Class B.

Titania is a particularly interesting semiconductor because there is no evidence in the literature of photodecomposition. To the contrary, Cun-ningham and his co-workers[22] have concluded that while ZnO photo-decomposes in vacuum, TiO_2 does not. If holes in TiO_2 cannot effectively oxidize and desorb lattice oxygen, then the photoadsorption/desorption reactions should be simpler to interpret than those for ZnO. It turns out that most observations suggest Class C effects, involving trapping of holes by foreign adsorbates.

In the usual case[34,35] it is concluded that the hole traps are OH^- groups on the titania surface (Figure 4.4) which become oxidized to OH radicals.[36] Thus, for example, Boonstra and Mutsaers[37] show that the coverage of oxygen photoadsorbed on TiO_2 is proportional to the coverage of hydroxyl groups on TiO_2. If the OH^- groups are exchanged for Cl^- groups by a simple anion exchange [Equation (4.22) and its discussion], the volume of oxygen photoadsorbable decreases. McLintock and Ritchie[38] also relate the photoadsorption of oxygen to the presence of water. Bickley and Jayantz[39] suggest that if there are no OH^- groups available, the donors in reduced rutile can act as hole traps in the oxygen chemisorption process.

Other species can be photoreduced as well as oxygen with the holes captured by the OH^- traps.[40,41] Silver and palladium ions from solution are deposited as metals on the TiO_2 surface by such a photoreduction process. Cunningham and Zainal[42] studied the photoreduction of several ions in solution and related the rate to the presence of other ions. If the other ions showed two-equivalent (hole trapping) behavior, the photoreduction was more efficient than when the hole traps were simply OH^-; ions of a class that could act as simple recombination centers decreased the rate of the photoreduction.

From these examples it is clear that in many cases Class C photoadsorption is difficult to distinguish from photodecomposition or photocatalysis. We will restrict the term "photocatalysis" to processes where a product is desorbed from the sample that is different from either of the species that have been photoadsorbed. But it is indeed a fine distinction. As described in Section 9.3.1, it is to be expected that if Class C photoadsorption continues, photocatalysis will almost always result.

9.4. Photocatalysis

Photocatalysis is the increase in a chemical reaction rate produced by optical excitation of the solid where the solid remains chemically unchanged. Detailed analyses and reviews of photocatalysis using a pure rigid band model have been made primarily by Volkenshtein[43] and Volkenshtein and Nagaev.[44]

A "theory" of photocatalysis is given in broad terms in Section 9.1, where it is indicated that if holes and electrons are captured by different species, a net chemical reaction can result. If both the electrons and the holes are directly involved in the reaction, steady state is reached when the rate of removal of electrons and holes by the reaction, plus any removal by simple recombination, equals the rate of generation by the illumination. If one type of carrier trapped at the surface creates an active site, upon which the reaction can proceed rapidly, steady state is reached when the rate of production of the site by the one carrier equals the rate of its removal by the other carrier.

We will not attempt a more general theory than Equations (9.4)–(9.6); there is so much variety possible in photocatalysis (as in catalysis itself) that a "general theory" of photocatalysis is as difficult as a "general theory" for catalysis. There are so many different possibilities of active species for hole capture and active species for electron capture, including lattice ions, reactants in various oxidation states, and intermediates in various oxidation states, not to speak of the variety (discussed in Chapter 10) in adsorption and chemical transformations not involving electrons or holes, that a general theory cannot be considered. As with Class C photoadsorption and photodesorption effects above, and as with catalysis itself as in Chapter 10, each type of reaction must be analyzed individually bearing in mind the simple rules developed above, e.g., that at steady state the flow of holes to the surface must equal the flow of electrons.

9.4.1. Photodecomposition of Adsorbed Species

Adsorbed N_2O photodecomposes on ZnO, forming gaseous nitrogen and adsorbed O^-. If the temperature is high enough that desorption of the O^- will occur (the conditions used by Tanaka and Blyholder,[45] who studied the reaction at about 400°C), then the photodecomposition is continuous. At this temperature the decomposition also occurs without illumination, and with no illumination (that is, in such a wide band gap n-type semiconductor, no holes) the rate is first order in the N_2O pressure, zero order in oxygen pressure. Illumination causes an increase in the rate, and the increase is found to be proportional to

$$p_{N_2O}/(k_1 + k_2 p_{N_2O} + k_3 p_{O_2}^{1/2})$$

where the p's are pressures, the k's are constants. The authors were able to explain such a law assuming hole and electron capture by surface species— the electrons being captured by N_2O and O_2 and the holes by N_2O^- and O^-.

Cunningham and his co-workers[46] and Lunsford's group[28] measured the N_2O decomposition on ZnO at low temperatures where the O^- could not desorb. Nitrogen was evolved. By ESR measurements it was concluded that O^- was indeed the species remaining even at temperatures as low as $77°K$, but the ESR signal observed was attributed to holes trapped in the bulk, not to the O^-.

In studies[47] of CD_3I decomposition over ZnO it was shown that unless the ZnO was dehydrated, hydrogen appeared in the photoproduced methane. Thus apparently the OH^- on the surface is involved in the photo-decomposition reaction.

The photolysis (decomposition) of water on TiO_2 has been the subject of considerable interest[48-51] because of the possible use of the effect in solar energy conversion, converting solar energy to the fuel, hydrogen. Although for commercial use TiO_2 would be impractical in solar energy conversion, as only about 10% of the solar energy is absorbed, direct water photolysis was first observed on this material,[48] so it is the material most studied. With a few other materials such as SnO_2[52] and $SrTiO_3$[53] it has also been shown that photocatalyzed water decomposition can occur.

In essence the mechanism of the photolysis is straightforward. With the TiO_2 in aqueous solution, the photoproduced holes are able to oxidize OH^- groups[36] to peroxide or to oxygen. If the TiO_2 is electrically connected to a platinum electrode, the photoproduced electrons will move to the Pt electrode and will be captured by protons to form hydrogen. Thus the overall reaction is water decomposition. Unfortunately, as was shown by Wrighton et al.,[49] the electrons do not have quite enough potential to reduce hydrogen, and a small voltage must be added to assist the photolytic process. However, this does represent one of the few cases in photocatalysis where a reasonable fraction of the photon energy is stored as chemical energy.

9.4.2. Photostimulated Catalytic Reactions

In photocatalysis probably the most extensively studied system is the photocatalyzed oxidation of CO on ZnO. Even this "simple" catalytic process is not well understood. Several authors[54-56] have determined that the rate of CO_2 production is first order in CO pressure, zero order in oxygen pressure. Others[57] find the reaction half-order in oxygen, zero order in CO. Volkenshtein[46] suggests that an observation of oxygen sensitivity should be attributed to low oxygen pressure, but in these latter experiments the oxygen pressure was moderately high (up to 60 Torr).

Doerffler and Hauffe[58] find that the photocatalytic CO oxidation reaction is extremely history sensitive, depending, for example, on the pretreatment with oxygen and the time of irradiation. When one considers that the holes from ZnO have been shown to oxidize lattice oxygen, O^-, O_2^-, and may also oxidize CO, CO_2^-, CO_2^{2-} (and, of course, the photoproduced electrons may also undergo many possible reactions), it is clear that the mechanism may change with the gaseous ambient. Certainly, transients lasting a long time may occur after a sample of large surface area is first illuminated. Thus it is understandable that different authors, differing in pretreatment details, obtain different results.

Photocatalytic oxidation of CO has been studied by Teichner and his group[59] on Ga_2O_3, ZnO, Sb_2O_5, SnO_2, and examined in detail on TiO_2 (anatase). These authors find Ga_2O_3 and anatase the most active photocatalysts.

The photocatalyzed oxidation or reduction of organic molecules is a particularly important example of photocatalysis. Such photocatalytic reactions are of technological importance, for example, on TiO_2, because in general we wish to prevent them. They occur in filled polymers or coatings (plastics or paints) in which TiO_2 pigment particles have been added to scatter the light and provide "hiding power" (in the paint case) or to absorb the solar ultraviolet and protect the plastic from direct photolysis (in the filled polymer case). Although TiO_2, particularly when coated with a thin layer of alumina or silica, performs well in the above roles, it is still a photocatalyst, and the absorbed solar uv produces electrons and holes which will slowly photocatalyze the oxidation of the polymeric binder.

We have discussed earlier the observations that holes from TiO_2 are sufficiently powerful oxidizing agents to oxidize OH^- to $OH\cdot$, and have discussed in relation to Figure 4.4 that unless given an exhaustive dehydrating treatment, the TiO_2 can be expected to have not only a complete monolayer of OH^- groups on the surface, but perhaps also adsorbed water on the surface of the hydrated solid. Thus in most cases it is not clear whether, on the one hand, the holes directly oxidize the organic species to a radical or, on the other, produce hydroxyl radicals or hydrogen peroxide and these radicals in turn oxidize the organic species. In either case the presence of the radical can initiate a chain oxidation reaction through the resin. In all probability there are many reaction routes,[60-64] and the prevalence of each depends on the hole reactivity (hole capture cross section) of the various groups (surface states) present on the surface.

It has been shown[38] that TiO_2 photocatalyzes the oxidation of ethylene and propylene, with CO_2, H_2O, and formaldehyde as the products. Other work has shown that even saturated linear hydrocarbons are oxidized,[65] which leads to ketones and aldehydes or CO_2 and H_2O. As would be expected, the reactivity of a primary carbon is lowest, with a secondary carbon (bonded to two other carbons) more easily oxidized by the holes, and a tertiary most easily oxidized. Alcohols are oxidized to aldehydes[22] or, in the case of isopropyl alcohol, to acetone.[35,39,66]

In many experiments the adsorbed OH groups on TiO_2 are clearly involved in the photocatalyzed oxidation reaction. The experiments by Cunningham *et al.*[22] suggest the participation of the OH^- groups, if only as adsorption centers, for with deuterated alcohols, a proton appears in the photodesorbed species. Bickley and co-workers[35] suggest the participation of OH^- groups as hole traps. Filimonov[67] concluded that oxygen associated with the TiO_2 appears in the product when photocatalytically oxidizing alcohols and alkanes, suggesting that the oxygen comes from the OH groups at the TiO_2 surface. The participation of the $OH \cdot$ or HO_2^- groups in studies of photo-oxidation (chalking) of paints has been determined by Volz *et al.*[36] by ESR measurements, and by Pappas and Fischer[68] using an enzymatic/spectroscopic detection method.

The role of the photoproduced electron in the reaction, of course, is to reduce oxygen from the gas phase. Bickley and his co-workers conclude that the photocatalysis occurs entirely by a conduction band mechanism, with the partially reduced oxygen as an active site and the OH^- only active as a trapping center for holes. Pappas and Fischer[68] propose excited neutral oxygen molecules, arising during hole/electron recombination via oxygen, are the active intermediates. However, on TiO_2, as discussed above, substantial evidence indicates that the OH radicals from the hole reaction can be the species that first attacks the hydrocarbon, with the reduced oxygen becoming attached subsequently or being used to restore the OH groups at the TiO_2 surface.

The holes from most other substances that have been extensively studied do not oxidize water. In ZnO the holes can provide active oxygen by decomposing the ZnO. Holes from ZnO seem to have insufficient oxidizing power to attack alkanes or alkenes, but sufficient to attack alcohols.[69]

Tanaka and Blyholder[70] have shown that the hydrogenation of ethylene to ethane on ZnO is unaffected by illumination. In this case the *lack* of photocatalytic effects provides valuable information about the surface sites involved in the reaction. One concludes that the dark reaction probably does not involve electrons or holes and does not involve sites that are produced by minority carriers.

Volkenshtein[43] gives an extensive review of photocatalysis at the solid/gas interface where he cites examples of apparent photoactivity in hydrogen deuterium exchange as well as in the redox reactions we have chosen as examples.

Photocatalysis in solution is not qualitatively different from photo-catalysis at the gas/solid surface. In fact we discussed the photocatalytic activity of TiO_2 in solution earlier. Studies in solution have the advantage that the flow of the photoproduced carriers normal to the surface can be measured as an electrode current. Thus, for example, the photocatalyzed oxidation of formic acid, HCOOH, on ZnO in aqueous solution[71]

$$HCOOH + O_2 \xrightarrow[\text{ZnO}]{h\nu} CO_2 + H_2O_2 \qquad (9.35)$$

has been studied step by step using electrochemical techniques. (a) By current doubling measurements it was found that the formic acid is oxidized by the photoproduced holes. (b) If no oxygen is present, it was determined that the resulting radical ($H\cdot$ or $HCOO\cdot$) injects an electron into the conduction band of the ZnO, but if oxygen is present the radical donates its electron to (reduces) the oxygen. (c) The photoproduced electrons moving to the surface reduce dissolved oxygen to peroxide or to water, but if the hole produced radicals ($H\cdot$ or $HCOO\cdot$) are present, the reduction stops at hydrogen peroxide. In summary, by studying these processes step by step using electrical measurements, a fairly complete description of the reaction steps could be developed.

9.5. Direct Excitation of Surface States by Photons

In this section we examine injection of electrons into a solid from its surface where the energy for the injection stems from absorption of light by a surface species. We will first discuss briefly the results where the surface species is an organic dye molecule, for which case "dye injection" is well documented. Then we will compare optical absorption leading to electron injection by other surface states.

Electron or hole injection by excited species at the surface of a semi-conductor or insulator has practical implications, particularly in photog-raphy. Both in electrophotography using ZnO and in standard silver based photography, dyes are used to extend the wavelength of the photoresponse of the material into the desired region of the visible. In the case of ZnO electrophotography, the process desired is the injection of an electron into the conduction band of the ZnO. The injected electron moves through the ZnO, helping to discharge the electrostatic field at the ZnO surface. In the case of normal photography, the injected electrons reduce the silver in the emulsion. As an example that is reasonably well documented, we will briefly discuss the case of dye injection into ZnO. Experiments where the dye is adsorbed at the gas/solid interface and experiments where the dye is suspended or dissolved in a solvent (permitting electrochemical studies) have both been extensively used and the results show no indication of a change in mechanism with these different media.

It is not known in most cases whether the electron injection into the ZnO conduction band occurs from the excited level of the dye molecule or from another surface state energy level. The latter mechanism requires an energy exchange where the dye molecule becomes de-excited, transferring its energy to and exciting the surface state, so the electron from the surface state can transfer to the ZnO conduction band. This question is discussed in many papers and reviews.[19,72-74] The two possibilities are termed the "direct injection" and "energy exchange" mechanisms, respectively. It is clear from many studies that the wavelength response follows the absorption spectrum of the dye used, although a slight shift in wavelength may occur, depending on adsorption. Thus direct excitation from the ground state level (the surface state) of the dye to the conduction band of the semiconductor can be eliminated immediately as a possible mechanism for dye injection: the electron is excited first to the excited state of the dye molecule.

There is a substantial body of evidence indirectly supporting direct electron transfer between the dye and the ZnO. For example, Heiland and Bauer[75] find that under certain conditions a net flow of electrons from the ZnO to the ground state of the excited dye can occur, as observed by the net loss of conduction electrons in the ZnO. Pettinger et al.[76] came to a similar finding. These workers used a heavily doped ZnO electrode so that tunneling (see Figure 8.5 and the associated discussion) could occur between the ZnO and the dye. They observed, as expected, electron injection from the excited dye while the sample was illuminated. Then when the light was first extinguished a pulse of cathodic current could be observed associated with cathodic tunneling. They suggest this cathodic current is associated with electron transfer to the now oxidized dye. If the energy exchange mechanism is operative, the dye molecule never should become oxidized, so the result seems to suggest the direct injection mechanism. In both these latter experiments, however, the electron can be given to an excited dye molecule by the reaction

$$e^- + D^* \rightarrow D^- \tag{9.36}$$

rather than necessarily reacting with an oxidized dye molecule. A reaction similar to Equation (9.36) was observed by Yoneyama *et al.*[77] on TiO_2 with the dye methylene blue.

A more indirect but perhaps more convincing argument for the direct electron injection is the observation of supersensitization. Here it is found that the addition of a reducing agent as a coadditive with the dye molecules increases the efficiency of the dye. The suggestion is that the reducing agent gives up an electron to the ground state of the excited dye molecule and the excited electron, no longer able to return to its ground state level, has a higher probability of being injected.[72,78–80] If this mechanism of super-sensitization is valid, it suggests strongly the direct injection mechanism.

Dyes used for such photographic sensitization are selected because they have an exceptionally high extinction coefficient, and in some cases are strongly adsorbed on the surface. Freund[81] has shown that at least one species (ferrocyanide ion) with a lower extinction coefficient, and presumably not adsorbed, can inject electrons in the same manner. The "dye injection" was observed with the ferrocyanide ion in solution, and the injection current observed as an electrochemical current. In the present case the magnitude of the current was only slightly above the dark current, and the wavelength dependence was very difficult to determine.

The other class of direct excitation that has been reported and is of great fundamental interest is that of direct electron excitation from surface states into the conduction band (or excitation of holes into the valence band) of semiconductors. Photovoltage and photoconductance[82–87] are the detection techniques usually used, measuring the changes in these signals as a function of the wavelength. These techniques were discussed in Section 3.1.5, and caution in interpretation was recommended.

However, it is of interest to discuss in more detail the expected wavelength dependence of a surface state absorption, inquiring as to what is the state to which the electron is being excited, assuming direct excitation to the excited state of the surface complex. The point is that the models of Chapter 5 suggest that if the sorbate/solid interaction is very strong, the excited state may meld into the band.

Consider Figure 5.3 and assume the LDS shown for the solid represents the conduction band of the semiconductor. By this model the shape of the excited state of the surface group that is absorbing the photon is represented in Figures 5.3a–5.3d. Figures 5a and 5b represent the expected states (ignoring their intrinsic broadening) of a dye—in other words, we expect the interaction between a large dye molecule and the solid to be weak, and therefore there will be little mix between the bands of the solid and the orbital of the excited state of the dye. In this case, the optical absorption spectrum is characteristic of the free dye molecule, or only slightly perturbed[88] from it. On the other hand, if the surface group is strongly affected by the bands of the solid, or even is associated with intrinsic surface states (such as a donor surface state associated with a surface sulfur ion on CdS), Figures 5.3c and 5.3d may apply, and the absorption spectrum observed by its photovoltage spectrum may indeed show a maximum for the photon energy needed to raise the electron to an energy near the band edge. In this latter case (or in the case of direct excitation to the conduction band) surface photoconductance or photovoltage measurements may therefore directly reflect surface state levels relative to the band edge.

We believe further work is needed to clarify exactly what absorption process is being measured in these complex but intriguing studies.

10 | Surface Sites in Heterogeneous Catalysis

10.1. General Concepts

10.1.1. The Role of the Catalyst

A particularly instructive application of the theory of surface states and surface sites is provided by the technology(ies) of heterogeneous catalysis. Although broad generalizations can be made,[1,2] in this discussion our interest is limited to reviewing the various forms of surface sites on a solid, reviewing adsorption on these sites, and in general relating the models that have been developed in the preceding chapters to catalytic observations. The objective is to show, on the one hand, that the surface site (state) concepts are and will be useful in the understanding of catalysis, and to show, on the other hand, that the extensive catalytic literature is a valuable source to provide a deeper understanding of the great variety of possible adsorbate/adsorbent reactions. With this objective in mind, the emphasis in this chapter will be on oxidation reactions. The active sites (acid and base, metal surface site) involved in bond rupture (cracking), bond shift (isomerization), and hydrogenation/dehydrogenation reactions will be reviewed. But the area most illustrative of the many types and actions of surface sites is found in the work on oxidation reactions.

The objective in catalysis is to increase the rate of a specific chemical reaction. The role of the catalyst may be fourfold: (a) to provide active sites (Chapter 4) that can dissociate molecules and present a chemically active species to the coreactant, (b) by adsorption or absorption (Chapters 5 and 7) to concentrate and provide a reservoir of active intermediate species,

(c) to provide a reservoir of electrons if needed (Chapters 7 and 8), particularly in the case of redox reactions, and (d) to provide a large density of active sites per gram of catalyst (Chapter 6). Catalysts with these characteristics will be "active," increasing the reaction rate. In addition there is the further desired function (d) to provide a uniform activation of the adsorbate, in a form such that only the specific reaction occurs that leads to the desired product. Catalysts with this attribute are termed "specific."

In Chapter 5 a sharp distinction was made, when describing an adsorbate on a solid, between a "surface molecule" picture and a "rigid band" model. In the surface molecule picture it was assumed that an entirely microscopic picture was appropriate, with chemisorption controlled by such variables as coordinatively unsaturated bonds available at the surface, with the bonds described by local atomic orbitals.[3,4] In the rigid band model it was assumed that an entirely macroscopic picture was appropriate, with chemisorption controlled by such variables as the Fermi energy, or the shape of the appropriate band.[5] It evolved that neither extreme view was entirely accurate, the real adsorbate/solid interaction is always a compromise, as illustrated by Figure 5.3 and the accompanying discussion. Unfortunately, in catalysis literature to date this compromise between the two extreme models has not been utilized. In the present discussion, therefore, following the literature, the models used will tend toward one or the other extreme. In intermediate cases where both models are useful, both will be presented and compared.

10.1.2. Some Correlations in Heterogeneous Catalysis

The development of new catalysts to accelerate specific reactions is approximately 10% theory, 50% experience and intuition, and 40% empirical optimization. As our knowledge of surface science grows, we can hope that these percentages will shift to the left in that tabulation, but at present heterogeneous catalysis is so complex that detailed analysis of even the simplest reaction is almost impossible.

Because of this lack of fundamental knowledge, broad correlations have been valuable to try to relate catalytic activity and selectivity to many properties of the solid. Some of these properties are undoubtedly independent variables directly related to catalytic activity. For example, it will be concluded that the acid strength of surface sites available has undoubtedly a direct relation to activity of a catalyst in certain hydrocarbon cracking reactions (carbon/carbon bond dissociation). The availability of active oxygen on the surface undoubtedly has a direct relation to the activity of many oxidation reactions (although in many cases the exact active oxygen species may not be identified). On the other hand, many properties of the solid may be related to its catalytic activity in an indirect way. For example, the electronic properties of a solid may indirectly determine the presence and activity of oxygen on its surface. At our present state of knowledge even indirect correlations are valuable.[6] They point the way to future, improved, theories, and in the meantime they can be included as guidelines helping in the "experience" and "intuition" aspects of catalyst design.

In this section we discuss example correlations, direct or indirect, between general properties of catalysts and their catalytic behavior.

10.1.2.1. The Volcano Relationship.

If a reactant is adsorbed too strongly, it forms a stable surface complex and is nonreactive with other species coadsorbing or impinging from the gas phase. If, on the other hand, the reactant is weakly adsorbed (e.g., physisorption, where the adsorption is so weak that the molecules are not dissociated), the reactant may not be presented in an active form. Thus it is to be expected that at some intermediate adsorption energy the adsorbed reactant will be at its most active. If we plot the activity against an appropriate energy parameter, we expect a maximum at some intermediate value (a "volcano" shaped curve). The Sabatier–Balandin volcano model[7,8] showed that for a metal catalyst if one plots the heat of formation of a metal oxide on the abscissa against the activity of that metal as a catalyst in oxidation reactions on the ordinate, one gets a volcano shaped curve.

The concept is reviewed by Emmett,[9] Appleby,[10] and Parsons[11] for electrocatalysis. Vijh[12] has suggested a relation between the Sabatier–Balandin concept and the electron theory (Section 10.1.2.3 below) through the relation between the heat of oxide formation and the semiconductor energy gap.

In later discussions we will be considering the redox character of the cation in the catalyst. This is a characteristic closely related to the metal/oxygen bond energy. It is of interest that Gale et al.[13] observe an optimum intermediate redox potential for ions used as catalysts in homogeneous oxidation reactions. Again a volcano effect appears; the catalyst should interact neither too strongly nor too weakly with the reactants.

10.1.2.2. Models Associated with d-Band Filling.

The so-called "electronic theory of catalysis" originated with Dowden,[14] who noted the possible importance of partially filled d bands in solids, and suggested that only with partially filled bands could covalent bonds be formed to adsorbates. The behavior of transition metals in providing active

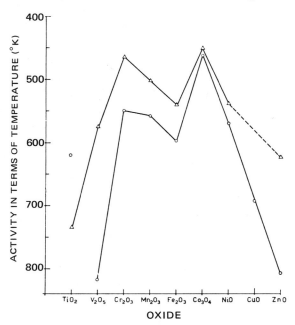

Figure 10.1. Activity of oxides of the first long period. The circles represent activity expressed as temperature (°K) required to oxidize 80% of hydrocarbons (the hydrocarbons considered were 2-pentene, n-pentene, isopentene, n-hexane, 2,3 dimethylbutane, and cyclohexane). The triangles represent activity expressed as the temperature required to oxidize 5% of NH_3.

sites is, however, not so simple, as discussed earlier in Section 5.2.2 and below in Section 10.4. Ligand field effects, bond direction and symmetry problems become involved in determining catalytic activity on transition metals and their oxides, and anion effects (Section 10.5.2 below) are involved in determining the activity of transition metal oxides. Nonetheless, there is still evidence that there may be in some cases a correlation between selectivity[15] or activity[16] and the occupancy of the d bands. Figure 10.1, from Giordano[17] is very suggestive of such a correlation. The activity is only high in the region of partly occupied d bands. The minimum in the middle of the transition metal series is associated with crystal field effects, to be discussed in Section 10.4 below.

10.1.2.3. The Rigid Band Model in Catalysis. By the "electronic factor" is sometimes meant the occupancy of the bands in the solid, and such theories of the "electronic factor" adhere strictly to the rigid band model of the solid surface. We will use the term "rigid band" as has been used in earlier chapters of the book, to describe effects in which local bonding is substantially ignored. Rigid band effects include, in the most general case, the partially filled d-band models described in the last subsection, but in the main its proponents are interested in catalysis on semiconductors.

Theories of catalysis based on the rigid band model[18] have been developed primarily for catalysis by Volkenshtein,[19,20] Hauffe,[21] Morrison,[22] Lee and his co-workers,[23,24] and Freund and Gomes.[25] As will be discussed below, the rigid band model is generally of most value in describing relatively inactive n-type catalysts such as zinc oxide, rather than the more active commercial catalysts.

However, if we consider the rigid band model in a slightly broader context, and relate the band properties of semiconductor solids to their catalytic activity, an interesting empirical correlation is found, suggested originally by Stone's group,[26] and reviewed by many authors, for example, Giordano[17] and Moro-Oka and Ozaki.[27] This is the observation that, in many reactions involving oxygen, p-type oxide semiconductors (where holes are readily available) are most active, insulating oxides intermediate, and n-type oxide semiconductors least effective. The correlation is not precise: the most active catalyst for one reaction will be active, but not necessarily the most active for others. Giordano however presents a long list of examples and references, showing the correlation follows quite generally for N_2O decomposition, for CO oxidation, for O_2 exchange or atom recombination, and for oxidation of many hydrocarbons.

The correlation may be related to another correlation to be described below (Section 10.5): the most active oxidation catalysts are in general those that can exchange lattice oxygen with the reactant. The probable connection between this property and the availability of holes will be discussed in Section 10.5.

10.2. Surface Sites Associated with Steps and Other Geometrical Factors

The influence of surface imperfections such as steps, edges, kinks, and terrace widths on adsorption was discussed in Section 6.3, where it was pointed out that such imperfections lead to coordinatively unsaturated surface atoms or groups of atoms that can form multiple bonds to adsorbates.

There are two classes of measurements that have been used to relate adsorption and catalytic activity to the presence of steps and the associated sites. One is the measurement of differences in activity for clusters, as was discussed in detail in Section 6.3. The other measurement is the measurement by the surface spectroscopies of adsorption and reaction on stepped surfaces of single crystals. Such measurements have received much attention since (as discussed in Section 3.2.7) the preparation of stepped surfaces and measurement of step density has become possible. It is found that if a single crystal is cut at a slight angle to a low index plane such as the (100) plane and annealed, the resulting structure, detected by LEED, is a series of steps and terraces. Thus the influence of crystalline steps can be deduced from the differences between stepped and nonstepped surfaces.

For example, Bottoms and Lidow[28] have shown stepped surfaces of ZnS adsorb CO and O_2 much more rapidly than do unstepped, and there is evidence that CO even dissociates at a surface step. Somorjai and his co-workers[29,30] have investigated both the adsorption of hydrocarbons and the adsorption of inorganic vapors on Pt, and shown that the presence of steps controls the rate in many cases. In catalytic studies, Saltsburg et al.[31] have shown the reaction

$$H_2 + D_2 = 2HD \tag{10.1}$$

proceeds at a negligible rate on a planar surface of platinum, but proceeds readily on a stepped surface.

For completeness, two other forms of heterogeneity should be mentioned: the edges of adsorbate islands[11,32,33] and the ends of dislocations. The edges of adsorbate islands (Section 7.1.3) are similar to a crystallographic step in offering attractive sites and also offer coordinatively unsaturated molecules of the reactant itself. Thus May[33] refers to the clean-off of oxygen at the edges of oxygen islands in the catalytic oxidation of hydrogen on nickel (110), providing an example of extra activity associated with the edge atoms of adsorbate islands. Giles and his co-workers[32] find a maximum activity with 1/2 monolayer of Pt on a mercury surface, with the activity decreasing as the monolayer is completed, suggesting active sites at the edge of adsorbate islands.

Thomas[34] has presented evidence that dislocations in solids can act as active centers. He cites work, for example, showing that the disproportionation of CO

$$2CO \rightarrow C + CO_2 \tag{10.2}$$

will proceed preferentially on dislocations. The effect is easily demonstrated: the carbon deposit at the dislocations can be observed after the reaction is completed. There is little doubt that attack of the crystal itself by a corrosive agent often preferentially occurs at the end of a dislocation, and such reactions are used to determine dislocation densities in crystals. However, the importance of dislocations in catalysis is probably not great, primarily because the density of sites is low, and indeed with catalyst clusters less than 100 Å in diameter, the stability of dislocations becomes questionable.

Finally, the reactivity on different crystal faces will often differ, and heterogeneity in reactions can be associated with various planes exposed. Sanders[35] has reviewed such effects expected for metal catalysts. In another review article, Siegel[36] describes hydrogen addition to (hydrogenation of) unsaturated hydrocarbons and relates the influence of the geometry of the adsorbate and the surface.

As an interesting specific example of crystal plane effects, Kolboe[37] in studies of dehydrogenation of isopropyl alcohol on powdered ZnO was able to analyze TPD curves and deduce that five different free energies were needed to explain the process. He concluded there are five different sites. In later work (the referenced article), he used acicular ZnO, dominated by one plane only, and found that now 75% of the reaction could be attributed to one particular site. Thus the heterogeneity observed earlier was shown to be associated with the exposure of several crystal planes.

10.3. The Role of Acid and Basic Sites in Catalytic Reactions

As reviewed in detail by Tanabe,[38] acid and base sites on an ionic solid are highly active in many catalytic reactions. We have discussed the origin of such sites in Section 4.2.3. Acid sites, with their high electron affinity or their reactive protons, can transform hydrocarbons into reactive forms by removing hydride ions, H^-, from the molecule, or conceivably donating protons, H^+ to the molecule. In either case the resulting positively charged hydrocarbon fragment is termed a carbonium ion.

Reactions that occur with acid sites present include skeletal isomerization (involving breaking and re-forming carbon/carbon bonds), cracking (decomposition of hydrocarbons), isomerization (movement of a double bond), and the dehydration of alcohols. Tanabe gives results for specific cases that suggest that in the above listing, the required acid strength of the centers decreases in the order listed. That is, dehydration of alcohols can be observed with weak acid sites, but skeletal isomerization of alkenes requires strong acid sites. A listing of the many acid initiated reactions is well beyond the scope of this book.

Both Brønsted and Lewis sites are active in catalytic reactions.[39–41] However, with the different mechanism between a Brønsted acid (possibly a proton donor) and a Lewis acid site (cannot be a proton donor), it is sometimes found that Lewis acids and Brønsted acids catalyze different reactions. For example, on a silica/alumina catalyst, only Brønsted acids will polymerize propylene,[42] whereas on the same catalyst the decomposition of isobutane correlates with the density of Lewis sites.[43] Ballivet *et al.*,[44] by selective poisoning of Brønsted and Lewis acid sites (using Na^+ exchange to remove Brønsted centers and using NH_3 to poison both Brønsted and Lewis centers), were able to show that both sites were active, but with different rates and different detailed mechanisms, in the isomerization of cis-2-butene on silica/alumina catalysts.

Basic sites, on the other hand, cause carbanion formation when a hydrocarbon molecule is adsorbed. The negative carbanion is obtained normally by extraction of a proton from the hydrocarbon. With very strong basic sites, particularly those associated with alkali atoms on an inert support,[38,45] the basic site will be in the form BH, where B is the alkali atom, and in this case the carbanion can form by the donation of the H^- to the hydrocarbon.

Activation of hydrocarbons by carbanion formation can lead to isomerization, or with strong basic sites can lead to polymerization or alkylation (the addition of extra carbon/hydrogen groups to the hydrocarbon). Both Tanabe[38] and Pines[46] provide excellent reviews of the details of the reactions of carbanions and the reaction steps leading to the polymerization or alkylation reactions. Pines, in particular, describes reactions on very active basic sites.

Baird and Lunsford[47] have identified the sites associated with basic activity in their studies as corner oxide ions, and made some generalizations regarding acid or base catalyzed isomerization reactions that are instructive. Consistent with the discussions (Chapters 4 and 6) of heterogeneity, these authors suggest O^{2-} ions on an edge or the corner of a cube is stable enough to exist (the Madelung potential is sufficiently high), but is relatively unstable. Therefore such corner O^{2-} ions provide strong basic sites, with the electron easily given up. They estimate the number of sites to be expected due to such edge or corner ions and find agreement with the number of very active sites in their isomerization reaction. The reaction they examined is the isomerization of 1-butene to form *cis*- or *trans*-2-butene. They, and others, find that basic catalysts tend to form the *cis*-2-butene. They relate the stereochemical preference to the fact that the shape of the cis butene should provide a better fit over the protruding corner O^{2-}. They suggest the cis molecule, adsorbed as a carbanion, "drapes" in a natural way over the corner oxygen ion. On the other hand, the trans form is less energetically favorable because, when adsorbed over a corner ion, one of the methyl groups must lose contact with the surface.

Double sites, with an acid and a base site closely adjacent, are to be expected, as discussed in Section 4.3.3, and are found to be catalytically active.[48–52] For example, such a double site can lead to a dehydrogenation as suggested by Kibley and Hall[50] where a basic site removes an H^+, and a neighboring strong acid site removes an H^-, so the neutral hydrocarbon, now dehydrogenated, can desorb. Pines and his co-workers[53] conclude that adsorbed alcohol molecules, one on a basic site and one on a neighboring acid site, can lead to ether formation. The oxidation of methanol on iron molybdate, reviewed in Section 10.6.2, provides another example where dual sites seem active. Such action is termed "acid/base bifunctional catalysis" and is reviewed by Tanabe.[38]

10.4. Covalent Bonding to Coordinatively Unsaturated Metal and Cationic Sites

Activation of reactants by local bonding to coordinatively unsaturated sites associated with metal atoms or cations at a surface will be reviewed very briefly in this section. A coordinatively unsaturated site (Section 5.2.2.3) is simply an unoccupied position oriented relative to a surface atom such that the lobes of dangling orbitals will facilitate bonding in a natural way. Thus the term "coordinatively unsaturated sites" is very broad and we will discuss sites where a simple covalent bond is expected, where π and multiple bonding can occur, and sites where the more complex ligand field effects will be anticipated.

On oxide catalysts covalent bonding is not uncommon, particularly on transition metal oxide catalysts that are not strongly acid or basic. In such cases olefins are activated by adsorption as covalently bonded radicals rather than by adsorption as carbonium ions or carbanions.[54,55] Thus, for example, the adsorption of propylene often occurs with hydrogen atom extraction:

$$\underset{\substack{| \\ H}}{\overset{\substack{H \quad H \\ | \quad |}}{H-C-C}}=\overset{\substack{H \\ \diagup}}{\underset{\substack{| \\ H}}{C}} \quad \rightleftharpoons \quad H\cdot \; + \; H-\underset{\substack{}}{\overset{\substack{H \quad H \quad H \\ | \quad | \quad |}}{C\cdots C\cdots C}}-H \qquad (10.3)$$

where the resulting hydrogen atom and the allyl radical are each bonded to a site on the surface. The $H\cdot$ may bond to an $O_L{}^-$ on the surface, as discussed in the next section, leading to an OH^- group, while the allyl radical is bonded to a transition metal ion that can provide d orbitals for covalent bonding.

Activation of reactants by heteropolar covalent bonding to metal ions at an oxide surface is also observed. Bonding of a strength similar to a metal formate was observed by Criado et al.[56] during the decomposition of formic acid on a series of metal oxides. Kubokawa et al.[54] report ir measurements that show propylene on ZnO at low temperature forms a π bonded allyl structure, but as the temperature is raised, the adsorbate converts to formate or acetate groups bonded as anions to the surface.

Similarly, on covalent and metallic solids activation of sorbates by such heteropolar covalent bonding can be observed. Schwartz and Madix[57] conclude such interaction occurs on germanium with oxygen or chlorine adsorbates. Frennet and Lienard[58] examine the model of the interaction of an adsorbate with a metal surface, emphasizing bonding to form a compound-like layer on the surface.

In general, however, the most active metal catalysts for hydrocarbon reactions are those that form a bond that is almost homopolar. Kemball[59] shows that in activation of hydrocarbons by covalent bonding to such a metal surface, many radical fragments can coexist. Experimentally he emphasized the simplest hydrocarbon, methane. He measured the deuteration of the CH_4, and found the product distribution showed an extremely high yield of CD_4. From this observation he was able to conclude that all carbon–hydrogen fragments coexist on the surface, with the dissociation steps occurring:

$$CH_4 \rightleftharpoons CH_3 + H \rightleftharpoons CH_2 + 2H \rightleftharpoons CH + 3H \qquad (10.4)$$

Transition metal catalysts provide highly active sites for hydrogenation/dehydrogenation or isomerization reactions. The mechanism will not be reviewed in detail. Broadly speaking it is reasonable that a hydrogen atom from the hydrocarbon can change its covalent bond from the carbon–hydrogen bond of the molecule to a metal–hydrogen bond, leaving a radical form for both the hydrogen and the hydrocarbon fragments.[60] The latter can perhaps be π bonded. Then isomerization (movement of double bonds) can easily follow, for example, during alternations of σ and π bonding. Such a role of π bonding in providing active adsorbates in hydrogenation reactions of olefins or aromatics on transition metal surfaces was introduced in Section 5.2.2.2. More details can be found in reviews by Siegel,[36] Bond and Wells,[61] Garnett and Sollich-Baumgartner,[60] and Garnett.[62]

Experimental studies of aromatics (ring compounds) using LEED and work function measurements has been made by Somorjai's group.[63] This work showed π bonding of aromatics to platinum is common. The possibility of π bonding of olefins to tungsten oxide is discussed, for example, by Pennella et al.[64]

Activation of adsorbates by ligand (crystal) field effects on transition metals can be expected when polar or charged molecules occupy an available coordination site. The adsorption of CO on tungsten was discussed in detail in Section 5.4.3, where it was shown that many forms of interaction were observed depending on the temperature, pressure, and history of the sample. Dowden[16] has reviewed crystal field effects at surface sites on transition metal oxides and sulfides, showing that crystal field effects may occur not only with highly polar adsorbates such as CO and OH$^-$, but perhaps even with hydrogen adsorption on oxide catalysts. He shows, in accordance with usual crystal field theories, that the contribution of crystal field effects to the bond energy has two maxima as the d levels are filled, with a minimum effect with five d electrons per atom. The results of such a correlation on catalytic activity are shown in Figure 10.1. Thus crystal field effects in bonding may, in some cases, have an important contribution to the activity of surface sites.

Finally, this section on local bonding to activate a reactant must include some comment on the activity of sites on transition metal atoms or ions in relaxing symmetry difficulties in a catalytic reaction. The problem is reviewed by Mango.[65] If a reaction is desired between two molecules that cannot interact because the occupied bonding orbitals in each molecule have directions that correspond to antibonding orbitals in the other, the interaction becomes forbidden due to symmetry problems. The action of the directed dangling orbitals on a transition metal catalyst can be to provide electron density to the antibonding orbitals in each of the reactant molecules. Then the reactants interact with each other; there can be bonding overlap of the now partially occupied orbitals. The symmetry problem is thus relaxed.

10.5. Sites in Oxidation Catalysis

10.5.1. Introduction

In oxidation catalysis the surface must provide sites for many functions. One function is oxygen exchange, by which we mean the adsorption of gaseous oxygen, its conversion to an active form, and its removal as part of the product. If oxygen is reduced when it is adsorbed, as expected on a highly ionic solid, the catalyst must provide electrons for this reduction. When the oxidized product is desorbed, the electrons must be recovered before the product can leave as a neutral molecule. Thus a second function is electron exchange. A third function is that of providing bonding sites for the reactants and intermediates, sites (as discussed relative to the volcano concept of Section 10.1.2.1) that bond strongly enough to activate the adsorbates but not so strongly that a stable complex forms.

Several of the functions may be provided by the same site. Anion sites are a good example of sites showing multiple functions. As will become obvious in Section 10.5.1, where anions and anion vacancies are discussed, the anionic site can act not only in anion exchange but also in adsorption and electron exchange. Some cationic sites also can show a dual function. It is observed that almost all good oxidation catalysts have sites available at the surface associated with single valency changes of the cation[55] (Cu^+/Cu^{2+}, Mo^{5+}/Mo^{6+}, etc.), and such "redox sites" are often considered highly active both as adsorption sites and in electron exchange.

In the subsections to follow, we will discuss the details of the various types of sites involved in oxidation catalysis and how the sites become involved in the reaction steps. However, before going into these details, it is appropriate to make a few comments about general models.

First, the role of the Fermi energy should be brought into perspective. The Fermi energy has been cited as a dominating feature in catalysis because it controls the availability of electrons. However, it should be noted that in general the Fermi energy is an effect, not a cause. If there is a large density of partially occupied energy levels at the surface or in the bulk, the Fermi energy will be located (pinned) *at the energy of these levels.* Such pinning is apparent for surface states from Figure 2.2 (where E_t is considered a partly occupied surface state) or from Equation (6.1), where it is clear that if both occupied and unoccupied states of a redox system are present, the Fermi energy is firmly pinned.

Thus the concept of a redox system controlling the exchange of unpaired electrons, and the concept of a Fermi energy at the surface of the solid controlling electron exchange, are simply two descriptions of the same process. At times the Fermi energy concept is particularly useful because it is a measureable property and its measurement permits determination of the energy of the levels of the redox system which dominates the catalyst surface.

A second general consideration of particular importance when discussing sites for electron exchange is the question of whether an adsorption process should best be pictured in terms of the surface molecule model or in terms of the rigid band model. As we move from Section 10.5.1.3 to Section 10.5.2 and then to Section 10.5.3, we will be progressing from cases where we are considering highly localized imperfections that normally are best described in terms of a surface molecule picture, through levels associated with the host cation that can be described either as localized surface state levels or as a narrow band, to electron exchange with levels that can only be described well by starting from a rigid band model.

The three cases are indicated in Figure 10.2. Figure 10.2a shows a band model that describes local imperfections or other sources of surface levels. The pinning of the Fermi energy is shown clearly by such a rigid band model, but otherwise it is wholly inappropriate as a good description of an adsorbate/surface state interaction. Localized levels as in Figure 10.2a can be associated with cations as described in Section 10.3.3.1—for example, surface states associated with surface Cr^{3+} on Cr_2O_3. Other examples would be ionic surface states such as surface O^- or foreign impurities at the surface. Figure 10.2b shows a narrow partially occupied band of levels, perhaps, for example, associated with a particular d band of the cation. The bonding to such cations at the surface will usually be described best by a surface molecule model that emphasizes directional characteristics of the d orbital. But to illustrate some of the characteristics, the rigid band diagram as in Figure 10.2b is useful. Examples where Figure

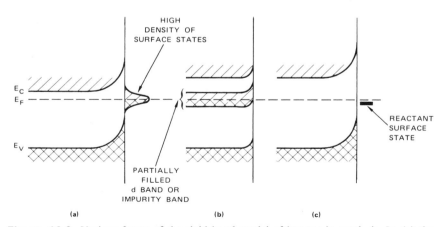

Figure 10.2. Various forms of the rigid band model of interest in catalysis. In (a) the surface Fermi energy is pinned by surface species (surface states) on the catalyst, in (b) by a partly filled narrow band such as a d band, and in (c) only by the reactant itself.

10.2b may provide a useful description are discussed in Section 10.5.3.2, and include many transition metal oxides such as Co_2O_3 and V_2O_4 and may include impurity bands associated with vacancies. Figure 10.2c shows a case where the electron exchange is with a broad conduction band of a solid, and there is little local interaction. In this case, as will be discussed in Section 10.5.4, the rigid band model is the best first approximation to a description of the system. Examples where Figure 10.2c provides the best description include primarily n-type semiconductors such as ZnO. The use of the rigid band model for other cases must be made with allowances for local effects.

In practice it will be found that very seldom can all the important sites on a given catalyst be described as clearly belonging to one of the three classifications of Figure 10.2, and the utility of Figure 10.2 is primarily to help classify the extremes in the reader's mind.

As a final point in the introduction, it is important to point out that there are two objectives associated with catalytic oxidation. In general, a given catalyst will be commercially utilized for one or the other objective but not both. One objective is the total combustion of the reducing agent, where, for example, an olefin is oxidized completely to CO_2 and H_2O. Such catalysts are needed, for example, as automobile exhaust catalysts. The other objective is the partial oxidation of a reactant, where the desired product is an intermediate species that is only partially oxidized. For example, one may wish to oxidize ethylene, C_2H_4, to the more expensive ethylene oxide, C_2H_4O. In the partial oxidation case we are particularly interested in the yield or the selectivity of the catalyst (respectively, the amount and the fraction of the desired product obtained) where in the complete combustion case we are only interested in the activity.

It is usually not clear why the oxidation process is able to stop at a single oxygen atom addition to an organic molecule. A probable cause in many cases may be that the number of active sites per unit area for oxidation is limited. Complete combustion results if the hydrocarbon is attacked at many points simultaneously, but if there is an isolated site offering a single oxygen atom, the reaction will stop after that atom is extracted from the catalyst. Another reason that may often contribute is the formation of a stable, desorbable intermediate, that desorbs as soon as it is formed. Another possible reason that has been offered and will be reported is that certain sites lead to complete combustion and other sites to partial oxidation. This may be related to a limitation in the density of active oxygen sites. Thus, for example, we will indicate the possibility that double-bonded oxygen bonded to the cation M as $M{=}O$ will, if it interacts with an olefin, lead to partial oxidation; whereas oxygen that has bonds to two cations will, if it interacts with an olefin, lead to complete combustion.

10.5.2. Oxygen Exchange Sites in Oxidation Catalysis

There are two main forms of oxygen that a metal or metal oxide catalyst can offer to an adsorbing reducing agent. One is best referred to as lattice oxygen, with the oxygen formally in the -2 oxidation state, and with the surface relocated into a structure resembling the bulk metal oxide. As discussed in Section 5.2.4, the distinction between adsorbed and lattice oxygen is not always clear. For the purpose of the present discussion we will consider the species lattice oxygen if anion vacancies result when the oxygen is removed. The second type is oxygen ionosorbed on the metal oxide surface. The ionosorbed oxygen will be formally in the oxidation state O^-, O_2^-, or perhaps O_3^- (see Section 7.2.1). Relocation in varying degrees will occur, but in general it is unrewarding with present knowledge to attempt a finer classification of oxygen sites depending on the chemical group formed by relocation.

In some cases it is not possible to determine the exact form of the oxygen that combines with the reducing agent in the reaction. In some cases, however, it can be shown that ionosorbed oxygen in some particular oxidation state is active, and this will be the subject in the first subsection below. In most cases it can be shown that oxygen that is extracted from lattice sites is the reactive species. These cases will be discussed in the second subsection and evidence will be presented regarding the reason for the activity of the lattice oxygen.

10.5.2.1. Ionosorbed Oxygen as the Active Form. On noble metal oxidation catalysts, one seldom considers the oxygen ionosorbed. In general on noble metal catalysts (Pt will be discussed in Section 10.6.1) the reactive oxygen can best be considered somewhere between lattice oxygen and covalently bonded adsorbed oxygen. There is at least one exception, and that is the case of oxygen on the catalyst silver, the oxygen that reacts with ethylene to form ethylene oxide. In this case the active species is the "ionosorbed" superoxide ion. The reaction

$$C_2H_4 + O_2^- \rightarrow C_2H_4O + O^- \qquad (10.5)$$

occurs. The residual O^- species resulting from reaction (10.5) are too reactive for ethylene oxide production and oxidize the ethylene all the way to CO_2 and water.[66–68]

On semiconductor catalysts, the high activity of the ionosorbed O^- species compared to the superoxide O_2^- species was discussed in Section 7.2.2.1. This much higher reactivity of the O^- species seems general. For example, the high reactivity of the adsorbed O^- species is observed again in the work of Herzog.[69] Herzog finds in propylene oxidation on Cu_2O that if N_2O instead of O_2 is used as a reactant, so that O^- is rapidly produced by N_2O decomposition, then CO_2 is produced rather than the desired acrolein. (Lattice oxygen is considered to produce acrolein.) Gravelle and Teichner[70] suggest O^- is often the active site in the CO oxidation on nickel oxide, and postulate intermediate groups such as CO_3^- can also be active. At high temperature they postulate the mechanism changes to a redox reaction where lattice oxygen must be involved, with the NiO being reduced by the CO and reoxidized by the O_2. Kubokawa and co-workers[71,72] suggest adsorbed oxygen on ZnO acts as the active center in the oxidation of hydrocarbons (from C_2 to C_4), with complexes between the preadsorbed oxygen and the impinging olefins acting as an intermediate. Nikisha et al.[73] find O^- on V_2O_5 provide active sites for the adsorption of O_2, leading to isotopic exchange. The O^- is also highly reactive with H_2 or CO, as indicated by the observation that adsorption of H_2 or CO stops the exchange reaction.

10.5.2.2. Lattice Oxygen Exchange.

On almost all good oxidation catalysts the lattice oxygen is given up to the reactant, to be replaced in another step of the reaction[74–79] by absorption of oxygen from the gas phase. For example, in the catalytic oxidation of NH_3 over Mn, Co, Cu, Fe, and V oxides, such a reduction then oxidation of the catalyst is observed.[80] Popovskii, Mamedov, and Boreskov[81] show for a long series of catalysts that the initial rate of catalyst reduction by the reducing agent (with no gaseous oxygen present) compares well with the rate of catalytic oxidation of the reducing agent. Again, Breysse et al.[82] show that CO oxidation over CeO_2 depends on such an exchange of lattice oxygen.

In this subsection models of the active site presented by the lattice oxygen, the oxidation state, and location of the lattice oxygen at the time it forms its bond with the reducing agent will be examined.

In the case of noble metal catalysts where the metal/oxygen bond is highly covalent and weak, the oxygen is easily released by the surface. The function of the noble metal is often simply to provide a bond just strong enough to dissociate the oxygen, forming an oxide that is on the verge of spontaneous dissociation at catalytic temperatures. Oxygen is thus "stored" in this weakly bonded reactive form until a CO or a hydrocarbon molecule impinges from the gas phase, forms a strong bond, and removes the oxygen (see Section 10.6.1).

In the case of metal oxides as catalysts, on the other hand, the oxygen is effectively charged and the disposition of the charge must be considered. With a strong ionic (heteropolar) catalyst, the partial charge on the lattice oxygen is high (or equivalently the cationic conduction band is at a high energy), and the extraction of lattice oxygen requires the breaking and reforming of heteropolar bonds. In particular, some of the steps must involve removal of valence electrons from the oxygen ion as it is extracted from its lattice site. For example, when a lattice oxide ion is added to CO to form CO_2, the electrons that were associated with the lattice O_L^{2-} must be absorbed by the solid at some stage in the reaction.

Three broad categories of active forms are possible: the lattice oxide ion (O_L^{2-}, the occupied surface state); an oxide ion with an electron missing, so that half the bonds are broken (O_L^-, an unoccupied ionic surface state); or an oxygen atom O unbonded to the ionic solid substrate. These three broad categories will be discussed in some detail. Each of these categories in turn leads to a spectrum of different possible sites, for the attractive site could be the oxygen in the appropriate formal oxidation state but relocated in many ways. The configuration could be anywhere between the one extreme with the oxygen occupying the normal lattice site and the other extreme with the oxygen relocated to the surface as an adsorbed species with the anion vacancy diffused away.

Lattice oxygen ions of formal valence -2 could form strong bonds to hydrocarbons (or to a hydrogen ion from a hydrocarbon) through an acid/base interaction, providing strong enough bonds that the oxygen atom can be removed from the solid. In some cases acid or basic sites have been correlated to redox activity,[83,84] but in general very strong acids or very strong bases are poor oxidation catalysts.

Tarama et al.[85] and Trifiró and Pasquon[86] have made an interesting suggestion that double-bonded lattice oxygen could provide an active adsorption site, forming covalent bonds even though the oxygen is in the formal -2 oxidation state. Trifiró and Pasquon view the interaction of a hydrocarbon RH with the double-bonded surface oxygen as follows:

$$[M{=}O] + R\text{H} \rightarrow M\begin{array}{c} \diagup\ O{-}H \\ \diagdown\ R \end{array} \qquad (10.6)$$

where M represents a cation of the oxide catalyst. However, it should be pointed out that the formation of such an adsorbed complex depends on the availability of unoccupied coordination sites. Whether the surface complex be of this form (Equation 10.6) or another, the double-bonded oxygen interacting with the hydrocarbon is considered to lead to partial oxidation. By this model there are two classes of oxide catalysts. The oxides MnO_2, Fe_2O_3, Co_3O_4, NiO, CuO, Cr_2O_3, and $CdMoO_4$ provide no double-bonded oxygen and in general oxidize hydrocarbons completely to CO_2 and H_2O. The oxides V_2O_5, MoO_3 and a series of molybdates (Fe, Co, Ni, Mn, Bi), all have ir absorption, which suggests they have double-bonded oxygen, and all are partial oxidation catalysts. The double-bond character of the oxygen is determined in these studies by the appearance of an ir band between 900 and 1000 cm^{-1} which is known to arise from double-bonded oxygen ions in solution complexes according to Bairaclough et al.[87] Akimoto and Echigoya[88] and Weiss et al.[89] also discuss the possibility that double-bonded oxide ions are active species in oxidation, although Akimoto and Echigoya consider the $M{=}O$ group to be an acid center. Boreskov[90] suggests an analogous model, relating the activity to the partial charge on the oxygen without specifying the nature of the bonds.

Oxygen in the formal -1 oxidation state is another attractive possibility for an active site. The unoccupied ionic surface states O_L^- are essentially radicals, and are expected to react strongly to form covalent bonds with impinging gas molecules. For example, one expects a strong bond to a hydrogen atom, perhaps extracted from an impinging hydrocarbon molecule, to form an OH^- group. The demonstrably high reactivity of the O^- form of oxygen where the O^- is ionosorbed has been discussed at length in the preceding subsection. By analogy, when a surface lattice oxygen ion is in the -1 oxidation state, it can also be expected to be highly reactive.

Depending on whether the solid is p type or n type, more or less O_L^- can be expected at equilibrium. As discussed in Chapter 4, the ionic surface states associated with anions will often be in the band gap, so statistics demands that a finite density

$$[O_L^-] = [O_L^{2-}] \exp[-(E_F - E_t)/kT] \qquad (10.7)$$

of O_L^- be present, where $[O_L^{2-}]$ is the concentration of lattice oxide ions at the surface (a half-monolayer or so), and E_t is the energy level of the ionic surface state. For a p-type oxide, E_F is near the valence band and hence near E_t.

There are also kinetic considerations that must be resolved. The -1 oxidation state must be generated by transfer of an electron[91,92] to levels in or on the solid. In the p-type semiconductor case, the valence band provides convenient levels for rapid electron exchange:

$$h^+ + O_L^{2-} \rightleftharpoons O_L^- \tag{10.8}$$

but there may be kinetic limitations on reaching equilibrium for many insulators or n-type materials because there are no partially occupied energy levels available near the energy level of the O_L^- species.

Thus if the -1 oxidation state is required to provide the active oxygen species, the general correlation of Section 10.1.2.3 (that p-type semiconductors are often the best, insulators intermediate, and n-type semiconductors the poorest oxidation catalysts) is immediately reconciled. For to a first approximation, the Fermi energy will be higher going from p-type, to insulating, to n-type oxides. Then from Equation (10.7), neglecting surface barrier effects (the density of surface states), the concentrations of O_L^- will decrease in that order.

Experimentally the chemical reactivity of lattice oxygen in the O^- oxidation state is often observed. For example, in electrochemical work on semiconductors the capture of holes at a surface oxide ion or at an ionic surface state often leads to decomposition of the material. Thus electrochemical etching of Ge, Si, and ZnO all apparently depend for rapid kinetics on the flow of holes to the surface. If the semiconductor is n type, optical excitation is required to form the holes.

The probability that a neutral oxygen atom could be available as the active surface oxygen for binding reducing agents on an ionic solid is low. In the lattice position such a zero-valent oxygen atom requires two electrons be removed from a lattice oxide ion. Equally unlikely is the possibility that an oxygen atom could move from its lattice site, leaving the two electrons on the remaining vacancy. The energy to form a neutral oxygen atom on the surface of a highly ionic solid is too high.

10.5.2.3. Anion Vacancies as Active Sites for Electron Exchange and Reoxidation of the Catalyst.

One form of local sites for electron exchange is the oxygen ion vacancy. When a lattice oxygen atom is removed, leaving a vacancy behind, the vacancy can be neutral ("neutral" means the two electrons that were associated with the oxygen remain on the vacancy) or one or both of the electrons can leave and the vacancy becomes "positively charged." Whether or not the electrons are retained by the vacancy makes a great difference in the catalytic behavior of the oxide.

In terms of the band model the anion vacancy behaves electronically as a donor. If the donor level is deep in the band, not too far from the valence band (as may be the case with many insulators), the valence electrons can simply be retained by the vacancy as the oxygen atom is plucked from the surface, and the vacancy is formed neutral. However, if the donor level is high in the band diagram, the energy of the neutral vacancy will be high, and oxygen removal to form neutral vacancies will be energetically unsound. For example, on ZnO it is believed that the donor level for oxygen ion vacancies is only a few hundredths of an electron volt from the conduction band, and the conduction band is 3 eV above the valence band (the energy of an electron on a lattice oxygen ion). Thus to form a neutral vacancy by extracting an oxygen atom, an electron from the valence band must be raised in energy almost to the conduction band (to the donor level of the vacancy). Such an energy expenditure (of 3 eV) is clearly difficult, and neutral vacancies will not form easily. However, in many cases the energy to form neutral vacancies can be reasonable, and neutral vacancies are sometimes assumed to result upon reduction of an oxide.

If the vacancy remains neutral, the electron and oxygen exchange is particularly easy to visualize. In essence the electron exchange problem is bypassed completely. The oxygen leaves the lattice sites as an atom, and the electrons stay with the vacancy. When an oxygen molecule is dissociated to reoxidize the crystal, the neutral atom returns to the vacancy without complicated reduction steps being necessary. Vacancies with sufficient electron affinity to remain neutral upon oxygen atom extraction were proposed by Dell et al.,[26] Winter,[93,94] and Boreskov and co-workers[95] to explain the catalytic activity of insulators in oxidation.

On the other hand, partially occupied (positively charged) vacancy levels can, if at an appropriate level, act as electron exchange centers, providing electrons to reduce gaseous oxygen and absorbing electrons in the oxidation step. Thus, for example, an anion vacancy acting as a donor could provide an electron to chemisorb O_2 as O_2^- and/or accept electrons from surface anions O_L^{2-} to form the active O_L^-. This possibility is particularly attractive if the vacancy density is sufficiently high that an "impurity band" forms[96] so the electrons can be conducted through the band to and from the appropriate sites. Electron exchange with narrow bands such as impurity bands will be discussed further in Section 10.5.3.2 below.

The mobility of vacancies is important in catalysis, and it is found that most good oxidation catalysts have mobile vacancies at reaction temperatures (see, for example, the discussions of Section 10.6.2). If vacancies are immobile, they will best be represented by a surface state picture of Figure 10.2a; if mobile, the impurity band will form and they will best be represented by a narrow band model of Figure 10.2b. When vacancies are mobile, they can be more active in the reoxidation process. If the vacancies can diffuse to the sites where oxygen is adsorbed, accelerating the reoxidation of the lattice, the catalyst can be much more active.

In many oxides, cation vacancies rather than anion vacancies are the dominant imperfection, so the fate of the anion vacancy is to be annihilated by a cation vacancy. This, for example, is to be expected for the excellent oxidation catalyst Cu_2O, where cation vacancies are highly mobile.[97,98] In this case the cation vacancy becomes the electron exchange center. If the cation vacancy is neutral (a hole is trapped), then when a lattice oxygen atom is removed, leaving an anion vacancy, the vacancy is promptly annihilated by two (for Cu_2O) cation vacancies at the surface, leading to a perfect lattice. The reoxidation of the Cu_2O involves the formation of new cation vacancies. Other cases where active catalysts are dominated by cation vacancies have been suggested.[99,100]

10.5.3. Dangling Bonds as Active Sites for Adsorption and Electron Exchange

The term "electron exchange" in this section is used in its broadest sense. In particular, the term will be used for all cases between the one extreme, heteropolar covalent bonding with a moderate partial charge on the adsorbate as provided by a particular surface atom, and the other extreme, ionosorption with essentially an integral partial charge on the adsorbate as provided by a narrow band in the solid. The reason for this broad definition is that there is no clear dividing line between cases where a heteropolar covalent bond (active redox centers on the catalyst surface) is the best description and cases where a rigid band model (ionosorption), as modified by a strong local interaction, is the best description. In fact, most oxidation catalysts of particular commercial interest fall right in the gray area, as will be described in Section 10.5.3.2 below.

10.5.3.1. Localized Surface States. The best example of a catalyst that apparently depends only on localized cationic surface states to provide both a redox electron exchange function and active adsorption sites is Cr_2O_3. The coordinatively unsaturated surface chromium ions, neutral in the $+3$ valence, readily assume a formal $+4$ valence, forming $(Cr-O_2)^{+3}$ upon adsorption of oxygen at room temperature [as per Equation (7.34)] or forming $(Cr-O)^{+3}$ at temperatures above 250°C.[101,102]

In either case they provide a high negative partial charge on the oxygen molecules or ion, as was discussed in Section 7.2.2.3. Other ligands can be added, bonding covalently with the surface chromium ions. Cr^{+3} can act in a redox capacity as indicated, or, the chromium being a transition metal, ligand field bonding can occur at the same site. Indeed, the surface Cr is found[103] to bond strongly to CO. Uptake of O_2 sharply saturates on a clean Cr_2O_3 at a value of about 10^{19} atoms/m², suggesting the uptake is limited to a monolayer and the redox states are strictly associated with the surface atoms. Shelef *et al.*[103] note also that unlike most oxidation catalysts, chromia is apparently "poisoned" by oxygen, for CO is oxidized far more rapidly on chromia by N_2O than by O_2. This observation seems consistent with the strong $Cr-O_2$ complex observed [as in Equation (7.34)] which can block the adsorption of CO or even the dissociation of O_2. A similar surface state, Mn^{+2}/Mn^{+3}, which can (at $T > 400°C$) be oxidized further to Mn^{+4} is suggested as an active center on MnO for oxygen adsorption.[104]

Other examples of active electron exchange centers that can best be considered as localized sites include highly dispersed additive surface states, such as used by Descamps *et al.*[105] and others[106] to control the Fermi energy at the ZnO surface. One-equivalent redox couples of varying oxidation potential were used. Similarly, some promoters can be associated with local electron exchange centers. For example, Spath and Torkar[107] suggest that the addition of BaO_2 to silver catalysts induces localized Ag donors at the surface of the catalyst.

10.5.3.2. Dangling Bonds Associated with the Host Cation.

Metal oxides where the oxidation state of the bulk cation can vary by one oxidation number (such as CuO/Cu_2O) are very common as active oxidation catalysts. When cations of both oxidation states are available on the surface, one of them will have unpaired electrons and can act as adsorption sites with a covalent bond formed between the adsorbate and the cation. Also the cation can change its formal valence, acting in electron exchange processes. The energy levels so offered at the surface of the solid can be considered either as partially occupied dangling bonds or as a narrow band of energy levels in the solid that is not surface localized. Thus such partially filled cationic levels are in the gray region between a cationic surface state, as described in the preceding subsection, and a narrow cationic band. For example, Hattori *et al.*[108] suggest that Ti^{3+} ions in TiO_2 act as electron donors, interacting with nitrobenzene to form a surface nitrobenzene radical. Without detailed information it is impossible to speculate whether the Ti^{3+} is an ionic surface state, and electron transfer from such a localized state to the nitrobenzene is the best description of the process, or whether an electron from the narrow Ti^{4+} "conduction band"* ionosorbs the nitrobenzene. A similar lack of knowledge applies to most narrow band systems as described below. The fact that many catalytic effects on such systems seem reconcilable to a local bonding model suggests that the ionic surface state model, or at least a model assuming an induced local bond, usually provides the most satisfactory picture. But it is useful at times to note the band character of the levels.

Independent of whether one uses a local bonding model or a narrow band model, one of the more important variables that affects the active cationic sites on the surface is the average oxidation state of the cation. By a strict local bonding model, the variable becomes the ratio of the number of surface cations in each oxidation state, or the ratio of "reducing agent" to "oxidizing agent" concentration present on the surface. By the narrow band model, the variable becomes the ratio of the density of unoccupied states in the band, capable of accepting electrons from reactants, to the density of occupied states in the band, capable of donating electrons to reactants. In many cases the optimum activity or selectivity (in partial oxidation) depends on reaching a certain ratio of the two oxidation states for the cation. For illustration in the hypothetical reaction:

$$Cu^+ + \tfrac{1}{2}O_2 \rightarrow Cu^{2+} + O^-$$

$$Cu^{2+} + R \rightarrow (Cu \cdot R)^{2+} \qquad (10.9)$$

$$(Cu \cdot R)^{2+} + O^- \rightarrow RO + Cu^+$$

a certain ratio of Cu^+/Cu^{2+} should be present on the surface of the catalyst to optimize simultaneously the first two steps of Equation (10.9).

The ratio of valence states of a cation in a catalyst can be varied in two ways. One is by interaction of the reactants with the catalyst. For example, there is evidence[109] that a certain Cu^+/Cu^{2+} stoichiometry must be attained for best results in the oxidation of propylene on Cu_xO, and this ratio is attained by adjusting the propylene/oxygen ratio (the propylene reduces the copper oxide, the oxygen reoxidizes it). The other method of varying the ratio of valence states is by adjusting the stoichiometry of the catalyst during its preparation, usually by introduction of another cation. Nakamura et al.[110] adjusted the ratio of V^{4+}/V^{5+} in a vanadia/phos-

---

* TiO_2 conducts by polarons, so even the concept of the electrons coming from a conduction band must be modified.

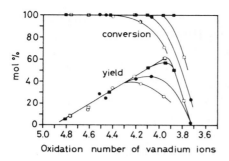

Figure 10.3. The relation between the oxidation number of the vanadium ions in V_2O_5/P_2O_5 catalysts, the conversion of 1-butene, and the yield of maleic anhydride. The various symbols represent different catalyst preparation methods.

phorus pentoxide catalyst by adjusting the phosphorus content and found an average oxidation state of the vanadium ion near +4 optimized the oxidation of butene to maleic anhydride. The results are shown in Figure 10.3. If the vanadium is so reduced that its oxidation state is driven below +4 (by addition of a large percentage of phosphorus), the rate of oxidation of butene and the selectivity both become poor, so the authors conclude that the V^{+4}/V^{+5} redox couple with a high concentration of V^{+4} is needed to provide active sites at the surface.

The local surface site model leads to a possible explanation for the observed data. It is known that the rate limiting step with vanadia catalysts is the reduction of the oxygen gas,[111–113] and for this step clearly a reduced catalyst will be most effective. Therefore, the most active catalyst should be that with the highest concentration of V^{+4}. However, when the olefin adsorbs on this surface, electrons are donated to the V^{+5}/V^{+4} band[111] (the olefin reduces V^{+5} to V^{+4}). Thus, all the vanadia should not be V^{+4}—there should be a few residual V^{+5} sites to accept the electrons. The reason that the catalyst produces CO_2 rather than maleic anhydride when the V^{+5} concentration is high (the conversion stays high in Figure 10.3 but the yield decreases) requires a more complex model. For example, if there are too many V^{+5} sites, the olefin may be attacked at several points simultaneously, as discussed above.

Many systems[114,99,109] have been shown to depend upon the ratio of oxidation states of such a cationic redox couple. In other systems strong maxima in activity are observed with variations in cation ratio, but it is not always clear that the maximum can be attributed to the optimization of the oxidation state of one of the cations involved. For example, the Fe/Mo oxide catalyst, used for oxidizing methanol to formaldehyde, provides a striking maximum when one plots the yield as a function of the Fe/Mo cation ratio,[115] which may be related to the average oxidation state of the Mo cation, but this relationship has not been proven.

As discussed, this class of catalysts that depend upon cationic redox sites can be described by a rigid narrow-band model. For some purposes this model has advantages, particularly when measurements of collective electron properties such as the Fermi energy are to be related to the electron exchange requirements on the catalyst. For example, in the vanadia case of Nakamura *et al.*, described above, where it was found that the most active catalyst is one with the vanadium ions in the V^{+4} state, the condition for maximum activity can equally well be defined in terms of the V^{+4}/V^{+5} band being almost completely occupied. The rate limiting step in vanadia catalysts being the reduction and absorption of lattice oxygen, the (almost) full band leads to the highest Fermi energy and the most rapid reduction of oxygen. However, empty levels must still be present in the band to permit electron injection from the olefin. In the case of V_2O_5 there is strong evidence[116,111] that olefins will donate electrons to the V^{+4}/V^{+5} band (or, by the other equivalent description, reduce V^{+5} surface ions).

In the V_2O_5 case oxygen reduction is rate limiting, so the Fermi energy should be high; in many other partial oxidation reactions, oxidation steps (electron transfer to the solid) are rate limiting, so the Fermi energy should be low. Thus, in studies of catalysts for the partial oxidation of propylene it was found[96] that the Fermi energy of the most active catalyst (relative to a common reference energy) is identical for bismuth molybdate, Cu_2O, and iron molybdate. The value was as low as possible while still high enough to induce oxygen reduction at a surface with no double layer. As will be discussed in Section 10.6.2, these catalysts all have either hydrogen abstraction from the olefin or the desorption of the product as the rate-limiting step, and a low Fermi energy accelerates this step as much as possible.

Other workers have related the Fermi energy to the optimum catalyst stoichiometry for oxidation of olefins. For example, Margolis *et al.*[117] showed the work function ϕ ($= E_e - E_F$ from Figure 2.2) is closely related to the selectivity for acrolein production on bismuth molybdate in the oxidation of propylene.

In summary, it appears from either the redox viewpoint or the rigid band viewpoint, one of the primary parameters in these catalysts with a variable valence cation is the ratio of concentration of the two valence states. The ratio can be described in terms of the redox character of the catalyst, whether the reducing agent (V^{+4}) or the oxidizing agent (V^{+5}) dominates at the surface, or (equivalently) can be described in terms of the surface Fermi energy.

10.5.4. Wide Bands as Electron Sources and Sinks: *n*-Type and *p*-Type Semiconductors

Electron exchange with wide bands, associated usually with ionosorption, has been discussed a great deal in theoretical analyses and in fundamental experimental catalytic studies. In most cases, however, catalysts that depend on electron transfer between the conduction band of a semiconductor and the adsorbate are too inactive catalytically for commercial interest.

ZnO is a particularly simple *n*-type material with a broad conduction band that has been studied extensively, and in general the results for redox reactions can be described best in terms of the rigid band model. The active surface species for oxidation reactions on ZnO has been assigned variously to ionosorbed O^- and O_2^-.[118,106] There is good evidence that as usual (Section 10.5.2.1) the O^- species is far more reactive than the O_2^- species.

Sancier[119] distinguishes the two species in their interaction with CO at room temperature, and finds a rapid reaction with O^-, a slow reaction with O_2^-. Tanaka and Miyahara[120] reacted O_2^- with CO at room temperature and found the CO desorbs unreacted at about 70°C. This confirms that the interaction of CO with the O_2^- is much weaker than its interaction with the O^-, but does not determine which form is the most active kinetically at the reaction temperature of about 225°C. It is necessary to compare the two possibilities of Equations (10.10) and (10.11):

$$CO + O^- \rightarrow CO_2^- \xrightarrow{k_1} CO_2 + e \tag{10.10}$$

and

$$CO + O_2^- \rightarrow CO_3^- \xrightarrow{k_2} CO_2 + O^- \tag{10.11}$$

where we assume

$$2O^- = O_2^- + e \tag{10.12}$$

is at equilibrium. One might expect k_1 to be less than k_2 as Equation (10.10) requires a highly activated electron injection step, but at high temperature the concentration of O^- will be much higher than O_2^-, from Equation (7.28). There is evidence for both viewpoints, and it is probable that both reactions occur, the relative rates depending on the detailed conditions.

The gas-phase kinetic measurements of CO oxidation on ZnO suggest a complex interaction, for different authors find different pressure dependencies. For example, Amigues and Teichner[121] find the reaction (at 260°C) independent of CO pressure, proportional to $p_{O_2}^{1/2}$. Deren and Mania,[122] on the other hand, find between 180 and 250°C that the rate is zero order in oxygen pressure, almost first order in CO pressure. The latter authors incidentally observe a "field effect" wherein a high field applied to the surface to induce extra negative charges causes a temporary increase in the oxidation rate, consistent with a (temporary) ionosorption of excess oxygen. The effect disappears after a few minutes, indicating perhaps that the induced charge becomes trapped in nonreactive surface states. For example, if O_2^- were the active species [Equation (10.11)], an induced charge would initially increase the density of these species, but with time, equilibrium would be regained, and most of the induced charge would reside in O^- species at the temperatures used, 223°C. This experiment may be construed as evidence for a high activity of the O_2^- species, but any deep surface states would lead to the same observation.

In studies of N_2O dissociation on ZnO, Cunningham and his co-workers[123] find doping effects that are in agreement with a simple ionosorption model for this reaction. The reaction (at room temperature) is

$$N_2O + e \rightarrow N_2\uparrow + O^- \qquad (10.13)$$

and the amount of nitrogen produced correlates qualitatively with the Weisz limitation (Section 7.2.2.2). The direction of the effect is correct: doping with the donor In gives the greatest adsorption, no doping intermediate, and doping with the acceptor Li gives the least adsorption. Quantitatively the correlation is not easily made without knowing the relative energy levels of the Li acceptor and the O^- "acceptor."

ZnO is of particular interest in reactions involving decomposition or oxidation of alcohols or synthesis of methanol, and although these are not all oxidation reactions, they may involve electron exchange. With methanol adsorbed at reaction temperatures, there is a substantial increase in surface conductance,[124,125] so there is undoubtedly electron exchange between the alcohols and the conduction band, although the electron transfer is not necessarily a step in the reactions.

There is some evidence that electron transfer is actually involved in the alcohol dehydrogenation reactions, for several authors have shown that bulk doping has an effect on the catalytic activity. For example, Sastri's group[126] in studies of isopropyl alcohol, McArthur et al.[127] in studies of the dehydrogenation of ethanol, and Miller and Wu[128] in the dehydrogenation of isobutyl alcohol suggest from doping experiments that electron transfer in ZnO is important, although the latter authors suggest valence band effects, a difficult concept in this strongly n-type material. Single-crystal work[129] has shown activation energies for electron injection from alcohols into the conduction band of ZnO range from 1.3 eV for methanol and ethanol to 0.5 eV for isopropyl alcohol. The 1.3-eV value may be related to the observed activation energy of that order for methanol and ethanol dehydrogenation on ZnO.[125–127] The fact that current doubling (Section 8.3.2) is observed with methanol on ZnO is considered[25] another indication that the species must react (oxidize or decompose) at a rate equal to the electron injection rate.

The active sites for adsorption and active intermediates on ZnO during alcohol reactions is not clear. There is evidence[130] that if the (0001̄) ZnO surface is covered with a monolayer of KOH, electron injection from methanol is dramatically reduced. Thus the active sites for methanol adsorption may be acidic. The importance of acid sites in methanol oxidation on iron molybdate will be described in Section 10.6.2. Tamaru's group[131] conclude methoxy groups are the active intermediate in methanol oxidation on ZnO, and this may correlate to the decomposition process.

Catalytic oxidation effects on p-type semiconductors are harder to analyze. With n-type materials we have oxides such as ZnO and TiO_2, where anion vacancies are not important, and where (particularly with ZnO) the conduction band is wide and uncomplicated. Thus a simple rigid band model is fruitful. On the other hand, with p-type materials, where in principle hole exchange between ionosorbed species and a broad valence band should lead to interesting correlations, the catalysts always seem complicated by other effects. For example the p-type catalyst Cu_2O has already been mentioned as an example of a catalyst with mobile vacancies (cationic), and has been mentioned as an example of a catalyst with a convenient cationic redox transition (Cu^{+1}/Cu^{+2}).

The most interesting active site for p-type oxide catalysts, common to all, is the anionic surface state, O_L^-, activated according to Equation (10.8). As discussed above, such an activated form of oxygen may be the key to the general higher activity of p-type catalysts (Section 10.1.2.3) compared to intrinsic or n-type catalysts. The interaction of such surface states with adsorbates is expected to be highly local, described best by Figure 10.2a, but the generation of the surface states by holes [Equation (10.8)] should be described best by the broad band model, Figure 10.2c.

Wood, Wise, and Yolles[109] have correlated the activity and selectivity of cuprous oxide in propene oxidation to acrolein with the stoichiometry, where the stoichiometry was determined by conductance measurements. They find that with varying stoichiometry (which leads to variations in not only the Cu^{+2}/Cu^{+1} ratio, but also the cation vacancy concentration and the hole concentration) the yield of acrolein passes through a maximum near the stoichiometric Cu_2O composition. They suggest the dominant variable is the oxidation state of active oxygen, as controlled by the Fermi energy [namely, the formation of O^- by Equation (10.8) or its equivalent for adsorbed species]. Supporting evidence for an interpretation that some such redox state dominates comes from studies[96] of the Fermi energy of the catalyst which show that when the catalytic selectivity is optimized, the Fermi energy is measured to be the same as that of bismuth molybdate and of iron molybdate partial oxidation catalysts. It was also shown that the primary action of the promoter methyl bromide, used in the propylene oxidation on cuprous oxide, is to stabilize the Fermi energy of the catalyst independent of the oxygen pressure. Thus the interpretation relates to the "redox character" of the catalyst, and because we cannot be sure whether it is the O_L^- or the Cu^+ surface states that dominate, the Fermi energy concept (which is valid independent of the chemical species) has an advantage.

10.6. Examples of Oxidation Catalysis

10.6.1. Platinum

In Section 7.3.2 the adsorption of gases on platinum was reviewed. It was shown that oxygen adsorption occurs when surface steps are available, with 50% coverage occurring even at low pressure on normal (heterogeneous) surfaces such as ribbons or supported catalysts. Three TPD peaks from oxygen adsorbed at low temperature are observed, two associated with molecular oxygen, desorbing at about 300°C, and one associated with atomic oxygen desorbing at 470°C.

Carbon monoxide, on the other hand, adsorbs strongly on the platinum (111) face,[132] with a strong linear carbon/platinum bond associated with overlap between the carbon orbitals and the d orbitals of a single platinum atom. On the (100) face there are three bonding states for CO on platinum, two with a linear structure as above, and one high-temperature form of CO that is assumed to bond to two platinum atoms in a bridged bond. On the (100) plane a favorable overlap configuration will develop between the carbon orbitals and the platinum d orbitals from neighboring atoms.

In the oxidation of CO on platinum, it is found that oxygen must be adsorbed for the reaction to proceed. If a monolayer of CO is adsorbed, the reaction is poisoned. Oxygen molecules apparently cannot dissociate rapidly over the CO monolayer. If a monolayer of O is adsorbed, the reaction proceeds with the impinging gaseous CO bonding directly with the adsorbed oxygen and the product CO_2 desorbing.[133,134] This mechanism (reactants from the gas phase interacting directly with adsorbates) is called a Rideal–Eley mechanism.

This very simple model (that only adsorbed oxygen can react) leads to many interesting observations in the CO oxidation reaction that without the model would appear complex. Consider variation in the CO/O_2 pressure ratio. If the ratio is very low, the desorbing CO_2 is replaced by oxygen, and the Rideal–Eley mechanism obtains as described above, characterized by a rate almost first order in CO pressure.[135] If, on the other hand, the CO/O_2 ratio is high, then the surface becomes mainly covered by CO, and the rate becomes limited by the rate at which oxygen can find empty sites on which to adsorb. The oxidation process becomes dominated by an interaction between coadsorbed species[133] (a Langmuir–Hinshelwood mechanism). At constant oxygen pressure, the rate becomes almost inversely proportional to the CO pressure.[135] Thus the oxidation rate shows a maximum when plotted as a function of CO pressure at constant O_2 pressure. Near the maximum, the system becomes quite unstable, and one observes oscillations with time in the reaction rate,[135,136] a rather unique phenomenon in catalysis. The temperature dependence of the CO oxidation rate is also a complicated function, with the two mechanisms available.[137]

In summary, it is the adsorbed oxygen atom, bonded weakly to the platinum atom, that is the active site for the oxidation reaction. The oxidation of CO on palladium is analogous.[138,139]

The oxidation of hydrogen on platinum is more complex[140,141] because both species must be dissociated to be reactive. There is evidence that oxygen can poison the surface,[140] and evidence that a layer of oxide must be present for an active catalyst.[141]

The oxidation of hydrocarbons on platinum shows important heterogeneity effects. In Chapter 5 the bonding of hydrocarbons to platinum was discussed, and in Chapter 6 the influence of heterogeneity on bonding was introduced. The implications of these two topics in catalysis on platinum is dramatically represented in the work of Anderson et al.[142] who studied the reactions of hydrocarbons, emphasizing n-hexane, on platinum. They used films evaporated in high vacuum to assure there were no complications due to contamination, and prepared films of varying thickness down to a "monolayer" of platinum. The "monolayer" actually sintered into crystallites of the order of 20 Å diameter, so cluster effects (Chapter 6) dominated. By depositing thicker films, catalysis on broad planar regions could be compared to catalysis on the clusters.

On platinum, hexane undergoes three types of reactions: skeletal isomerization (to methyl pentanes, for example), hydrogenolysis (the formation of CH_4, for example), and cyclization (the formation of ring compounds). From earlier work using ^{13}C-labeled hexanes, reviewed in the referenced paper,[142] it was concluded that the active surface intermediate is a C_5 carbocyclic group.

In the film work under discussion, it was shown that of the three reactions cyclization is highly favored while skeletal isomerization and hydrogenolysis are suppressed on the clusters, relative to the product distribution with larger particles.

The model is as follows. In order for skeletal isomerization to be observed, the adsorbed carbocyclic intermediate must open at a different carbon atom than that of the original adsorption site, so this means that the intermediate must be bonded to two or more platinum atoms. With the small 20 Å particles, the adsorption will be almost entirely on corner or edge atoms (Chapter 6), so adsorption will be dominantly on one site only, and the ring must reopen at the same position as it was formed. Thus desorption as a ring compound is possible, and desorption as the original hexane is possible, but skeletal isomerization is unlikely. On the broader area films where two or more points of adsorption are probable for the intermediate, isomerization is correspondingly favored.

Kahn et al.[143] review and list facile and demanding reactions of hydrocarbons on platinum, showing, for example, hydrogenation or dehydrogenation of cyclic compounds is facile but hydrogenolysis and isomerization of alkanes is, as Anderson et al. found, demanding. Egghart[144] reviews other fragments that are active intermediates in hydrocarbon reactions on platinum, emphasizing oxidation. For example, he suggests that if an acetic acid group forms, the oxidation will be slow because the group poisons the surface, but if a formate group (a C_1 fragment) forms, it is rapidly oxidized further.

10.6.2. Partial Oxidation Catalysts: Bismuth and Iron Molybdate

10.6.2.1. Introduction. Molybdena based catalysts provide an excellent example for discussion because such a variety of active site configurations have been proposed by one author or another. The remarkable characteristics of this family of catalysts have been explained based on ionosorbed oxygen, lattice oxygen exchange, electronic effects, redox sites, acid sites, and doubly bonded oxygen. At reaction temperature the role of ionosorbed oxygen has been discounted recently, but the other effects are all currently still considered possible, and there is evidence for each viewpoint.

The molybdena based catalysts are partial oxidation catalysts. Thus with bismuth molybdate, propylene (C_3H_6) is oxidized to acrolein (C_3H_6O), and in the presence of ammonia, to acrylonitrile. Similarly, oxidation of methanol over iron molybdate produces formaldehyde and water, and so on. Such partial oxidation reactions are of great commercial significance.

For many of these molybdena based catalysts[66] the oxidation starts with the adsorption of the olefin by hydrogen abstraction. An H atom is split off, leaving an allylic radical, as in Equation (10.3). Adams and Jennings[145] show the hydrogen extraction is the rate limiting step in the oxidation of propylene to acrolein on bismuth molybdate, by showing a strong isotope effect. The rate is slower if the hydrogen is replaced by deuterium. They also show an allylic intermediate results from the hydrogen abstraction for they find the position of a deuterium atom in the propylene [on the left side of Equation (10.3)] has no effect on the deuterium position in the product. This observation requires a symmetrical intermediate, the allyl intermediate.

Adams and his co-workers[146] also report that the oxidation of propylene in the 400–500°C temperature range of commercial interest is first order in propylene pressure, zero order in oxygen, and unaffected by product pressure.

In the following subsections we will discuss the various models for active sites in the oxidation of propylene on bismuth molybdate, then briefly discuss the oxidation of methanol on iron molybdate.

10.6.2.2. Lattice Oxygen Exchange.

There is little doubt that the oxidation of propylene on bismuth molybdate occurs by a mechanism wherein the lattice is reduced by the propylene and reoxidized by the gaseous oxygen. In fact, the initial propylene oxidation rate is the same whether oxygen is present or not.[147] Keulks,[148] and Wragg, Ashmore, and Hockey[149] have shown that in the oxidation process if $^{18}O_2$ is used as the gaseous reactant, it does not appear in the product, except in accordance with the ratio of ^{16}O to ^{18}O in the lattice. This implies a rapid oxygen (vacancy) diffusion. Otsubo et al.[150] have carried such isotope experiments a step further, using isotopic oxygen in the initial preparation of the bismuth oxide or the molybdena. In other words, they prepared catalysts that in effect could be characterized as $Bi_2O_3 \cdot Mo^{18}O_3$ and compared these catalysts to those that could be called $Bi_2{}^{18}O_3 \cdot MoO_3$. They made the interesting observation that the lattice oxygen comes from the bismuth oxide, not the molybdenum oxide (which incidentally implies that the oxygen is not mixed during the initial calcining of the catalyst before its use in the reaction). After the catalyst is substantially reduced, the oxygen from the molybdena starts to transfer to the bismuth oxide layer and then appears in the product.

10.6.2.3. Conductance and Electron Exchange.

Changes in conductance of bismuth molybdate have been observed by many authors, for example, Peacock et al.,[151] who observed that the conductance rises with propylene present (at 400°C or so) and is restored by reoxidation in oxygen. The direction of conductance change indicates an n-type semiconductor. The rate of the reduction or oxidation process was found linear in gas pressure. With a 1:1 ratio of propylene and oxygen, the observed conductance corresponds to 10% reduction of the catalyst. The conductance of the catalyst is, however, quite insensitive to the propylene to oxygen ratio between 9:1 and 1:9.

Morrison[96] points out that bismuth molybdate has, when optimized, a Fermi energy as low as possible consistent with maintaining the ability to reduce gaseous oxygen. This provides the optimum condition to remove electrons in the appropriate oxidation step. It is suggested that the active center for the rate limiting step, hydrogen abstraction, may be the O_L^- ion, the oxidized ionic surface state associated with surface oxygen, and for optimum activity the Fermi energy should be low to maximize the concentration of O_L^- ions [Equation (10.7)].

If O_L^- ions are active, there must be energy levels near the ionic surface state O_L^- to absorb the electrons, or the rate would be limited by electron transfer kinetics. As bismuth molybdate is an n-type semiconductor, it is not expected to be an active oxidation catalyst (Section 10.1.2.3), so its remarkable activity must be reconciled, and, in particular, sites for electron exchange must be identified. It was suggested that the large density of oxygen vacancies in the active catalyst could be partially occupied donors and provide an impurity band that acts as a source and sink of electrons. The other possibility, that the Mo^{+6}/Mo^{+5} redox surface state acts as an electron exchange center, will be discussed below.

10.6.2.4. Phases in the Bismuth Oxide/Molybdena System.

In the discussions above, the term "bismuth molybdate" has been used loosely for convenience. Actually, there has been a substantial effort to identify the phases present as a function of the Bi/Mo ratio, and to relate the activity of the catalyst to the phases present. The work has been reported primarily by Schuit's group.[152-154]

Three dominant compounds are found, with bismuth molybdate itself, $Bi_2O_3 \cdot MoO_3$ becoming when calcined $(BiO)_2MoO_4$, the so-called koechlinite structure, consisting of alternating layers of the bismuth group and the molybdenum group. This is the material that Otsubo et al.[150] studied and, as mentioned above, they found that the active oxygen comes from the bismuth group. With a Bi:Mo ratio of 2:3, a scheelite structure appears, $Bi_2(MoO_4)_3$, also a layer structure. With a Bi:Mo ratio of 6:1, a calcium fluoride structure, derived from the Bi_2O_3 structure, dominates. A highly active catalyst results from any ratio in the region of Bi:Mo of 2:3 to 2:1, but, as pointed out by Batist et al.[153] no single phase gives as high an activity as a binary system.

The layer structure of the active forms is undoubtedly important in providing the high mobility of the anion vacancies. In fact, as pointed out by these authors, it is possible that the active sites for oxygen reduction could be on one crystal face, and the active sites for the abstraction of lattice oxygen could be on another crystal face, with the oxygen vacancies moving easily from one face to the other.

10.6.2.5. Cation Sites.

10.6.2.5a. *Redox Sites.* The role of the Mo^{+6}/Mo^{+5} ionic surface state is not at all clear in the bismuth molybdate catalyst. Sancier et al.[155] note that as the Mo^{+5} ESR signal increases, the activity of the catalyst decreases. In these experiments the pressure ratio of propylene to oxygen was increased to increase the Mo^{+5} signal. They, therefore, suggest that the formation of the Mo^{+5} oxidation state of the molybdenum is undesirable. Peacock et al.,[156] on the other hand, conclude that active O^- states are formed (to extract the hydrogen from the propylene) by the electron being transferred to the Mo^{+6} surface ion, reducing it to Mo^{+5}. If the Mo^{+6}/Mo^{+5} band is indeed the active electron exchange center, the observed high ratio of Mo^{+6}/Mo^{+5} for an active catalyst is consistent with the fact that the rate limiting step is hydrogen abstraction. The higher the Mo^{+6} concentration, the more surface O_L^- will form (by either a surface molecule or a Fermi energy model) which will result in more sites for hydrogen abstraction.

The other action of the Mo^{+6} site may be as a site for the adsorption of the allyl radical. Peacock and his co-workers propose this model. The fact that ammonia is a strong poison for the propylene oxidation[157] may be regarded as indirect evidence that Mo^{+6} sites are important, for the ammonia could be expected to interact with acid Mo^{+6} sites.

10.6.2.5b. *Acid Sites.* Ai and Suzuki[158] suggest the bismuth molybdate catalyst, with phosphorus added, as used for the partial oxidation of butene, butadiene, and furan, shows an activity that is related directly to the acidity of the surface. They measured the dehydration of isopropyl alcohol as a measure of acidity, and found the acidity depends on the Bi/Mo ratio. They found a maximum in the acidity that corresponded to the maximum in oxidation activity.

In oxidation by bismuth molybdate it should, however, be noted that Adams[159] pointed out that an oxidation mechanism involving carbonium ions (that is, involving strong acid centers), could not be reconciled with the behavior. A carbonium ion mechanism would lead to rates for the oxidation of different alkenes that span many orders of magnitude, whereas the observed rates are remarkably insensitive to the exact alkene used. Thus, the more usual model is that the olefins are adsorbed as radicals, rather than as ions.

10.6.2.5c. *Multiple Sites.* Haber and Gryzybowska[147] suggest that the rate limiting step of propylene oxidation on bismuth molybdate is hydrogen abstraction onto bismuth centers; this is followed by another hydrogen abstraction onto Mo centers, and finally the remaining hydrocarbon desorbs, taking a lattice oxygen with it.

Matsuura and Schuit[160] also suggest two centers, defining them initially as A sites and B sites. They find the adsorption isotherm for butadiene on bismuth molybdate shows two distinct characteristics, and define the site for strong adsorption, where the isotherm indicates a single site, as the A site. The A site seems well defined: butadiene can adsorb on it, but the site can be removed by reduction of the sample (the removal being observed by a lower capacity for butadiene adsorption). Readsorption of oxygen will occur in the amount of one atom per site removed. The site is thus considered an oxygen species of a special nature, which, when removed, can be regenerated by adsorption of an equal number of oxygen atoms. From this evidence the A center is designated as an O^{2-} site in a special position. The B site is much more complex. It is considered to be a combination of three sites, and arguments are presented, similar to those for the A site, that the B center is a combination of a vacancy and two O^{2-} ions. In the actual oxidation (here of butene) it is concluded that dehydrogenation of butene uses both A and B sites, and reoxidation of the lattice occurs by consumption of oxygen in vacant A sites.

10.6.2.6. The Oxidation of Methanol on Iron Molybdate.

As a final example, we will briefly describe the sites proposed for iron molybdate, a catalyst effective in the oxidation of methanol to formaldehyde and water. The interesting difference between this system and the preceding one is that there is evidence that acid centers are active in this reaction involving alcohol.

Indeed, Pernicone and his co-workers[161-163] feel that the oxidation of methanol occurs essentially by the mechanism

$$O_L^{2-} + V_0 + CH_3OH \rightarrow (V_0 \cdot CH_3O)^- + (H \cdot O_L)^- \qquad (10.14)$$

$$(V_0 \cdot CH_3O)^- + Mo^{+6} + O_L^{2-} \rightarrow Mo^{+5} + (H \cdot O_L)^- + CH_2O\uparrow + V_0 \quad (10.15)$$

$$Mo^{+5} + 2(H \cdot O_L)^- \rightarrow H_2O\uparrow + O_L^{2-} + Mo^{+6} + V_0 \qquad (10.16)$$

$$\tfrac{1}{2}O_2 + V_0 \rightarrow O_L^{2-} \qquad (10.17)$$

Thus three types of active sites are involved in the proposed model. The methanol adsorbs as a methoxy group in Equation (10.14), complexed with an anion vacancy as the active center. The Mo^{+6}/Mo^{+5} redox state acts as an electron exchange site, and the basic lattice oxide ion acts in proton extraction.

The reaction is different from the propylene oxidation on bismuth molybdate also in that here the rate limiting step is the desorption of the product. This is determined by noting that the CH_2O and the H_2O desorb at different rates, whereas if the initial steps in the reaction were rate limiting, the two products would desorb at the same time. Thus one of the oxidation steps (10.15) or (10.16) is rate limiting. It is found in this case, as with the bismuth molybdate, that a single phase is not desirable. Although $Fe_2(MoO_4)_3$ is the optimum ratio of Fe:Mo (in fact Boreskov's group[115] find a sharp maximum at this ratio), Pernicone's group find that a small excess of molybdenum is required to optimize the catalyst. This changes the lattice constants slightly and introduces defects.

These examples show clearly the variety in defects, electrical properties, and cationic redox sites, mobile anion vacancies, and acid/base centers that all seem to be important in providing very active catalysts. There seems little doubt that many of these catalysts must meet at least two and perhaps three requirements with respect to site properties before they are at the point of becoming of commercial interest. In modern surface science we are reaching the point of being able to define and to measure each of the types of sites individually, and in the foreseeable future we will be able to put catalysts together by theoretical prediction, rather than by empirical testing of innumerable combinations.

✱ | References

References for Chapter 1

1. F. F. Volkenshtein, *The Electronic Theory of Catalysis on Semiconductors* (MacMillan, New York, 1963).
2. S. G. Davison and J. P. Levine, in *Solid State Physics*, Vol. 25, eds. F. Seitz and D. Turnbull (Academic Press, New York, 1970).
3. W. M. H. Sachtler and P. Van der Plank, *Surface Sci.* **18**, 62 (1969).
4. O. Johnson, *J. Catal.* **28**, 503 (1973).
5. Z. Knor, *Adv. Catal.* **22**, 51 (1972).
6. J. R. Schrieffer and P. Soven, *Phys. Today* **28**, 24 (1975).
7. J. W. Gadzuk, *Surface Sci.* **43**, 44 (1974).
8. M. J. Kelly, *Surface Sci.* **43**, 587 (1974).
9. D. R. Penn, *Surface Sci.* **39**, 333 (1973).
10. J. C. Slater and K. H. Johnson, *Phys. Today* **27**, 34 (1974).
11. H. H. Ibach, K. Horn, R. Dorn, and H. Lüth, *Surface Sci.* **38**, 433 (1973).
12. F. Beck and H. Gerischer, *Z. Elektrochem.* **63**, 943 (1959).
13. S. R. Morrison, *Prog. Surface Sci.* **1**, 105 (1971).
14. H. Gerischer, *Surface Sci.* **18**, 97 (1969).
15. M. Green and M. J. Lee, in *Solid State Surface Science*, Vol. 1, ed. M. Green (Marcel Dekker, New York, 1969).
16. P. Mark, *J. Phys. Chem. Solids* **29**, 689 (1968).
17. D. A. Dowden, *Catal. Rev.* **5**, 1 (1971).
18. S. Siegel, *Adv. Catal.* **16**, 123 (1966).
19. R. T. Sanderson, *Chemical Periodicity* (Reinhold, New York, 1971).
20. F. Seel, *Atomic Structure and Chemical Bonding*, p. 38 (John Wiley and Sons, New York, 1963).
21. J. C. Phillips, *Surface Sci.* **37**, 24 (1973).
22. H. Pines and J. Manassen, *Adv. Catal.* **16**, 49 (1966).
23. K. Tanabe, *Solid Acids and Bases* (Academic Press, New York, 1970).
24. S. R. Morrison, *Surface Sci.* **50**, 329 (1975).

25. J. W. May, *Adv. Catal.* **21**, 151 (1970).
26. M. Prettre and B. Claudel, *Elements of Chemical Kinetics* (Gordon and Breach, London, 1970).
27. J. C. Phillips, *Surface Sci.* **44**, 290 (1974).
28. G. Blyholder and R. W. Sheets, *J. Catal.* **39**, 152 (1975).
29. L. Pauling, *The Nature of the Chemical Bond* (Cornell U. Press, Ithaca, 1960).
30. H. P. Boehm, *Disc. Faraday Soc.* **52**, 264 (1971).
31. J. T. Kummer and Y. Y. Yao, *Can. J. Chem.* **45**, 421 (1967).
32. R. Nosker, P. Mark, and J. D. Levine, *Surface Sci.* **19**, 291 (1970).
33. G. I. Young, *J. Colloid Sci.* **13**, 67 (1958).
34. H. S. Taylor, *J. Am. Chem. Soc.* **53**, 578 (1931).
35. L. D. Schmidt, *Catal. Rev. Sci..Eng.* **9**, 115 (1974).
36. R. D. Giles, J. A. Harrison, and H. R. Thrisk, *J. Electroanal. Chem.* **20**, 47 (1969).
37. J. M. Thomas, *Adv. Catal.* **19**, 293 (1969).
38. B. Lang, R. W. Joyner, and G. A. Somorjai, *Surface Sci.* **30**, 454 (1972).

References for Chapter 2

1. A. Many, Y. Goldstein, and N. B. Grover, *Semiconductor Surfaces* (Interscience, New York, 1965).
2. D. R. Frankl, *Electrical Properties of Semiconductor Surfaces* (Pergamon, London, 1967).
3. P. Mark, *Surface Sci.* **25**, 192 (1971).
4. P. B. Weisz, *J. Chem. Phys.* **21**, 1531 (1953).
5. J. A. Stratton, *Electromagnetic Theory* (McGraw-Hill, New York, 1971).
6. J. N. Zemel and M. Kaplett, *Surface Sci.* **13**, 17 (1969).
7. S. R. Morrison, *Adv. Catal.* **7**, 259 (1955).
8. D. H. Lindley and P. C. Banbury, *J. Phys. Chem. Solids* **14**, 200 (1959).
9. A. E. Yunovich, in *Surface Properties of Semiconductors*, ed. E. N. Frumkin, p. 88 (Consultants Bureau, New York, 1964).
10. S. R. Morrison, *Phys. Rev.* **114**, 437 (1959).
11. S. R. Morrison, *J. Vac. Sci. Tech.* **7**, 84 (1970).
12. S. R. Morrison, *Phys. Rev.* **102**, 1297 (1956).
13. H. Gerischer, *Surface Sci.* **18**, 97 (1969).
14. R. R. Dogonadze, A. M. Kuznetsoo, and A. A. Chernenko, *Russ. Chem. Rev.* **34** 759 (1965).
15. R. A. Marcus, *J. Chem. Phys.* **24**, 966 (1956).
16. S. G. Christov, *Ber. Bunsenges. Phys. Chem.* **79**, 357 (1975).
17. H. Gerischer, in *Advances in Electrochemistry and Electrochemical Engineering*, Vol. I, p. 139, ed. P. Delahay (Interscience, New York, 1961).
18. J. M. Hale, in *Reactions of Molecules at Electrodes*, p. 229, ed. N. S. Hush (Wiley, London, 1971).
19. R. Memming and F. Möllers, *Ber. Bunsenges. Phys. Chem.* **76**, 475 (1972).
20. R. A. Marcus, *J. Chem. Phys.* **43**, 679 (1965).

References for Chapter 3

1. J. P. Rynd and A. K. Rastogi, *Surface Sci.* **48**, 22 (1975).
2. A. Many, Y. Goldstein, and N. B. Grover, *Semiconductor Surfaces* (Interscience, New York, 1965).
3. D. R. Frankl, *Electrical Properties of Semiconductor Surfaces* (Pergamon, London, 1967).
4. D. R. Frankl, *CRC Crit. Rev. Solid State Sci.* **4**, 455 (1974).
5. C. E. Reed and C. G. Scott, in *Surface Physics of Phosphors and Semiconductors*, ed. C. G. Scott and C. E. Reed (Academic Press, New York, 1975).
6. J. C. Riviere, in *Solid State Surface Science*, ed. M. Green (M. Dekker, New York, 1969).
7. J. C. Tracy and J. M. Blakely, *Surface Sci.* **13**, 313 (1969).
8. B. O. Seraphin, *Surface Sci.* **8**, 399 (1967).
9. D. E. Aspnes and A. Frova, *Phys. Rev. B* **2**, 1037 (1970).
10. B. Hoffmann, *Z. Physik* **219**, 354 (1969).
11. H. Gobrecht and R. Thull, *Ber. Bunsenges. Phys. Chem.* **74**, 1234 (1970).
12. J. Shappir, *Surface Sci.* **26**, 545 (1971).
13. K. J. Haas, D. C. Fox, and M. J. Katz, *J. Phys. Chem. Solids* **26**, 1779 (1965).
14. H. C. Gatos and J. Lagowski, *J. Vac. Sci. Tech.* **10**, 130 (1973).
15. G. Heiland and W. Mönch, *Surface Sci.* **37**, 30 (1973).
16. H. Lüth, *Surface Sci.* **37**, 90 (1973).
17. J. R. Maltby, C. E. Reed, and C. G. Scott, *Surface Sci.* **51**, 89 (1975).
18. L. J. Brillson, *J. Vac. Sci. Tech.* **12**, 249 (1975).
19. G. Heiland, Proc. 2nd Int. Conf. on Electrophotography, Soc. Phot. Sci. Engr. p. 117, 1973.
20. H. Harreis and G. Heiland, *Surface Sci.* **24**, 643 (1971).
21. J. F. Dewald, *J. Phys. Chem. Solids* **14**, 155 (1960).
22. W. H. Laflere, F. Cardon, and W. P. Gomes, *Surface Sci.* **44**, 541 (1974).
23. R. Degrupe, W. P. Gomes, F. Cardon, and J. Vennik, *J. Electrochem. Soc.* **122**, 711 (1975).
24. W. L. Brown, *Phys. Rev.* **100**, 590 (1955).
25. W. L. Brown, W. H. Brattain, C. G. B. Garrett, and H. C. Montgomery, *Semiconductor Surface Physics*, p. 117, ed. R. H. Kingston (U. Pennsylvania Press, Philadelphia, 1957).
26. H. Statz, G. A. deMars, L. Davis, and A. Adams, Jr., *Semiconductor Surface Physics*, p. 139, ed. R. H. Kingston (U. Pennsylvania Press, Philadelphia, 1957).
27. S. R. Morrison, *Surface Sci.* **27**, 586 (1971); *J. Catal.* **34**, 462 (1974).
28. F. Meyer, E. G. de Kluizenaar, and G. A. Bootsma, *Surface Sci.* **27**, 88 (1971).
29. G. A. Bootsma and F. Meyer, *Surface Sci.* **14**, 52 (1969).
30. F. Meyer and A. Kroes, *Surface Sci.* **47**, 124 (1975).
31. F. Meyer, *Phys. Rev. B* **9**, 3622 (1974).
32. R. C. O'Handley and D. K. Burge, *Surface Sci.* **48**, 214 (1975).
33. R. Dorn, H. Lüth, and G. J. Russel, *Phys. Rev. B* **9**, 1951 (1974).
34. F. Lukes, *Surface Sci.* **49**, 344 (1975).
35. G. Petermann, H. Tributsch, and R. Bogomolni, *J. Chem. Phys.* **57**, 1026 (1972).
36. A. Amith, *J. Phys. Chem. Solids* **14**, 271 (1960).

37. M. H. Brodsley and J. N. Zemel, *Phys. Rev.* **155**, 780 (1967).

38. R. L. Petritz, *Phys. Rev.* **110**, 1254 (1958).

39. J. J. Harris and A. J. Crocker, *Surface Sci.* **30**, 692 (1972).

40. B. Feuerbacher, *Surface Sci.* **47**, 115 (1975).

41. C. R. Brundle, *Surface Sci.* **48**, 99 (1975).

42. D. E. Eastman and M. I. Nathan, *Phys. Today* **28**, 44 (1975).

43. T. E. Fischer, *J. Vac. Sci. Tech.* **9**, 860 (1972).

44. J. E. Rowe, H. Ibach, and H. Froitzheim, *Surface Sci.* **48**, 44 (1975).

45. R. Ludeke and A. Koma, *CRC Crit. Rev. Solid State Sci.* **5**, 259 (1975).

46. K. N. Ramachendran and C. C. Cox, *J. Vac. Sci. Tech.* **10**, 1068 (1973).

47. R. L. Park, *Surface Sci.* **48**, 80 (1975).

48. J. E. Holliday, *Surface Sci.* **48**, 137 (1975).

49. E. Plummer, J. W. Gadzuk and D. R. Penn, *Phys. Today* **28**, 63 (1975).

50. E. W. Plummer and R. D. Young, *Phys. Rev. B* **1**, 2088 (1970).

51. J. W. Gadzuk and E. W. Plummer, *Rev. Mod. Phys.* **45**, 487 (1973).

52. P. J. Estrup, *Phys. Today* **28**, 33 (1975).

53. H. D. Beckey and F. W. Röllgen, *J. Vac. Sci. Tech.* **9**, 471 (1972).

54. L. Ernst and J. H. Block, *Surface Sci.* **49**, 293 (1975).

55. E. W. Müller and T. T. Tsong, *Field Ion Microscopy* (Elsevier, New York, 1969).

56. J. E. Goell, *J. Appl. Opt.* **12**, 729 (1973).

57. H. D. Hagstrum and G. E. Becker, *J. Chem. Phys.* **54**, 1015 (1971).

58. H. D. Hagstrum, *J. Vac. Sci. Tech.* **12**, 7 (1975).

59. G. A. Somorjai and H. H. Farrell, *Adv. Chem. Phys.* **20**, 215 (1972).

60. M. B. Webb and M. G. Lagally, *Solid State Phys.* **28**, 301 (1973).

61. J. C. Tracy and J. M. Burkstrand, *CRC Crit. Rev. Solid State Sci.* **4**, 380 (1974).

62. H. Ibach, *J. Vac. Sci. Tech.* **9**, 713 (1972).

63. G. E. Laramore, *J. Vac. Sci. Tech.* **9**, 625 (1972).

64. J. W. May, *Adv. Catal.* **21**, 151 (1970).

65. R. L. Park, *Phys. Today* **28**, 52 (1975).

66. D. F. Stein, *J. Vac. Sci. Tech.* **12**, 268 (1975).

67. L. Fiermans, R. Hoogervijs, and J. Vennick, *Surface Sci.* **47**, 1 (1975).

68. T. A. Carlson, *Phys. Today* **25**, 31 (1972).

69. D. A. Shirley, *J. Vac. Sci. Tech.* **12**, 280 (1975).

70. G. Slodzian, *Surface Sci.* **48**, 161 (1975).

71. B. F. Phillips, *J. Vac. Sci. Tech.* **11**, 1093 (1974).

72. H. W. Werner, *Surface Sci.* **47**, 301 (1975).

73. R. F. Goff, *J. Vac. Sci. Tech.* **10**, 355 (1973).

74. D. P. Smith, *Surface Sci.* **25**, 171 (1971).

75. H. Niehus and E. Bauer, *Surface Sci.* **47**, 222 (1975).

76. D. J. Ball, T. M. Buck, D. Macnair, and G. H. Wheatley, *Surface Sci.* **30**, 69 (1972).

77. H. Yakowitz, *J. Vac. Sci. Tech.* **11**, 1100 (1974).

78. S. L. Bernasek and G. A. Somorjai, *Surface Sci.* **48**, 204 (1975).

79. J. A. Schwartz and R. J. Madix, *Surface Sci.* **46**, 317 (1974).

80. G. A. Somorjai and S. B. Brumbach, *CRC Crit. Rev. Solid State Sci.* **4**, 429 (1974).

81. D. Menzel, *Surface Sci.* **47**, 370 (1975).

82. J. H. Leck and B. P. Stimpson, *J. Vac. Sci. Tech.* **9**, 293 (1972).

83. D. Lichtman, *CRC Crit. Rev. Solid State Sci.* **4**, 395 (1974).

84. J. L. Freeouf and D. E. Eastman, *CRC Crit. Rev. Solid State Sci.* **5**, 245 (1975).

85. B. F. Lewis and T. E. Fischer, *Surface Sci.* **41**, 371 (1974).
86. B. Feuerbacher and B. Fitton, *Phys. Rev. B* **8**, 4890 (1973).
87. F. C. Brown, *Solid State Phys.* **29**, 1 (1974).
88. J. W. Gadzuk, *J. Vac. Sci. Tech.* **12**, 289 (1975).
89. R. Ludeke and L. Esaki, *Surface Sci.* **47**, 132 (1975).
90. E. Sickafus and F. Steinrisser, *Phys. Rev. B* **6**, 3714 (1972).
91. R. L. Park, J. E. Houston, and D. G. Schreiner, *Rev. Sci. Instr.* **41**, 1810 (1970).
92. R. G. Musket and S. W. Taatjes, *J. Vac. Sci. Tech.* **9**, 1041 (1972).
93. R. G. Musket, *J. Vac. Sci. Tech.* **9**, 603 (1972).
94. E. W. Müller, *Z. Phys.* **106**, 541 (1937).
95. J. W. Gadzuk, *Phys. Rev.* **182**, 416 (1969).
96. B. A. Politzer and P. H. Cutler, *Surface Sci.* **22**, 277 (1970).
97. E. W. Plummer and J. W. Gadzuk, *Phys. Rev. Lett.* **25**, 1493 (1970).
98. D. R. Penn, *Phys. Rev. B* **11**, 3208 (1975).
99. W. B. Shepherd and W. T. Peria, *Surface Sci.* **49**, 461 (1973).
100. R. Leysen, H. van Hove, J. Marien, and J. Loosveldt, *Phys. Stat. Solidi A* **11**, 539 (1972).
101. E. W. Müller, *Z. Phys.* **131**, 36 (1951).
102. T. Adachi and S. Nakamura, *Jap. J. Appl. Phys.* **11**, 275 (1972).
103. L. Ernst and J. H. Block, *Surface Sci.* **49**, 293 (1975).
104. T. Utsumi and O. Nishikawa, *J. Vac. Sci. Tech.* **9**, 477 (1972).
105. M. G. Inghram and R. Gomer, *J. Chem. Phys.* **22**, 1279 (1954).
106. E. W. Müller and T. Sakurai, *J. Vac. Sci. Tech.* **11**, 885 (1974).
107. T. Sakurai and E. W. Müller, *Surface Sci.* **50**, 38 (1975).
108. A. van Oostrum, *CRC Crit. Rev. Solid State Sci.* **4**, 353 (1974).
109. D. Tabor and J. M. Wilson, *J. Vac. Sci. Tech.* **9**, 695 (1972).
110. R. E. Schlier and H. E. Farnsworth, *Semiconductor Surface Physics*, p. 3, ed. R. H. Kingston (U. Pennsylvania Press, Philadelphia, 1957).
111. T. W. Haas, G. J. Dooley, III, J. T. Grant, A. J. Jackson, and M. P. Hooker, *Progress in Surface Science*, Vol. I, p. 155, ed. S. G. Davison (Pergamon, New York, 1972).
112. B. A. Joyce and J. H. Neave, *Surface Sci.* **34**, 401 (1973).
113. E. A. Wood, *J. Appl. Phys.* **35**, 1306 (1963).
114. J. A. Strozier, Jr., D. W. Jepsen, and F. Jona, in *Surface Physics of Materials*, Vol. I, ed. J. M. Blakely (Academic Press, New York, 1975).
115. L. L. Kesmodel and G. A. Somorjai, *Phys. Rev. B* **11**, 630 (1975).
116. J. E. Demuth, D. W. Jepson, and P. M. Marcus, *Phys. Rev. Lett.* **31**, 540 (1973).
117. M. van Hove and D. S. Y. Tong, *J. Vac. Sci. Tech.* **12**, 230 (1975).
118. T. N. Rhodin and D. S. Y. Tong, *Phys. Today* **28**, 23 (1975).
119. W. P. Ellis and R. L. Schwoebel, *Surface Sci.* **11**, 82 (1968).
120. M. Henzler and J. Klabes, *Jap. J. Appl. Phys.*, Suppl. 2, Pt. 2, 389 (1974).
121. J. Perferau and G. E. Rhead, *Surface Sci.* **24**, 555 (1971).
122. R. E. Weber and W. T. Peria, *J. Appl. Phys.* **38**, 4355 (1967).
123. P. W. Palmberg and T. N. Rhodin, *J. Appl. Phys.* **39**, 2425 (1968).
124. L. A. Harris, *J. Appl. Phys.* **39**, 1419 (1968).
125. T. Kawai, K. Kunimori, T. Kondow, T. Orishi, and K. Tamaru, *Jap. J. Appl. Phys.*, Suppl. 2, Pt. 2, 513 (1974).
126. L. H. Jenkins, D. M. Zehner, and M. F. Chung, *Surface Sci.* **38**, 327 (1973).
127. D. P. Smith, *J. Appl. Phys.* **38**, 340 (1967).

128. T. E. Madey, J. J. Czyzewski, and J. T. Yates, Jr., *Surface Sci.* **49**, 465 (1975).
129. R. H. Jones, D. R. Olander, W. J. Siekhaus, and J. A. Schwartz, *J. Vac. Sci. Tech.* **9**, 1429 (1972).
130. A. West and G. A. Somorjai, *J. Vac. Sci. Tech.* **9**, 668 (1972).
131. R. J. Madix and A. A. Susu, *J. Vac. Sci. Tech.* **9**, 915 (1972).
132. D. L. Smith, Ph. D. Thesis, U. California, Berkeley, 1969.
133. M. L. Hair, *Infrared Spectroscopy in Surface Chemistry* (Marcel Dekker, New York, 1967).
134. M. L. Kottke, R. G. Greenler, and H. G. Tomkins, *Surface Sci.* **32**, 231 (1972).
135. N. J. Harrick, *Ann. N. Y. Acad. Sci.* **101**, 928 (1963).
136. P. A. Redhead, *Vacuum* **12**, 203 (1962).
137. G. Ehrlich, *Adv. Catal.* **14**, 255 (1963).
138. J. L. Falconer and R. J. Madix, *Surface Sci.* **48**, 393 (1975).
139. D. A. King, *Surface Sci.* **47**, 384 (1975).
140. K. Tanabe, *Solid Acids and Bases* (Academic Press, New York, 1970).
141. T. Yamanaka and K. Tanabe, *J. Phys. Chem.* **79**, 2409 (1975).
142. E. P. Parry, *J. Catal.* **2**, 371 (1963).
143. H. Knözinger, *Surface Sci.* **41**, 339 (1974).
144. A. Terenin, *Adv. Catal.* **15**, 227 (1964).
145. Y. Kageyama, T. Yotsuyanagi, and K. Aumura, *J. Catal.* **36**, 1 (1975).
146. P. Soven, E. W. Plummer, and N. Kar, *CRC Crit. Rev. Solid State Sci.* **6**, 109 (1976).
147. J. H. Block, *CRC Crit. Rev. Solid State Sci.* **6**, 133 (1976).
148. E. W. Müller, *CRC Crit. Rev. Solid State Sci.* **6**, 85 (1976).
149. R. L. Palmer and J. N. Smith, *Catal. Rev.-Sci. Eng.* **12**, 279 (1975).
150. M. Smutek, S. Cerny, and F. Buzek, Adv. Catal. **24**, 343 (1975).

References for Chapter 4

1. S. G. Davison and J. D. Levine, *Solid State Physics*, Vol. 25, eds. H. Ehrenreich. F. Seitz and D. Turnbull (Academic Press, New York, 1970).
2. A. K. Vijh, *J. Mat. Sci.* **5**, 379 (1970).
3. F. Seitz, *Modern Theory of Solids*, pp. 408ff (McGraw-Hill, New York, 1940).
4. J. C. Phillips, *Rev. Mod. Phys.* **42**, 317 (1970).
5. P. Manca, *J. Phys. Chem. Solids* **20**, 268 (1961).
6. J. D. Levine and P. Mark, *Phys. Rev.* **144**, 751 (1966).
7. R. K. Swank, *Phys. Rev.* **153**, 884 (1967).
8. P. Mark, *Catal. Rev. Sci. Eng.* **12**, 71 (1975).
9. A. Kurtin, T. C. McGill, and C. A. Mead, *Phys. Rev. Lett.* **22**, 1433 (1969).
10. J. C. Phillips, *Surface Sci.* **37**, 24 (1973).
11. J. van Laar and J. J. Scheer, *Surface Sci.* **8**, 342 (1967).
12. T. Smith, *Surface Sci.* **25**, 45 (1971).
13. J. M. Chen, *Surface Sci.* **25**, 305 (1971).
14. L. J. Brillson, *J. Vac. Sci. Tech.* **12**, 249 (1975).
15. M. Henzler, *Surface Sci.* **25**, 650 (1971).
16. J. A. Appelbaum and D. R. Hamann, *Phys. Rev. Lett.* **31**, 106 (1973).
17. J. A. Appelbaum and D. R. Hamann, *Phys. Rev. Lett.* **32**, 225 (1974).

18. E. Louis and F. Yndurain, *Phys. Stat. Solidi B* **57**, 175 (1973).
19. R. O. Jones, in *Structure and Chemistry of Solid Surfaces*, p. 141, ed. G. Somorjai (Wiley, New York, 1969).
20. F. Yndurain and M. Elices, *Surface Sci.* **29**, 540 (1972).
21. J. A. Appelbaum and D. R. Hamann, *Phys. Rev. B* **6**, 2166 (1972).
22. J. A. Appelbaum, in *Surface Physics of Materials*, Vol. I, ed. J. M. Blakely (Academic Press, New York, 1975).
23. F. Yndurain and L. M. Falicov, *J. Phys. C* **8**, 1571 (1975).
24. J. D. Joannopoulos and M. L. Cohen, *Phys. Rev. B* **10**, 5075 (1974).
25. D. J. Chadi and M. L. Cohen, *Phys. Rev. B* **11**, 732 (1975).
26. D. J. Chadi and M. L. Cohen, *Solid State Comm.* **16**, 691 (1975).
27. R. Dorn, H. Lüth and G. J. Russell, *Phys. Rev. B* **9**, 1951 (1974).
28. K. Hirabayashi, *J. Phys. Soc. Jap.* **27**, 1475 (1969).
29. K. C. Pandy and J. C. Phillips, *Phys. Rev. Lett.* **32**, 1433 (1974).
30. K. C. Pandy and J. C. Phillips, *Phys. Rev. B* **13**, 750 (1976).
31. J. R. Schrieffer and P. Soven, *Phys. Today* **28**, 24 (1975).
32. J. W. Davenport, T. L. Einstein, and J. R. Schrieffer, in Proc. 2nd Int. Conf. on Solid Surfaces, Kyoto, 1974.
33. J. B. Pendry and F. Forstman, *J. Phys. C* **3**, 59 (1970).
34. F. Forstman, *Z. Phys.* **235**, 69 (1970).
35. F. Forstman and J. B. Pendry, *Z. Phys.* **235**, 75 (1970).
36. F. Forstman and V. Heine, *Phys. Rev. Lett.* **24**, 1419 (1970).
37. B. J. Waclawski and E. W. Plummer, *Phys. Rev. Lett.* **29**, 783 (1972).
38. B. Feuerbacher and B. Fitton, *Phys. Rev. Lett.* **29**, 786 (1972).
39. K. Tanaka and K. Tamaru, *Bull. Chem. Soc. Jap.* **37**, 1862 (1964).
40. R. T. Sanderson, *Chemical Periodicity* (Reinhold, New York, 1960).
41. J. K. Wilmshurst, *Can. J. Chem.* **38**, 467 (1960).
42. V. N. Filimov, Yu. N. Lopatin, and D. A. Sukhov, *Kinet. Katal.* **10**, 458 (1969).
43. P. Mars, in *The Mechanism of Heterogeneous Catalysis*, ed. J. H. de Boer (Elsevier, Amsterdam, 1960).
44. K. Tanabe, T. Sumijoshi, K. Shibata, T. Kiyoura, and J. Kitagawa, *Bull. Chem. Soc. Jap.* **47**, 1064 (1974).
45. A. J. Leonard, P. Ratnassamy, F. D. Declerk, and J. J. Fripiat, *Disc. Faraday Soc.* **52**, 98 (1971).
46. J. Koutecky, *Adv. Chem. Phys.* **9**, 85 (1965).
47. J. Levine, *Surface Sci.* **34**, 90 (1973).
48. V. Heine, *Jap. J. Appl. Phys.*, Suppl. 2, Pt. 2, 679 (1974).
49. R. E. Schlier and H. E. Farnsworth, *J. Chem. Phys.* **30**, 917 (1959).
50. E. Tosatti and P. W. Anderson, *Solid State Comm.* **14**, 773 (1974).
51. J. A. Appelbaum, G. A. Baraff, and D. R. Hamann, *Phys. Rev. B* **11**, 3822 (1975).
52. M. F. Chung and H. E. Farnsworth, *Surface Sci.* **22**, 93 (1970).
53. G. Heiland, Proc. 2nd Int. Conf. on Electrophotography, October, 1973, p. 117.
54. R. Leysen, G. van Orshaegan, H. van Hove, and A. Neyens, *Phys. Stat. Solidi A* **18**, 613 (1973).
55. S.-C. Chang and P. Mark, *Surface Sci.* **46**, 293 (1974).
56. A. U. MacRae and G. W. Gobeli, *J. Appl. Phys.* **35**, 1629 (1964).
57. J. Derrien, F. A. d'Avitaya, and A. Gelachant, *Surface Sci.* **47**, 162 (1975).

58. J. D. Levine and S. G. Davison, *Phys. Rev.* **174**, 911 (1968).
59. G. Heiland, P. Kunstmann, and H. Pfister, *Z. Phys.* **176**, 485 (1963).
60. R. Nosker, P. Mark, and J. D. Levine, *Surface Sci.* **19**, 291 (1970).
61. P. Mark, *J. Vac. Sci. Tech.* **10**, 893 (1973).
62. Y. Margoninski and Z. H. Kalman, *Phys. Lett.* **47A**, 201 (1974).
63. H. van Hove and R. Leysen, *Phys. Stat. Solidi A* **9**, 361 (1972).
64. B. J. Hopkins, R. Leysen, and P. A. Taylor, *Surface Sci.* **48**, 486 (1975).
65. W. A. Harrison and S. Ciraci, *Phys. Rev. B* **10**, 1516 (1974).
66. J. Marien, R. Leysen, and H. van Hove, *Phys. Stat. Solidi A* **5**, 121 (1971).
67. R. Leysen, H. van Hove, J. Marien, and J. Loosveldt, *Phys. Stat. Solidi A* **11**, 539 (1972).
68. H. Lüth and G. Heiland, *Phys. Stat. Solidi A* **14**, 573 (1972).
69. H. Lüth, *Surface Sci.* **37**, 90 (1973).
70. L. J. Brillson, *Surface Sci.* **51**, 45 (1975).
71. R. H. Williams and A. J. McEvoy, *Phys. Stat. Solidi A* **12**, 277 (1972).
72. T. E. Fischer, *Phys. Rev.* **139A**, 1228 (1965).
73. D. J. Miller and D. Haneman, *Phys. Rev. B* **3**, 2918 (1971).
74. J. Higinbotham and D. Haneman, *Surface Sci.* **32**, 466 (1972).
75. D. Haneman, *Jap. J. Appl. Phys.*, Suppl. 2, Pt. 2, 371 (1974).
76. H. Gatos and M. C. Lavine, *J. Electrochem. Soc.* **107**, 427 (1960).
77. G. Hochstrasser and J. F. Antonini, *Surface Sci.* **32**, 644 (1972).
78. R. L. Nelson and J. W. Hale, *Disc. Faraday Soc.* **52**, 77 (1971).
79. J. H. Dinan, L. K. Galbraith, and T. E. Fischer, *Surface Sci.* **26**, 587 (1971).
80. D. E. Eastman and W. D. Grobman, *Phys. Rev. Lett.* **28**, 1378 (1972).
81. P. E. Gregory and W. E. Spicer, *Phys. Rev. B* **13**, 725 (1976).
82. P. E. Gregory, W. E. Spicer, S. Ciraci, and W. A. Harrison, *Appl. Phys. Lett.* **25**, 511 (1974).
83. D. E. Eastman and J. L. Freeouf, *Phys. Rev. Lett.* **33**, 1601 (1974).
84. J. van Laar and J. J. Sheer, *Surface Sci.* **8**, 342 (1967).
85. H. Froitzheim and H. Ibach, *Surface Sci.* **47**, 713 (1975).
86. A. Huijser and J. van Laar, *Surface Sci.* **52**, 202 (1975).
87. W. Ranke and K. Jacobi, *Solid State Comm.* **13**, 705 (1973).
88. R. Ludeke and L. Esaki, *Surface Sci.* **47**, 132 (1975).
89. R. Ludeke and A. Koma, *CRC Crit. Rev. Solid State Sci.* **5**, 245 (1975).
90. K. Jacobi, *Surface Sci.* **51**, 29 (1975).
91. J. L. Freeouf and D. E. Eastman, *CRC Crit. Rev. Solid State Sci.* **5**, 245 (1975).
92. H. P. Boehm, *Disc. Faraday Soc.* **52**, 264 (1971).
93. K. Tanabe, *Solid Acids and Bases* (Academic Press, New York, 1970).
94. J. B. Peri, *J. Phys. Chem.* **69**, 211, 220, 231 (1965).
95. J. K. Lee and S. W. Weller, *Anal. Chem.* **30**, 1057 (1958).
96. E. Borello, G. D. Gatta, B. Fubini, C. Mortiera, and G. Venturello, *J. Catal.* **35**, 1 (1974).
97. E. P. Parry, *J. Catal.* **2**, 371 (1963).
98. H. P. Boehm, *Adv. Catal.* **16**, 179 (1966).
99. S. G. Hindin and S. W. Weller, *J. Phys. Chem.* **60**, 1501 (1956).
100. N. W. Cant and L. H. Little, *Can. J. Chem.* **43**, 1252 (1965).
101. M. L. Hair and W. Hertl, *J. Phys. Chem.* **77**, 1965 (1973).
102. P. B. West, G. R. Haller, and R. L. Burwell, Jr., *J. Catal.* **29**, 486 (1973).

103. M. Yamadayo, K. Shimomura, T. Konoshita, and H. Uchida, *Shokubai* (*Tokyo*) **7**, 313 (1965) (as reviewed in Ref. 93).
104. J. Koubek, J. Volf, and J. Pasek, *J. Catal.* **38**, 385 (1975).
105. K. Shibata, T. Kyoura, J. Kitagawa, T. Sumiyoshi, and K. Tanabe, *Bull. Chem. Soc. Jap.* **46**, 2985 (1973).
106. J. A. Schwartz, *J. Vac. Sci. Tech.* **12**, 321 (1975).
107. C. L. Thomas, *Ind. Eng. Chem.* **41**, 2564 (1949).
108. A. J. Tench and R. L. Nelson, *Trans. Faraday Soc.* **63**, 2254 (1967).
109. B. D. Flockhart, K. Y. Liew, and R. C. Pink, *J. Catal.* **32**, 20 (1974).
110. S. R. Morrison, *Surface Sci.* **34**, 462 (1975).
111. J. J. Lander and J. Morrison, *J. Appl. Phys.* **34**, 1403 (1963).
112. J. B. Marsh and H. E. Farnsworth, *Surface Sci.* **1**, 3 (1964).
113. J. E. Florio and W. D. Robertson, *Surface Sci.* **24**, 17 (1971).
114. P. P. Auer and W. Mönch, *Jap. J. Appl. Phys.*, Suppl. 2, Pt. 2, 397 (1974).
115. J. W. May, *Adv. Catal.* **21**, 151 (1970).
116. L. L. Kesmodel and G. A. Somorjai, *Phys. Rev. B* **11**, 630 (1975).
117. D. Aberdan, R. Badoing, C. Gaubert, and Y. Gauthien, *Surface Sci.* **47**, 181 (1975).
118. J. C. Phillips, *Surface Sci.* **44**, 290 (1974).
119. D. R. Palmer, S. R. Morrison, and C. E. Dauenbaugh, *J. Phys. Chem. Solids* **14**, 27 (1960).
120. D. R. Palmer, S. R. Morrison, and C. E. Dauenbaugh, *Phys. Rev.* **129**, 608 (1963).
121. Y. Margoninski, *J. Vac. Sci. Tech.* **9**, 920 (1972).
122. F. G. Allen and G. W. Gobeli, *Phys. Rev.* **127**, 150 (1962).
123. P. Handler, *Appl. Phys. Lett.* **3**, 96 (1963).
124. M. Henzler, *Phys. Stat. Solidi* **19**, 833 (1967).
125. M. Henzler and G. Heiland, *Solid State Comm.* **4**, 499 (1966).
126. B. F. Lewis and T. E. Fischer, *Surface Sci.* **41**, 371 (1974).
127. J. E. Rowe and H. Ibach, *Phys. Rev. Lett.* **32**, 421 (1974).
128. L. F. Wagner and W. E. Spicer, *Phys. Rev. Lett.* **28**, 1381 (1972).
129. T. Murotani, K. Fujiwara, and M. Nishijima, *Jap. J. Appl. Phys.*, Suppl. 2, Pt. 2, 409 (1974).
130. A. N. Boonstra and J. van Ruler, *Surface Sci.* **4**, 141 (1969).
131. D. R. Frankl, *Electrical Properties of Semiconductor Surfaces*, Chap. 7 (Pergamon Press, New York, 1967).
132. C. Sebenne, D. Bolmont, G. Guichan, and M. Balkanski, *Jap. J. Appl. Phys.*, Supp. 2, Pt. 2, 405 (1974); *Phys. Rev. B* **12**, 3280 (1975).
133. C. Chiarotti, P. Chiaradia, and S. Nannarone, *Surface Sci.* **49**, 315 (1975).
134. J. E. Rowe, H. Ibach, and H. Froitzheim, *Surface Sci.* **48**, 44 (1975).
135. W. Müller and W. Mönch, *Phys. Rev. Lett.* **27**, 250 (1971).
136. M. M. Traum, J. E. Rowe, and N. E. Smith, *J. Vac. Sci. Tech.* **12**, 298 (1975).
137. W. B. Shepherd and W. T. Peria, *Surface Sci.* **38**, 461 (1973).
138. L. Ernst and J. H. Block, *Surface Sci.* **49**, 293 (1975).
139. R. Ludeke and A. Koma, *Phys. Rev. B* **13**, 739 (1976).
140. I. Petroff and C. R. Viswanathan, *Phys. Rev. B* **4**, 799 (1971).
141. B. J. Waclawski, T. V. Vorburger, and R. J. Stein, *J. Vac. Sci. Tech.* **12**, 301 (1975).
142. B. Feuerbacher and N. E. Christensen, *Phys. Rev. B* **10**, 2373 (1974).
143. B. Feuerbacher, *Surface Sci.* **47**, 115 (1975).
144. E. W. Plummer and J. W. Gadzuk, *Phys. Rev. Lett.* **25**, 1493 (1970).

145. R. Kasowski, *Phys. Rev. Lett.* **33**, 83 (1974).
146. P. J. Feibelman and D. E. Eastman, *Phys. Rev. B* **10**, 4932 (1974).
147. V. E. Henrich, G. Dresselhaus, and H. J. Zeiger, *Phys. Rev. Lett.* **36**, 1335 (1976).
148. J. R. Chelikowsky and M. L. Cohen, *Phys. Rev. B* **13**, 826 (1976).
149. A. R. Lubinsky, C. B. Duke, B. W. Lee, and P. Mark, *Phys. Rev. Lett.* **36**, 1058 (1976).
150. R. A. Powell and W. E. Spicer, *Phys. Rev. B* **13**, 2601 (1976).
151. P. E. Gregory and W. E. Spicer, *Phys. Rev. B* **13**, 725 (1976).
152. W. A. Harrison, *Surface Sci.* **55**, 1 (1976).
153. J. B. Peri, *J. Catal.* **41**, 227 (1976).

References for Chapter 5

1. C. B. Duke and N. O. Lipari, *J. Vac. Sci. Tech.* **12**, 222 (1975).
2. A. Ignatiev, F. Jona, D. W. Depsa, and P. M. Marcus, *J. Vac. Sci. Tech.* **12**, 226 (1975).
3. J. W. May, *Adv. Catal.* **21**, 151 (1970).
4. M. Henzler, *Surface Sci.* **24**, 209 (1971).
5. W. M. H. Sachtler, *Surface Sci.* **22**, 468 (1970).
6. A. A. Holscher and W. M. H. Sachtler, *Disc. Faraday Soc.* **41**, 29 (1966).
7. R. Riwan, C. Guillot, and J. Paigne, *Surface Sci.* **47**, 183 (1975).
8. J. E. Demuth, D. W. Jepsen, and P. M. Marcus, *Phys. Rev. Lett.* **32**, 1182 (1974).
9. T. E. Madey, J. J. Czyzewski, and J. T. Yates, Jr., *Surface Sci.* **49**, 465 (1975); *Phys. Rev. Lett.* **32**, 777 (1974).
10. J. N. Butler, J. Giner, and J. M. Parry, *Surface Sci.* **18**, 140 (1969).
11. D. A. Rand and R. Woods, *Surface Sci.* **41**, 611 (1974).
12. V. Ponec, *Catal. Rev.* **11**, 41 (1975).
13. W. M. H. Sachtler and P. van der Plank, *Surface Sci.* **18**, 62 (1969).
14. Y. Soma and W. M. H. Sachtler, *Jap. J. Appl. Phys.*, Suppl. 2, Pt. 2, 241 (1974).
15. L. Pauling, *The Nature of the Chemical Bond* (Cornell U. Press, Ithaca, 1960).
16. R. T. Sanderson, *Chemical Periodicity* (Reinhold, New York, 1971).
17. R. Ludeke and A. Koma, *Phys. Rev. Lett.* **34**, 1170 (1975).
18. G. A. Bootsma and F. Meyer, *Surface Sci.* **18**, 123 (1969).
19. J. Fahrenfort, L. L. van Rijen, and W. M. H. Sachtler, Proceedings of Symposium on Mechanics of Heterogeneous Catalysis, p. 23, Amsterdam, 1959.
20. G. C. Bond, *Surface Sci.* **18**, 11 (1969).
21. Z. Knor, *Adv. Catal.* **22**, 51 (1972).
22. O. Johnson, *J. Catal.* **28**, 503 (1973).
23. G. C. Bond and P. B. Wells, *Adv. Catal.* **15**, 92 (1964).
24. J. L. Garnett, *Catal. Rev.* **5**, 229 (1971).
25. R. B. Moyes and P. B. Wells, *Adv. Catal.* **23**, 121 (1973).
26. M. Orchin, *Adv. Catal.* **16**, 1 (1966).
27. W. H. Weinberg, H. A. Deans, and R. P. Merrill, *Surface Sci.* **41**, 312 (1974).
28. J. L. Gland and G. A. Somorjai, *Surface Sci.* **41**, 387 (1974).
29. R. Queau and R. Poilblanc, *J. Catal.* **27**, 200 (1972).
30. A. J. Goodsel and G. Blyholder, *Chem. Comm.* **17**, 1132 (1970).

31. A. J. Sargood, C. W. Jowett, and B. J. Hopkins, *Surface Sci.* **22**, 343 (1970).
32. W. H. Weinberg and R. P. Merrill, *Surface Sci.* **39**, 206 (1973).
33. W. H. Weinberg, *J. Catal.* **28**, 459 (1973).
34. G. C. Bond, *Surface Sci.* **18**, 11 (1969).
35. P. Mark, *J. Phys. Chem. Solids* **29**, 689 (1968).
36. R. O. James and T. W. Healey, *J. Coll. Interf. Sci.* **40**, 42, 53, 65 (1972).
37. J. van Laar and J. J. Scheer, *Surface Sci.* **8**, 347 (1967).
38. R. Dorn, H. Lüth, and G. J. Russell, *Phys. Rev. B* **9**, 1951 (1974).
39. F. S. Stone, *Adv. Catal.* **13**, 1 (1962).
40. H. P. Boehm, *Adv. Catal.* **16**, 179 (1966).
41. R. Kh. Burshtein, L. A. Larin, and S. I. Sergeev, in *Surface Properties of Semiconductors*, ed. A. N. Frumkin (Consultants Bureau, New York, 1964).
42. G. J. Russell and D. Haneman, *Surface Sci.* **27**, 362 (1971).
43. G. A. Somorjai, *J. Phys. Chem. Solids* **24**, 175 (1963).
44. E. Guesne, C. Sebenne, and M. Balkanski, *Surface Sci.* **24**, 18 (1971).
45. A. M. Morgan and D. A. King, *Surface Sci.* **23**, 259 (1970).
46. G. K. Hall and C. H. B. Mee, *Surface Sci.* **28**, 598 (1971).
47. M. W. Roberts and B. R. Wells, *Surface Sci.* **15**, 325 (1969).
48. A. Benninghoven and L. Wiedmann, *Surface Sci.* **41**, 483 (1974).
49. T. Robert, M. Bartel, and G. Offergeld, *Surface Sci.* **33**, 123 (1972).
50. J. W. Gadzuk, in *Surface Physics of Materials*, Vol. II, ed. J. M. Blakely (Academic Press, New York, 1975).
51. T. B. Grimley, *Ber. Bunsenges. Phys. Chem.* **75**, 1003 (1971).
52. J. R. Schrieffer and P. Soven, *Phys. Today* **28**, 24 (1975).
53. T. B. Grimley, *Adv. Catal.* **12**, 1 (1960).
54. J. C. Robertson and C. W. Wilmsen, *J. Vac. Sci. Tech.* **8**, 53 (1971).
55. L. W. Anders, R. S. Hansen, and L. S. Bartell, *J. Chem. Phys.* **10**, 5277 (1973).
56. R. P. Messmer, B. McCarroll, and C. M. Singal, *J. Vac. Sci. Tech.* **9**, 891 (1972).
57. A. J. Bennett, B. McCarroll, and R. B. Messmer, *Surface Sci.* **24**, 191 (1971).
58. A. van der Avoird, S. P. Liebmann, and D. J. M. Fassaert, *Phys. Rev. B* **10**, 1230 (1974).
59. J. C. Slater and K. H. Johnson, *Phys. Today* **27**, 34 (1974).
60. I. P. Batra and O. Robaux, *Surface Sci.* **49**, 653 (1975).
61. I. P. Batra and P. S. Bagus, *Solid State Comm.* **16**, 1097 (1975).
62. D. E. Eastman and J. K. Cashion, *Phys. Rev. Lett.* **27**, 1520 (1971).
63. N. Rösch and T. N. Rhodin, *Phys. Rev. Lett.* **32**, 1189 (1974); *Disc. Faraday Soc.* **58**, 28 (1974).
64. J. E. Demuth and D. E. Eastman, *Phys. Rev. Lett.* **32**, 1123 (1974); *Jap. J. Appl. Phys.*, Suppl. 2, Pt. 2, 827 (1974).
65. T. B. Grimley, *J. Vac. Sci. Tech.* **8**, 31 (1971).
66. D. R. Penn, *Surface Sci.* **39**, 333 (1973).
67. J. R. Schrieffer, *J. Vac. Sci. Tech.* **9**, 561 (1971).
68. D. M. Newns, *Phys. Rev.* **178**, 1123 (1969).
69. E. A. Hyman, *Phys. Rev. B* **11**, 3739 (1975).
70. S. K. Lyo and R. Gomer, *Phys. Rev. B* **10**, 4161 (1974).
71. D. J. M. Fassaert, H. Verbeeck, and A. van der Avoird, *Surface Sci.* **29**, 501 (1972).
72. S. P. Liebmann, A. van der Avoird, and D. J. M. Fassaert, *Phys. Rev. B* **11**, 1503 (1975).

73. T. L. Einstein, *Phys. Rev. B* **11**, 577 (1975).
74. S. C. Ying, J. R. Smith, and W. Kohn, Phys. Rev. *B* **11**, 1483 (1975).
75. H. B. Huntington, L. A. Turk, and W. W. White, *Surface Sci.* **48**, 187 (1975).
76. J. W. Gadzuk, *J. Vac. Sci. Tech.* **12**, 289 (1975).
77. P. J. Feibelman and D. E. Eastman, *Phys. Rev. B* **10**, 4932 (1974).
78. D. Menzel, *J. Vac. Sci. Tech.* **12**, 313 (1975).
79. C. R. Brundle, *Surface Sci.* **48**, 99 (1975).
80. K. Y. Yu, J. C. McMenamin, and W. E. Spicer, *J. Vac. Sci. Tech.* **12**, 286 (1975).
81. E. Sickafus and F. Steinrisser, *Phys. Rev. B* **6**, 3714 (1972).
82. A. Liebsch, *Phys. Rev. Lett.* **32**, 1203 (1974).
83. A. Liebsch and G. W. Plummer, *Disc. Faraday Soc.* **58**, 19 (1974).
84. J. W. Gadzuk, *Phys. Rev. B* **10**, 5030 (1974); *Surface Sci.* **43**, 44 (1974).
85. M. M. Traum, J. E. Rowe, and N. E. Smith, *J. Vac. Sci. Tech.* **12**, 298 (1975).
86. M. M. Traum, N. V. Smith, and I. J. DiSalvo, *Phys. Rev. Lett.* **32**, 1241 (1974).
87. J. Anderson and G. J. Lapeyre, *Phys. Rev. Lett.* **36**, 376 (1976).
88. B. Feuerbacher and B. Fitton, *Solid State Comm.* **15**, 295 (1974).
89. J. E. Rowe and S. B. Christman, *J. Vac. Sci. Tech.* **12**, 293 (1975).
90. B. J. Waclawski, T. V. Vorburger, and R. J. Stein, *J. Vac. Sci. Tech.* **12**, 301 (1975).
91. R. Gomer, *Jap. J. Appl. Phys.*, Suppl. 2, Pt. 2, 213 (1974).
92. B. Feuerbacher, *Surface Sci.* **47**, 115 (1975).
93. P. L. Young and R. Gomer, *Surface Sci.* **44**, 277 (1974).
94. P. J. Estrup and J. Anderson, *J. Chem. Phys.* **45**, 2254 (1966).
95. P. W. Tamm and L. D. Schmidt, *J. Chem. Phys.* **55**, 425 (1971).
96. J. Anderson, G. W. Rubloff, M. A. Passler, and P. J. Stiles, *Phys. Rev. B* **10**, 2401 (1974).
97. E. W. Plummer and A. E. Bell, *J. Vac. Sci. Tech.* **9**, 583 (1972).
98. J. W. Gadzuk and E. W. Plummer, *Rev. Mod. Phys.* **45**, 487 (1973).
99. A. M. Bradshaw, D. Menzel, and M. Steinkelberg, *Disc. Faraday Soc.* **58**, 46 (1974).
100. K. A. Kress and G. J. Lapeyre, *Phys. Rev. Lett.* **28**, 1639 (1972).
101. C. R. Helms and W. E. Spicer, *Phys. Rev. Lett.* **28**, 565 (1972).
102. P. R. Norton and P. J. Richards, *Surface Sci.* **49**, 567 (1975).
103. S. R. Morrison, *Surface Sci.* **20**, 110 (1971).
104. S. R. Morrison, *Surface Sci.* **10**, 459 (1968).
105. H. G. Völz, G. Kämpf, and H. G. Fitzky, *Farbe u. Lack.* **78**, 1037 (1972).
106. A. Fujishima and K. Honda, *Nature* **238**, 37 (1972).
107. M. S. Wrighton, D. S. Ginley, P. T. Wolczanski, A. B. Ellis, D. L. Morse, and A. Linz, *Proc. Natl. Acad. Sci. USA* **72**, 1518 (1975).
108. P. E. Gregory and W. E. Spicer, *Surface Sci.* **54**, 229 (1976).
109. J. E. Demuth and D. E. Eastman, *Phys. Rev. B* **13**, 1523 (1976).
110. R. P. Messmer, S. K. Knudson, K. H. Johnson, J. B. Diamond, and C. Y. Yang, *Phys. Rev. B* **13**, 1396 (1976).
111. S. G. Davison and Y. S. Huang, *Solid State Commun.* **15**, 863 (1974).
112. E. N. Foo and S. G. Davison, *Surface Sci.* (1976), in press.
113. B. Feuerbacher and R. F. Willis, *Phys. Rev. Lett.* **36**, 1339 (1976).

References for Chapter 6

1. D. R. Frankl, *J. Electrochem. Soc.* **109**, 238 (1962).
2. P. J. Boddy and W. H. Brattain, *J. Electrochem. Soc.* **109**, 812 (1962).
3. S. R. Morrison, *J. Phys. Chem. Solids* **14**, 214 (1960).
4. S. R. Morrison, *J. Vac. Sci. Tech.* **7**, 84 (1969).
5. S. R. Morrison, *Surface Sci.* **27**, 586 (1971).
6. J. R. Anderson, R. J. MacDonald, and Y. Shimoyama, *J. Catal.* **20**, 147 (1971).
7. G. C. Bond, *Plat. Met. Rev.* **19**, 126 (1975).
8. J. T. Yates and C. W. Garland, *J. Phys. Chem.* **65**, 617 (1961).
9. R. van Hardeveld and F. Hartog, *Adv. Catal.* **22**, 75 (1972).
10. V. Haensel, U.S. Patent No. 2,566,521 (1949) Alumina Platinum Oxysulfide and its Preparation.
11. L. Heard and M. J. Herder, U.S. Patent No. 2,659,701 (1953), Solubilized Platinum Sulfide Reforming Catalyst.
12. E. S. Zhmud, V. S. Boronin and O. M. Poltorak, *Russ. J. Phys. Chem.* **39**, 431 (1965).
13. O. M. Poltorak and V. S. Boronin, *Russ. J. Phys. Chem.* **39**, 781 (1965).
14. H. A. Benesi and R. M. Curtis, *J. Catal.* **10**, 328 (1968).
15. K. Morikawa, T. Shirasaki, and M. Okada, *Adv. Catal.* **20**, 97 (1969).
16. Yu. I. Yermakov, B. N. Kuznetsov, and Yu. A. Rynden, *Reaction Kinet. Catal. Lett.* **2**, 151 (1975).
17. T. E. Whyte, Jr. *Catal. Rev.* **8**, 117 (1973).
18. H. Chon, R. A. Fisher, E. Tomeszko, and J. G. Aston, Proc. 2nd Int. Cong. on Catalysis, p. 217, Paris, 1961.
19. H. L. Gruber, *J. Phys. Chem.* **66**, 48 (1962).
20. J. E. Benson and M. Boudart, *J. Catal.* **4**, 704 (1965).
21. G. R. Wilson and W. K. Hall, *J. Catal.* **24**, 306 (1972).
22. N. Giordano and E. Moretti, *J. Catal.* **18**, 228 (1970).
23. P. Wentrcek, K. Kimoto, and H. Wise, *J. Catal.* **33**, 279 (1974); **36**, 247 (1975).
24. J. Freel, *J. Catal.* **25**, 139, 149 (1972).
25. J. W. Sprys, L. Bartosiewicz, R. McCune, and H. K. Plummer, *J. Catal.* **39**, 91 (1975).
26. S. E. Wanke and P. C. Flynn, *Catal. Rev. Sci. Eng.* **12**, 93 (1975).
27. R. A. Herrmann, *J. Phys. Chem.* **65**, 2189 (1960).
28. R. T. K. Baker, C. Thomas and R. B. Thomas, *J. Catal.* **38**, 510 (1975).
29. J. W. Geus, *Chemisorption and Reactions on Metallic Films*, Vol. 1, ed. J. R. Anderson (Academic Press, New York, 1971).
30. R. Klein, *Surface Sci.* **29**, 309 (1972).
31. R. Gomer, R. Wortman, and R. Lundy, *J. Chem. Phys.* **26**, 1147 (1957).
32. K. Tanaka and K. Miyahara, *J. Chem. Soc. Chem. Commun.* **1973**, 877 (1973).
33. J. C. Slater and K. H. Johnson, *Phys. Today* **27**, 34 (1974).
34. R. O. Jones, P. J. Jennings, and G. S. Painter, *Surface Sci.* **53**, 409 (1975).
35. R. C. Baetzold, *Comments Solid State Phys.* **4**, 62 (1972).
36. E. G. Schlosser, *Ber. Bunsenges. Phys. Chem.* **73**, 358 (1969).
37. W. Romanowski, *Surface Sci.* **18**, 373 (1969).
38. E. Ruckenstein, *J. Catal.* **35**, 441 (1974).

39. J. H. Sinfelt, *J. Catal.* **29**, 308 (1973).
40. D. B. Tanner and A. J. Sievers, *Phys. Rev. B* **11**, 1330 (1975).
41. R. van Hardeveld and A. van Montfoort, *Surface Sci.* **4**, 396 (1966); **17**, 90 (1969).
42. J. J. Ostermaier, J. R. Katzer, and W. H. Manogue, *J. Catal.* **33**, 457 (1974).
43. J. C. Wu and P. Harriott, *J. Catal.* **39**, 395 (1975).
44. S. R. Morrison, *Surface Sci.* **10**, 459 (1968).
45. J.-P. Bonnelle and M. Guelton, *Surface Sci.* **27**, 375 (1971).
46. M. Green and M. J. Lee, in *Solid State Surface Science*, Vol. I, ed. M. Green (Marcel Dekker, New York, 1969).
47. S. R. Morrison, *Surface Sci.* **15**, 363 (1969).
48. J. A. Jackson, J. R. Szedon, and T. A. Temofonte, *J. Electrochem. Soc.* **119**, 1424 (1972).
49. K. Tanabe, T. Sumiyoshi, K. Shibata, T. Kiyoura, and J. Kitagawa, *Bull. Chem. Soc. Jap.* **47**, 1064 (1974).
50. K. Tanabe, *Solid Acids and Bases* (Academic Press, New York, 1970).
51. K. Shibata, T. Kiyoura, J. Kitagawa, T. Sumiyoshi, and K. Tanabe, *Bull. Chem. Soc. Jap.* **46**, 2985 (1973).
52. M. L. Hair and W. Hertl, *J. Phys. Chem.* **77**, 1965 (1973).
53. S. Szczepanska and S. Malinowski, *J. Catal.* **27**, 1 (1972).
54. R. J. Tedeschi, *Corrosion* **31**, 130 (1975).
55. W. W. Arbogast, Jr., *Corrosion* **30**, 179 (1974).
56. R. Walker, *Corrosion* **31**, 97 (1975); **29**, 290 (1973).
57. P. Fugassi and E. G. Haney, Proc. 2nd Int. Conf. on Titanium, 1972; *Corrosion* **29**, 140 (1973).
58. S. R. Morrison, *J. Vac. Sci. Tech.* **7**, 84 (1970).
59. K. M. Sancier, *Surface Sci.* **21**, 1 (1970).
60. I. N. Putilova, S. A. Balezin, and V. P. Barannik, *Metallic Corrosion Inhibitors* (Pergamon Press, New York, 1960).
61. M. Guelton, J. P. Bonnelle, and J. P. Beaufils, *J. Chim. Phys.* **7–8**, 1122 (1971).
62. J. P. Bonnelle, J. M. Balois, and J. P. Beaufils, *J. Chim. Phys.* **6**, 1045 (1972).
63. P. Descamps, M. Guelton, J. P. Bonnelle, and J. P. Beaufils, *C. R. Acad. Sci. (Paris)* **279c**, 813 (1974).
64. S. R. Morrison and J. P. Bonnelle, *J. Catal.* **25**, 416 (1972).
65. K. Prelec and Th. Sluyters, *Rev. Sci. Instr.* **44**, 1451 (1973).
66. L. Holbrook and H. Wise, *J. Catal.* **20**, 367 (1971).
67. S. R. Morrison, *J. Catal.* **34**, 462 (1974).
68. A. N. Webb, *Ind. Eng. Chem.* **49**, 261 (1957).
69. H. R. Gerberich, F. E. Lutinski, and W. K. Hall, *J. Catal.* **6**, 209 (1966).
70. H. Pines, *Adv. Catal.* **12**, 117 (1960).
71. N. W. Cant and L. H. Little, *Can. J. Chem.* **43**, 1252 (1965).
72. A. V. Kiselev, *Disc. Faraday Soc.* **52**, 14 (1971).
73. S. Khoobiar, *J. Phys. Chem.* **68**, 411 (1964).
74. D. Bianchi, G. E. E. Garder, G. M. Pajonk, and S. J. Teichner, *J. Catal.* **38**, 135 (1975).
75. K. M. Sancier, *J. Catal.* **20**, 106 (1971).
76. G. E. Batley, A. Ekstrom, and D. A. Johnson, *J. Catal.* **36**, 285 (1975).
77. P. A. Sermon and G. C. Bond, *Catal. Rev.* **8**, 211 (1973).
78. G. M. Schwab, *Surface Sci.* **13**, 198 (1969).

79. S. S. Kilchitskaya and V. I. Strikha, *Surface Sci.* **38**, 149 (1973).
80. A. B. Kuper, *Surface Sci.* **13**, 172 (1969).
81. S. Bacyaroy and P. Mark, *Surface Sci.* **30**, 53 (1972).
82. P. Mark, *Surface Sci.* **25**, 192 (1971).
83. A. N. Frumkin, ed., *Surface Properties of Semiconductors* (Consultants Bureau, New York, 1962) (several papers on the real surface).
84. V. F. Kiselev, S. N. Kozlov, Yu. F. Novotskii-Vlasov, and R. V. Pruknikov, *Surface Sci.* **11**, 111 (1968).
85. S. N. Kozlov, V. F. Kiselev, and Yu. F. Novototskii-Vlasov, *Surface Sci.* **28**, 395 (1971).
86. J. Lagowski, I. Baltos, and H. Gatos, *Surface Sci.* **40**, 216 (1973).
87. J. Lagowski, E. S. Sproles, Jr. and H. Gatos, *Surface Sci.* **30**, 653 (1972) and references therein.
88. T. M. Valahas, J. Sochanski, and H. C. Gatos, *Surface Sci.* **26**, 41 (1971).
89. H. P. Bonzel, *CRC Crit. Rev. Solid State Sci.* **6**, 171 (1976).
90. D. W. Bassett, *Surface Sci.* **53**, 74 (1975).

References for Chapter 7

1. J. E. Lennard-Jones, *Proc. Roy. Soc.* **A106**, 463 (1924).
2. S. Brunauer, P. H. Emmett, and E. Teller, *J. Am. Chem. Soc.* **60**, 309 (1938).
3. S. Brunauer, L. S. Deming, W. E. Deming, and E. Teller, *J. Am. Chem. Soc.* **62**, 1723 (1940).
4. J. H. de Boer, *The Dynamical Character of Adsorption* (Oxford, London, 1968).
5. K. Hauffe and S. R. Morrison, *Adsorption* (de Gruyter, Berlin, 1974) (in German).
6. H. E. Neustadter and R. J. Bacigalupe, *Surface Sci.* **6**, 246 (1967).
7. S. Rees and J. P. Olivier, *On Physical Adsorption* (Interscience, New York, 1964).
8. A. A. Adamson, *Physical Chemistry of Surfaces* (Interscience, New York, 1963).
9. R. Aveyard and D. A. Haydon, *An Introduction to the Principles of Surface Chemistry* (Cambridge Univ. Press, London, 1973).
10. M. F. Chung and H. E. Farnsworth, *Surface Sci.* **24**, 635 (1971).
11. H. Ibach, K. Horn, R. Dorn, and H. Lüth, *Surface Sci.* **38**, 433 (1973).
12. K. Baron, D. W. Blakely, and G. A. Somorjai, *Surface Sci.* **41**, 45 (1974).
13. A. J. Sargood, C. W. Jowett, and B. J. Hopkins, *Surface Sci.* **22**, 343 (1970).
14. R. Gomer, R. Wortman, and R. Lundy, *J. Chem. Phys.* **26**, 1147 (1957).
15. R. Klein, *Surface Sci.* **29**, 309 (1972).
16. S. R. Morrison, *Adv. Catal.* **7**, 259 (1955).
17. N. H. Turner, *J. Catal.* **36**, 262 (1975).
18. C. Aharoni and F. C. Tompkins, *Adv. Catal.* **21**, 1 (1970).
19. T. Takaishi, *JCS Faraday Trans. I* **68**, 801 (1972).
20. J. Oudar, *J. Vac. Sci. Tech.* **9**, 657 (1972).
21. J. R. Schrieffer and P. Soven, *Phys. Today* **28**, 24 (1975).
22. G. L. Price and J. A. Venables, *Surface Sci.* **49**, 264 (1975).
23. R. J. H. Voorhoeve and R. S. Wagner, *Met. Trans.* **2**, 3421 (1971).
24. R. C. Henderson and R. F. Helm, *Surface Sci.* **30**, 310 (1972).
25. M. Kaburagi and J. Kanamori, *Jap. J. Appl. Phys.*, Suppl. 2, Pt. 2, 145 (1974).

26. J. W. May, *Adv. Catal.* **21**, 151 (1970).
27. D. L. Adams, *Surface Sci.* **42**, 12 (1974).
28. P. J. Estrup, *Phys. Today* **28**, 33 (1975).
29. J. C. Tracy and P. W. Palmberg, *J. Chem. Phys.* **51**, 4852 (1969).
30. J. C. Tracy and P. W. Palmberg, *Surface Sci.* **14**, 274 (1969).
31. J. C. Tracy, *J. Chem. Phys.* **56**, 2736, 2748 (1972).
32. J. C. Tracy and J. M. Burkstrand, *CRC Crit. Rev. Solid State Sci.* **4**, 381 (1974).
33. A. Ignatiev and F. Jona, *Surface Sci.* **49**, 189 (1975).
34. W. H. Weinberg, H. A. Deans, and R. P. Merrill, *Surface Sci.* **41**, 312 (1974).
35. W. Heiland and E. Taglauer, *J. Vac. Sci. Tech.* **9**, 620 (1972).
36. T. E. Madey, H. A. Engelhardt, and D. Menzel, *Surface Sci.* **48**, 304 (1975).
37. R. Riwan, C. Guillot, and J. Paigne, *Surface Sci.* **47**, 183 (1975).
38. A. M. Bradshaw, D. Menzel, and M. Steinkilberg, *Jap. J. Appl. Phys.*, Suppl. 2, Pt. 2, 841 (1974).
39. M. van Hove and S. Y. Tong, *J. Vac. Sci. Tech.* **12**, 230 (1975).
40. R. M. Lambert and C. M. Comrie, *Surface Sci.* **46**, 61 (1974).
41. A. Ignatiev, F. Jona, D. W. Depson, and P. M. Marcus, *J. Vac. Sci. Tech.* **12**, 226 (1975).
42. G. Rovida, F. Pratesi, M. Maglietta, and E. Ferroni, *J. Vac. Sci. Tech.* **9**, 796 (1972).
43. P. B. Weisz, *J. Chem. Phys.* **21**, 1531 (1953).
44. A. Many, Y. Goldstein and N. B. Grover, *Semiconductor Surfaces* (Interscience, New York, 1965).
45. Th. Volkenshtein and O. Peshev, *J. Catal.* **4**, 301 (1965).
46. D. A. Melnick, *J. Chem. Phys.* **26**, 1136 (1957).
47. S. R. Morrison, in *Current Problems in Electrophotography*, eds. W. F. Berg and K. Hauffe (de Gruyter, Berlin, 1972).
48. S. R. Morrison and P. H. Miller, Jr., *J. Chem. Phys.* **25**, 1064 (1956).
49. M. A. Seitz and T. O. Sokolz, *J. Electrochem. Soc.* **121**, 162 (1974).
50. K. Hauffe, *Z. Elektrochem.* **65**, 321 (1961).
51. J. H. Lunsford, *Catal. Rev.* **8**, 135 (1973).
52. A. J. Tench and P. J. Holroyd, *Chem. Comm.* **1968**, 471 (1968).
53. N.-B. Wong and J. H. Lunsford, *J. Chem. Phys.* **56**, 2664 (1972).
54. N.-B. Wong, Y. B. Taarit, and J. H. Lunsford, *J. Chem. Phys.* **60**, 2148 (1974).
55. B. N. Shelimov, C. Naccache and M. Che, *J. Catal.* **37**, 279 (1975).
56. V. A. Schvets, V. M. Vorotyntsev, and V. B. Kazanski, Kinet. Katal. **10**, 356 (1969).
57. C. Naccache, *Chem. Phys. Lett.* **11**, 323 (1971).
58. S. Yoshida, T. Matsuzaki, T. Kashiwazaki, K. Mori, and K. Tarama, *Bull. Chem. Soc. Jap.* **47**, 1564 (1974).
59. E. Arijs, F. Cardon, and W. Maenhout-Van der Vorst, *J. Solid State Chem.* **6**, 326 (1973).
60. E. Kh. Enikeev, L. Ya. Margolis, and S. Z. Roginskii, *Dokl. Akad. Nauk. S.S.S.R.* **129**, 372 (1959).
61. T. I. Barry and F. S. Stone, *Proc. Roy. Soc.* **A255**, 124 (1960).
62. E. Kh. Enikeev, *Problemy Kinet. Katal.* **10**, 88 (1960).
63. K. Sancier, *J. Catal.* **9**, 331 (1967).
64. J. H. Lunsford and J. P. Jayne, *J. Chem. Phys.* **44**, 1487 (1966).
65. M. Codell, J. Weisberg, H. Gisser, and R. D. Iyengar, *J. Am. Chem. Soc.* **91**, 7762 (1969).

66. A. Tench and T. Lawson, *Chem. Phys. Lett.* **8**, 177 (1971).
67. J. Cunningham, J. J. Kelly, and A. L. Penny, *J. Phys. Chem.* **74**, 1992 (1970).
68. H. Chon and J. Pajares, *J. Catal.* **14**, 257 (1969).
69. J. Volger, *Phys. Rev.* **79**, 1023 (1950).
70. E. E. Hahn, *J. Appl. Phys.* **22**, 855 (1951).
71. K. Tanaka and G. Blyholder, *J. Chem. Soc. Chem. Commun.* **1971**, 1343 (1971).
72. K. Tanaka and K. Miyahara, *J. Chem. Soc. Chem. Commun.* **1973**, 877 (1973).
73. S. R. Morrison, *Surface Sci.* **27**, 586 (1971).
74. V. A. Schvets and V. B. Kazanski, *J. Catal.* **25**, 123 (1972).
75. M. Komuro, *Bull. Chem. Soc. Jap.* **48**, 756 (1975).
76. W. P. Gomes, *Surface Sci.* **19**, 172 (1970).
77. H. Clark and D. J. Berets, *Adv. Catal.* **9**, 204 (1957).
78. M. J. Katz and K. J. Haas, *Surface Sci.* **19**, 380 (1970).
79. T. Robert, M. Bartel and G. Offergeld, *Surface Sci.* **33**, 123 (1972).
80. S. R. Morrison, to be published.
81. R. L. Burwell, G. L. Haller, K. C. Taylor, and J. F. Read, *Adv. Catal.* **20**, 1 (1969).
82. M. P. McDaniel and R. L. Burwell, Jr., *J. Catal.* **36**, 394 (1975).
83. A. A. Davydov, Yu. M. Shchekochikhin, and N. P. Keier, *Kinet. Katal.* **13**, 1088 (1972).
84. A. A. Davydov, Yu. M. Shchekochikhin, N. P. Keier, and A. P. Zeif, *Kinet. Katal.* **10**, 1125 (1969).
85. D. J. Miller and D. Haneman, *Phys. Rev.* **133**, 2918 (1971).
86. G. J. Russell and D. Haneman, *Surface Sci.* **27**, 362 (1971).
87. A. U. MacRae and G. W. Gobeli, *J. Appl. Phys.* **35**, 1629 (1964).
88. R. Dorn, H. Lüth and G. J. Russell, *Phys. Rev. B* **9**, 1951 (1974).
89. A. N. Webb, *Ind. Eng. Chem.* **49**, 261 (1957).
90. R. M. Cornell, A. M. Posner, and J. P. Quick, *J. Coll. Int. Sci.* **53**, 6 (1975).
91. G. A. Parks, *Chem. Rev.* **65**, 177 (1965).
92. R. O. James and T. W. Healy, *J. Coll. Int. Sci.* **40**, 42, 53, 65 (1972).
93. W. H. Weinberg, R. M. Lambert, C. M. Comrie, and J. W. Linnett, *Surface Sci.* **30**, 299 (1972).
94. B. J. Wood, N. Endow, and H. Wise, *J. Catal.* **18**, 70 (1970).
95. J. A. Joebstl. *J. Vac. Sci. Tech.* **12**, 347 (1975).
96. B. Lang, R. W. Joyner, and G. A. Somorjai, *Surface Sci.* **30**, 454 (1972).
97. S. L. Bernasek and G. A. Somorjai, *Surface Sci.* **48**, 204 (1975).
98. G. A. Somorjai, *Catal. Rev.* **7**, 87 (1972).
99. M. Alnot, A. Cassuto, J. Fusy, and A. Pentenero, *Jap. J. Appl. Phys.*, Suppl. 2, Pt. 2, 79 (1974).
100. P. Mark, *Catal. Rev. Sci. Eng.* **12**, 71 (1975).
101. J. Völter, M. Procop, and H. Bernet, *Surface Sci.* **39**, 453 (1973).
102. A. West and G. A. Somorjai, *J. Vac. Sci. Tech.* **9**, 668 (1972).
103. W. H. Weinberg, H. A. Deans, and R. P. Merrill, *Surface Sci.* **41**, 312 (1974).
104. J. G. McCarty and R. J. Madix, *Surface Sci.* **54**, 121 (1976).
105. R. M. Lambert and C. M. Comrie, *Surface Sci.* **46**, 61 (1974).
106. E. N. Sickafus and H. P. Bonzel, *Prog. Surf. Membrane Sci.* **4c**, 116 (1971).
107. P. R. Norton and P. J. Richards, *Surface Sci.* **44**, 129 (1974).
108. J. K. Roberts, *Some Problems in Adsorption* (Cambridge University Press, 1939).
109. M. Procop and J. Völter, *Z. Phys. Chem. Leipzig* **250**, 387 (1972).

110. M. Procop and J. Völter, *Surface Sci.* **33**, 69 (1972).
111. Y. Kubokawa, S. Jakashina, and O. Toyama, *J. Phys. Chem.* **68**, 1244 (1964).
112. L. T. Dixon, R. Bartl, and J. W. Gryder, *J. Catal.* **37**, 368 (1975).
113. J. D. Clewley, J. F. Lynch, and T. B. Flanagan, *J. Catal.* **36**, 291 (1975).
114. B. Weber, J. Fusy, and A. Cassuto, *J. Chim. Phys.* **66**, 708 (1968).
115. P. R. Norton, *J. Catal.* **36**, 211 (1975).
116. E. Zaremba and W. Kohn, *Phys. Rev. B* **13**, 2270 (1976).
117. J. G. McCarty and R. J. Madix, *Surface Sci.* **54**, 210 (1976).
118. R. Ducros and R. P. Merrill, *Surface Sci.* **55**, 227 (1976).

References for Chapter 8

1. M. Hofman-Perez and H. Gerischer, *Z. Elektrochem.* **65**, 771 (1961).
2. J. F. Dewald, *Proceedings of Second Conference on Semiconductor Surfaces*, ed. J. Zemel (Pergamon Press, New York, 1960).
3. P. J. Boddy, *Surface Sci.* **13**, 52 (1969).
4. S. R. Morrison, *Prog. Surface Sci.* **1**, 105 (1971).
5. R. A. Marcus, *J. Chem. Phys.* **24**, 966 (1956).
6. R. A. Marcus, *Can. J. Chem.* **37**, 155 (1959).
7. R. A. Marcus, *J. Chem. Phys.* **43**, 679 (1965).
8. R. R. Dogonadze, A. M. Kuznetsov, and Yu. A. Chizmadzhev, *Russ. J. Phys. Chem.* **38**, 652 (1964).
9. R. R. Dogonadze, A. M. Kuznetsov, and A. A. Cherenko, *Russ. Chem. Rev.* **34**, 759 (1965).
10. J. M. Hale, in *Reactions of Molecules at Electrodes*, ed. N. S. Hush (Wiley Interscience, London, 1971).
11. T. Erdey-Gruz, *Kinetics of Electrode Processes* (Wiley Interscience, New York, 1972).
12. P. van Rysselberghe, *Electrochim. Acta* **8**, 583, 709 (1963); **9**, 1547 (1964); **10**, 107 (1965).
13. H. Gerischer, in *Advances in Electrochemistry and Electrochemical Engineering*, Vol. I, ed. P. Delahay (Interscience, New York, 1961).
14. D. M. Tench and E. Yeager, *J. Electrochem. Soc.* **121**, 318 (1974).
15. T. Freund and S. R. Morrison, *Surface Sci.* **9**, 119 (1968).
16. S. R. Morrison, *Surface Sci.* **15**, 363 (1969).
17. S. R. Morrison and T. Freund, Proc. NBS Workshop on Electrocatalysis on Non-Metallic Surfaces, ed. A. D. Franklin, 1976.
18. H. Gerischer, NATO Summer Course on Fundamental Processes on Semiconductor Surfaces, U. of Ghent, 1968.
19. H. Gerischer, *Surface Sci.* **18**, 97 (1969).
20. W. M. Latimer, *Oxidation Potentials* (Prentice–Hall, New York, 1952).
21. F. Beck and H. Gerischer, *Z. Elektrochem.* **63**, 943 (1959).
22. H. Gerischer and E. Meyer, *Z. Phys. Chem. NF* **74**, 302 (1971).
23. R. A. L. Vanden Berghe, and W. P. Gomes, *Ber. Bunsenges. Phys. Chem.* **76**, 481 (1972).
24. Yu. V. Pleskov, *Croatica Chem. Acta* **44**, 179 (1972).

25. V. A. Myamlin and Y. V. Pleskov, *Electrochemistry of Semiconductors* (Plenum Press, New York, 1967).
26. D. R. Turner, *The Electrochemistry of Semiconductors*, ed. P. J. Holms (Academic Press, London, 1962).
27. W. H. Brattain and C. G. B. Garrett, *B. S. T. J.* **34**, 129 (1955).
28. R. Memming and F. Möllers, *Ber. Bunsenges. Phys. Chem.* **76**, 475 (1972).
29. F. Cardon and W. P. Gomes, *Ber. Bunsenges. Phys. Chem.* **74**, 436 (1970).
30. S. R. Morrison and T. Freund, *J. Chem. Phys.* **47**, 1543 (1967).
31. M. C. Markham and M. C. Upreti, *J. Catal.* **4**, 229 (1965).
32. S. R. Morrison and T. Freund, *Electrochim. Acta* **13**, 1343 (1968).
33. W. P. Gomes, T. Freund, and S. R. Morrison, *J. Electrochem. Soc.* **115**, 818 (1968).
34. W. P. Gomes, T. Freund, and S. R. Morrison, *Surface Sci.* **13**, 201 (1969).
35. T. Freund and W. P. Gomes, *Catal. Rev.* **3**, 1 (1969).
36. H. Gerischer, in the discussion following Ref. 32.
37. H. Gerischer and H. Rösler, *Chem. Ing. Tech.* **4**, 176 (1970).
38. R. A. L. Vanden Berghe, W. P. Gomes, and F. Cardon, *Z. Phys. Chem. NF* **92**, 91 (1974).
39. R. Memming, *J. Electrochem. Soc.* **116**, 785 (1969).
40. K. Micka and H. Gerischer, *J. Electroanal. Chem.* **38**, 397 (1972).
41. F. Cardon and W. P. Gomes, *Surface Sci.* **27**, 286 (1971).
42. F. Lohmann, *Ber. Bunsenges. Phys. Chem.* **70**, 428 (1966).
43. W. H. Laflere, F. Cardon, and W. P. Gomes, *Surface Sci.* **44**, 541 (1974).
44. F. Möllers and R. Memming, *Ber. Bunsenges. Phys. Chem.* **76**, 469 (1972).
45. A. Fujishima, A. Sakamoto, and K. Honda, *Seisan-Kenkyu* **21**, 450 (1969).
46. R. A. L. Vanden Berghe, F. Cardon, and W. P. Gomes, *Surface Sci.* **39**, 368 (1973).
47. R. de Gryse, W. P. Gomes, F. Cardon, and J. Vennik, *J. Electrochem. Soc.* **122**, 711 (1975).
48. W. P. Gomes and F. Cardon, *Z. Phys. Chem. NF* **86**, 330 (1973).
49. R. Memming and F. Möllers, *Ber. Bunsenges. Phys. Chem.* **76**, 609 (1972).
50. R. A. L. Vanden Berghe and W. P. Gomes, *Ber. Bunsenges. Phys. Chem.* **76**, 481 (1972).
51. T. Freund, *J. Phys. Chem.* **73**, 468 (1969).
52. W. P. Gomes, *Surface Sci.* **19**, 172 (1970).
53. B. Pettinger, H.-R. Shöppel, and H. Gerischer, *Ber. Bunsenges. Phys. Chem.* **78**, 450 (1974).
54. T. Freund, in "Effect of Environment on Thermal Control Coatings" Final Report, Jet Propulsion Lab., NASA, Contract 951522, October, 1969, by S. R. Morrison and T. Freund.
55. H. Gerischer, *Surface Sci.* **13**, 265 (1969).
56. Y. V. Pleskov, *Proceedings of the International Conference on Semiconductor Physics, Prague* (Academic Press, London, 1962) (in Russian).
57. Y. V. Pleskov, *Dokl. Akad. Nauk. S.S.S.R.* **126**, 111 (1959).
58. W. H. Brattain and P. J. Boddy, *J. Electrochem. Soc.* **109**, 574 (1962).
59. V. A. Tyagai and G. Ya. Kolbasov, *Surface Sci.* **28**, 423 (1971).
60. R. A. L. Vanden Berghe, F. Cardon, and W. P. Gomes, *Ber. Bunsenges. Phys. Chem.* **78**, 331 (1974).
61. H. Gerischer, A. Maurer, and W. Mindt, *Surface Sci.* **4**, 431 (1966).
62. R. Memming and G. Neumann, *Surface Sci.* **10**, 1 (1968).

63. P. J. Boddy and W. H. Brattain, *J. Electrochem. Soc.* **109**, 812 (1962).
64. R. Memming, *Surface Sci.* **2**, 436 (1964).
65. D. J. Bernard and P. Handler, *Surface Sci.* **40**, 141 (1973).
66. V. A. Benderskii, Ya. M. Zolotovitskii, Ya. L. Kogan, M. L. Khidekel, and B. M. Shub, *Dokl. Akad. Nauk. S.S.S.R.* **222**, 606 (1975).
67. E. C. Dutoit, F. Cardon, and W. P. Gomes, to be published.
68. M. Gleria and R. Memming, *J. Electroanal. Chem.* **65**, 163 (1975).

References for Chapter 9

1. W. Shockley and W. T. Read, Jr., *Phys. Rev.* **87**, 835 (1952).
2. A. Many, Y. Goldstein, and N. B. Grover, *Semiconductor Surfaces* (Interscience, New York, 1965).
3. D. R. Frankl, *Electrical Properties of Semiconductor Surfaces* (Pergamon, London, 1967).
4. A. Many and D. Gerlich, *Phys. Rev.* **107**, 404 (1957).
5. Reference 3, p. 237.
6. S. R. Morrison, *J. Vac. Sci. Tech.* **7**, 84 (1970).
7. R. J. Collins and D. G. Thomas, *Phys. Rev.* **112**, 388 (1958).
8. K. M. Sancier, *Surface Sci.* **21**, 1 (1970).
9. S. R. Morrison, in *Surface Physics of Phosphors and Semiconductors*, eds. C. G. Scott and C. E. Reed (Academic Press, New York, 1975).
10. F. F. Volkenshtein and I. V. Karpenko, in *Surface Properties of Semiconductors*, p. 79, ed. A. N. Frumkin (Consultants Bureau, New York, 1964).
11. P. Mark, *Catal. Rev.* **1**, 165 (1967).
12. L. K. Galbraith and T. E. Fischer, *Surface Sci.* **30**, 185 (1972).
13. S. Baidyaroz, W. R. Bottoms, and P. Mark, *Surface Sci.* **28**, 517 (1971).
14. S. R. Morrison, *Adv. Catal.* **7**, 259 (1955).
15. D. A. Melnick, *J. Chem. Phys.* **26**, 1136 (1957).
16. D. B. Medved, *J. Chem. Phys.* **28**, 870 (1958); *J. Phys. Chem. Sol.* **20**, 255 (1961).
17. E. Arijs, F. Cardon, and W. Maenhout-Van der Vorst, *Z. Phys. Chem. NF* **94**, 255 (1975).
18. D. Eger, Y. Goldstein, and A. Many, *RCA Rev.* **36**, 508 (1975).
19. G. Heiland, 2nd Int. Conf. on Electrophotography, p. 117, Soc. Phot. Sci. and Engrs., Oct., 1973.
20. W. Hirschwald and E. Thull, *Disc. Faraday Soc.* **58**, 176 (1974).
21. D. G. Thomas, *J. Phys. Chem. Solids* **3**, 229 (1957).
22. J. Cunningham, E. Finn, and N. Samman, *Disc. Faraday Soc.* **58**, 160 (1974).
23. F. Steinbach and R. Harborth, *Disc. Faraday Soc.* **58**, 143 (1974).
24. K. Hauffe and H. Volz, *Ber. Bunsenges. Phys. Chem.* **77**, 967 (1973).
25. Y. Shapira, S. M. Cox, and D. Lichtman, *Surface Sci.* **54**, 43 (1976).
26. M. F. Chung and H. E. Farnsworth, *Surface Sci.* **25**, 321 (1972).
27. K. Tanaka and G. Blyholder, *J. Phys. Chem.* **76**, 3184 (1972).
28. N. Wong, Y. B. Taarit, and J. H. Lunsford, *J. Chem. Phys.* **60**, 2148 (1974).
29. C. L. Balestra and H. C. Gatos, *Surface Sci.* **28**, 563 (1971).
30. C. E. Weitzel and L. K. Monteith, *Surface Sci.* **44**, 438 (1974).

31. C. E. Weitzel and L. K. Monteith, *Surface Sci.* **40**, 555 (1973).
32. R. L. Nelson and J. W. Hale, *Disc. Faraday Soc.* **52**, 77 (1971).
33. M. Petrera, F. Trifiro, and G. Benedek, *Jap. J. Appl. Phys.*, Suppl. 2, Pt. 2, 315 (1974).
34. M. Hughes, *FATIPEC Congr.* **10**, 67 (1970).
35. R. I. Bickley, G. Munuera, and F. S. Stone, *J. Catal.* **31**, 398 (1973).
36. H. G. Völz, G. Kämpf, and H. G. Fitzky, *Farbe u. Lack.* **78**, 1037 (1972).
37. A. H. Boonstra and C. A. H. A. Mutsaers, *J. Phys. Chem.* **79**, 1694 (1975).
38. I. S. McLintock and M. Ritchie, *Trans. Faraday Soc.* **61**, 1007 (1965).
39. R. I. Bickley and R. K. M. Jayantz, *Disc. Faraday Soc.* **58**, 194 (1974).
40. J. J. Kelly and J. K. Vondeling, *J. Electrochem. Soc.* **122**, 1103 (1975).
41. F. Möllers, H. J. Tolle, and R. Memming, *J. Electrochem. Soc.* **121**, 1160 (1974).
42. J. Cunningham and H. Zainal, *J. Phys. Chem.* **76**, 2362 (1972).
43. Th. Volkenshtein, *Adv. Catal.* **23**, 157 (1973).
44. F. F. Volkenshtein and V. B. Nagaev, *Kinet. Catal.* **14**, 325, 1291 (1974); **16**, 320 (1975).
45. K. Tanaka and G. Blyholder, *J. Phys. Chem.* **75**, 1037 (1971).
46. J. Cunningham, J. J. Kelly, and A. L. Penny, *J. Phys. Chem.* **75**, 617 (1971).
47. J. Cunningham and A. L. Penny, *J. Phys. Chem.* **76**, 2353 (1972).
48. A. Fujishima and K. Honda, *Nature* **238**, 37 (1972).
49. M. S. Wrighton, D. S. Ginley, P. T. Wolczanski, A. B. Ellis, D. L. Morse, and A. Linz, *Proc. Natl. Acad. Sci. USA* **72**, 1518 (1975).
50. T. Ohnishi, Y. Nakato, and H. Tsubomura, *Ber. Bunsenges. Phys. Chem.* **79**, 523 (1975).
51. K. L. Hardee and A. J. Bard, *J. Electrochem. Soc.* **122**, 739 (1975).
52. H. Kim and H. A. Laitenen, *J. Electrochem. Soc.* **122**, 53 (1975).
53. J. G. Mavroides, D. I. Tchernev, J. A. Kafalas, and D. F. Kolesar, *Mat. Res. Bull.* **10**, 1023 (1975).
54. F. Romero-Rossi and F. S. Stone, 2nd Int. Congr. on Catalysis, p. 1486, Paris, 1960.
55. V. S. Zakharenko, A. E. Cherkashin, and N. P. Keier, *Dokl. Akad. Nauk. S.S.S.R.* **211**, 628 (1973).
56. F. Steinbach and R. Barth, *Ber. Bunsenges. Phys. Chem.* **73**, 884 (1969).
57. K. Tanaka and G. Blyholder, *J. Phys. Chem.* **76**, 1807 (1972).
58. W. Doerffler and K. Hauffe, *J. Catal.* **3**, 17 (1964).
59. A. Thevenet, F. Juillet and S. J. Teichner, *Jap. J. Appl. Phys.*, Suppl. 2, Pt. 2, 529 (1974).
60. G. Irick, Jr., *J. Appl. Polym. Sci.* **16**, 2387 (1972).
61. H. Knoll and U. Kühnhold, *Angew. Chem. Int. Ed.* **6**, 978 (1967).
62. W. A. Weyl and T. Forland, *Ind. Eng. Chem.* **42**, 257 (1950).
63. P. F. Cornaz and G. C. A. Schuit, *Disc. Faraday Soc.* **41**, 290 (1966).
64. A. J. Tench and T. Lawson, *Chem. Phys. Lett.* **7**, 459 (1970); **8**, 177 (1971).
65. N. Djeghri, M. Formenti, F. Juillet, and S. J. Teichner, *Disc. Faraday Soc.* **58**, 185 (1974).
66. R. I. Bickley and F. S. Stone, *J. Catal.* **31**, 389 (1973).
67. V. N. Filimonov, *Kinet. Katal.* **7**, 512 (1966).
68. S. P. Pappas and R. M. Fischer, *J. Paint Tech.* **46**, 65 (1974).
69. K. Sancier and S. R. Morrison, *Surface Sci.* **36**, 622 (1973).

70. K. Tanaka and G. Blyholder, *J. Phys. Chem.* **76**, 1394 (1972).
71. S. R. Morrison and T. Freund, *J. Chem. Phys.* **47**, 1543 (1967).
72. E. Inoue, in *Current Problems in Electrophotography*, p. 146, eds. W. F. Berg and K. Hauffe (deGruyter, Berlin, 1972).
73. H. Meier, *Spectral Sensitization* (Focal Press, London, 1968).
74. M. Vodenicharova and G. H. Jensen, *J. Phys. Chem. Solids* **36**, 1241 (1975).
75. G. Heiland and W. Bauer, *Tappi* **56**, 83 (1973).
76. B. Pettinger, H.-R. Schöppel, and H. Gerischer, *Ber. Bunsenges. Phys. Chem.* **77**, 960 (1973).
77. H. Yoneyama, Y. Toyoguchi, and H. Tamura, *J. Phys. Chem.* **76**, 3460 (1972).
78. H. Tributsch and H. Gerischer, *Ber. Bunsenges. Phys. Chem.* **73**, 582 (1969).
79. H. Gerischer and H. Tributsch, *Ber. Bunsenges. Phys. Chem.* **72**, 437 (1968).
80. U. Bode, K. Hauffe, Y. Ishikawa, and H. Putsch, *Z. Phys. Chem. NF* **85**, 144 (1973).
81. T. Freund, *Surface Sci.* **33**, 295 (1972).
82. J. C. Lagowski, E. S. Sproles, Jr., and H. C. Gatos, *Surface Sci.* **30**, 653 (1972).
83. C. L. Balestra, J. Lagowski, and H. C. Gatos, *Surface Sci.* **26**, 317 (1971).
84. J. Lagowski, C. L. Balestra, and H. C. Gatos, *Surface Sci.* **29**, 203, 213 (1972).
85. H. Lüth, *Surface Sci.* **37**, 90 (1973).
86. G. Heiland and W. Mönch, *Surface Sci.* **37**, 30 (1973).
87. H. Lüth and G. Heiland, *Phys. Stat. Solidi A* **14**, 573 (1972).
88. H. Gerischer, *Disc. Faraday Soc.* **58**, 219 (1974).

References for Chapter 10

1. G. Gati, *J. Catal.* **34**, 203 (1974).
2. J. Deren, B. Russer, and J. Nowotny, *J. Catal.* **34**, 124 (1974).
3. Z. Knor, *Adv. Catal.* **22**, 51 (1972).
4. W. M. H. Sachtler and P. Van der Plank, *Surface Sci.* **18**, 62 (1969).
5. J. N. Butler, J. Giner, and J. M. Parry, *Surface Sci.* **18**, 140 (1969).
6. G. C. Bond, *Surface Sci.* **18**, 11 (1969).
7. P. Sabatier, *Ber. Bunsenges. Phys. Chem.* **44**, 2001 (1911).
8. A. A. Balandin, *Adv. Catal.* **10**, 120 (1958).
9. P. H. Emmett, "Chemisorption and Catalysis" in Symp. on Properties of Surfaces, ASTM Mat. Sci. Series No. 4, 1963.
10. A. J. Appleby, *Catal. Rev.* **4**, 221 (1970).
11. R. Parsons, *Surface Sci.* **18**, 28 (1969).
12. A. K. Vijh, *J. Catal.* **31**, 51 (1973).
13. L. H. Gale, D. C. Olson, and J. Vasilevski, *J. Catal.* **24**, 548 (1972).
14. D. A. Dowden, *J. Chem. Soc.* **1950**, 242 (1950).
15. O. Johnson, *J. Catal.* **28**, 503 (1973).
16. D. A. Dowden, *Catal. Rev.* **5**, 1 (1971).
17. N. Giordano, *Chim. L'Industria* **51**, 1189 (1969).
18. O. Peshev, W. Malakhov, and Th. Volkenshtein, *Prog. Surface Sci.* **6** (1976).
19. F. F. Volkenshtein, *The Electronic Theory of Catalysis on Semiconductors* (Mac Millan, New York, 1963).
20. F. F. Volkenshtein, *Russ. Chem. Rev.* **35**, 537 (1966).

21. K. Hauffe, *Rev. Pure Appl. Chem.* **18**, 79 (1968).
22. S. R. Morrison, *J. Catal.* **20**, 110 (1971).
23. V. J. Lee and D. R. Mason, Proc. 3rd Int. Cong. on Catalysis, p. 556, Amsterdam, 1964.
24. V. J. Lee, *J. Catal.* **17**, 178 (1970).
25. T. Freund and W. P. Gomes, *Catal. Rev.* **3**, 1 (1969).
26. R. M. Dell, F. S. Stone, and P. F. Tiley, *Trans. Faraday Soc.* **49**, 201 (1953).
27. Y. Moro-Oka and A. Ozaki, *J. Catal.* **5**, 116 (1966).
28. W. R. Bottoms and D. B. Lidow, *J. Electrochem. Soc.* **122**, 119 (1975).
29. K. Baron, D. W. Blakely, and G. A. Somorjai, *Surface Sci.* **41**, 45 (1974).
30. B. Lang, R. W. Joyner, and G. A. Somorjai, *Surface Sci.* **30**, 454 (1972).
31. H. Saltsburg, J. N. Smith, and R. L. Palmer, in "Surface Science of Evaporation and Effusion," Am. Vac. Soc. Symp., Los Alamos, 1969.
32. R. D. Giles, J. A. Harrison, and H. R. Thirsk, *J. Electroanal. Chem.* **20**, 47 (1969).
33. J. W. May, *Adv. Catal.* **21**, 151 (1970).
34. J. M. Thomas, *Adv. Catal.* **19**, 293 (1969).
35. J. V. Sanders, *Jap. J. Appl. Phys.*, Suppl. 2, Pt. 2, 479 (1974).
36. S. Siegel, *Adv. Catal.* **16**, 123 (1966).
37. S. Kolboe, *J. Catal.* **27**, 379 (1972).
38. K. Tanabe, *Solid Acids and Bases* (Academic Press, New York, 1970).
39. H. Pines and C. N. Pillai, *J. Am. Chem. Soc.* **83**, 3270 (1961).
40. H. Pines and F. G. Schapnell, *J. Org. Chem.* **31**, 1375 (1966).
41. A. J. Leonard, P. Ratnasamy, F. D. Declerck, and J. J. Fripiat, *Disc. Faraday Soc.* **52**, 98 (1971).
42. V. C. F. Holm, G. C. Bailey, and A. Clark, *J. Phys. Chem.* **63**, 129 (1959).
43. M. Sato, T. Aonuma, and T. Shiba, Proc. 3rd Int. Cong. on Catalysis, Vol. I, Paper 17, Amsterdam, 1964.
44. D. Ballivet, D. Barthomeuf, and Y. Trambouze, *J. Catal.* **26**, 34 (1972); **34**, 423 (1974).
45. M. Ishizura, K. Aika, and A. Ozaki, *J. Catal.* **38**, 189 (1975).
46. H. Pines, *Adv. Catal.* **12**, 117 (1960).
47. M. J. Baird and J. H. Lunsford, *J. Catal.* **26**, 440 (1972).
48. Z. G. Szabo, B. Jover, and R. Ohmacht, *J. Catal.* **39**, 225 (1975).
49. K. Morishige, H. Hattori, and K. Tanabe, *Bull. Chem. Soc. Jap.* **48**, 3088 (1975).
50. C. L. Kibley and W. K. Hall, *J. Catal.* **31**, 65 (1973).
51. F. Figueras, A. Nohl, L. de Mourgues, and Y. Trambouze, *Trans. Faraday Soc.* **67**, 155 (1971).
52. H. Niiyama, S. Moiri, and E. Etchigoya, *Bull. Chem. Soc. Jap.* **45**, 655 (1972).
53. E. Licht, Y. Schächter, and H. Pines, *J. Catal.* **34**, 338 (1974).
54. Y. Kubokawa, H. Miyata, T. Ono, and S. Kawasaki, *J. Chem. Soc. Chem. Commun.* 655 (1974).
55. L. Ya. Margolis, *Catal. Rev.* **8**, 241 (1973).
56. J. M. Criado, F. Gonzalez, and J. M. Trillo, *J. Catal.* **23**, 11 (1971).
57. J. A. Schwarz and R. J. Madix, *Surface Sci.* **46**, 317 (1974).
58. A. Frennet and G. Lienard, *Surface Sci.* **18**, 80 (1969).
59. C. Kemball, *Catal. Rev.* **5**, 32 (1971).
60. J. L. Garnett and W. A. Sollich-Baumgartner, *Adv. Catal.* **16**, 95 (1966).
61. G. C. Bond and P. B. Wells, *Adv. Catal.* **15**, 92 (1964).

62. J. L. Garnett, *Catal. Rev.* **5**, 229 (1971).
63. J. L. Gland and G. A. Somorjai, *Surface Sci.* **41**, 387 (1974).
64. F. Pennella, R. B. Regier, and R. L. Banks, *J. Catal.* **34**, 52 (1974).
65. F. D. Mango, *Adv. Catal.* **20**, 291 (1969).
66. W. M. H. Sachtler, *Catal. Rev.* **4**, 27 (1970).
67. A. W. Czanderna, *J. Phys. Chem.* **68**, 2765 (1964).
68. P. A. Kilty and W. M. H. Sachtler, *Catal. Rev.* **10**, 1 (1974).
69. W. Herzog, *Ber. Bunsenges. Phys. Chem.* **76**, 64 (1972).
70. P. C. Gravelle and S. J. Teichner, *J. Catal.* **20**, 168 (1970).
71. Y. Kubokawa, T. Ono, and N. Yano, *J. Catal.* **28**; 471 (1973).
72. T. Ono, T. Tomino, and Y. Kubokawa, *J. Catal.* **31**, 167 (1973).
73. V. V. Nikisha, B. N. Shelimov, V. A. Schvets, A. P. Griva, and V. B. Kazanski, *J. Catal.* **28**, 230 (1973).
74. P. Mars and D. W. Van Krevelan, *Chem. Eng. Sci.* **3**, 41 (1954).
75. M. J. Fuller and M. E. Warwick, *J. Chem. Soc. Chem. Commun.* **1973**, 210 (1973).
76. B. Viswanathan, V. Ramalingam, and M. V. C. Sastri, *Ind. J. Chem.* **12**, 205 (1974).
77. G. K. Boreskov, *Kinet. Catal.* **14**, 2 (1973).
78. W. M. H. Sachtler, *Catal. Rev.* **4**, 27 (1970).
79. G. K. Boreskov, *Adv. Catal.* **15**, 285 (1964).
80. N. I. Il'Chenko and G. I. Golodets, *J. Catal.* **39**, 57 (1975).
81. V. V. Popovskii, E. A. Mamedov, and G. K. Boreskov, *Kinet. Katal.* **13**, 145 (1972).
82. M. Breysse, M. Guenin, B. Claudel, and H. Latreille, *J. Catal.* **27**, 275 (1972).
83. M. Akimoto and E. Echigoya, *J. Catal.* **29**, 191 (1973).
84. M. Ai and T. Ikawa, *J. Catal.* **40**, 203 (1975).
85. K. Tarama, S. Teranishi, S. Yoshida, and N. Tamura, Proc. 3rd Int. Cong. on Catalysis, Vol. II, p. 24, Amsterdam, 1965.
86. F. Trifirò and I. Pasquon, *J. Catal.* **12**, 412 (1968).
87. G. C. Bairaclough, J. Lewis, and R. S. Nyholm, *J. Chem. Soc.* **1959**, 3552 (1959).
88. M. Akimoto and E. Echigoya, *J. Catal.* **35**, 278 (1974).
89. F. Weiss, J. Marion, J. Metzger, and J.-M. Cognion, *Kinet. Catal.* **14**, 32 (1973).
90. G. K. Boreskov, *Accounts Chem. Res.* **2**, 981 (1973).
91. V. B. Kazanski, *Kinet. Catal.* **14**, 72 (1973).
92. V. V. Nikisha, B. N. Shelimov, V. A. Svets, A. P. Griva, and V. B. Kazanski, *J. Catal.* **28**, 230 (1973).
93. E. R. S. Winter, *J. Catal.* **15**, 144 (1969).
94. E. R. S. Winter, *J. Catal.* **22**, 158 (1971).
95. V. I. Gorgoraki, G. K. Boreskov, and L. A. Kisatkina, *Kinet. Katal.* **7**, 266 (1966).
96. S. R. Morrison, *J. Catal.* **34**, 462 (1974).
97. M. O'Keefe, Y. Ebisuzaki, and W. J. Moore, *J. Phys. Soc. Jap.* **18**, Suppl. 2, 131 (1963).
98. R. Frerichs and I. Liberman, *Phys. Rev.* **121**, 991 (1961).
99. K. Aykan, A. W. Sleight, and D. B. Rogers, *J. Catal.* **29**, 185 (1973).
100. L.-R. le Coustumer, A. d'Huyssen, and J.-P. Bonnelle, *C.R. Acad. Sci.* (*Paris*) **278c**, 989 (1974).
101. R. L. Burwell, Jr., G. L. Haller, K. C. Taylor, and J. F. Read, *Adv. Catal.* **20**, 1 (1969).
102. M. P. McDaniel and R. L. Burwell, Jr., *J. Catal.* **36**, 394 (1975).

103. M. Shelef, K. Otto, and H. Gandhi, *J. Catal.* **12**, 361 (1968).
104. H. C. Yao and M. Shelef, *J. Catal.* **31**, 377 (1973).
105. P. Descamps, M. Guelton, J.-P. Bonnelle, and J.-P. Beaufils, *C.R. Acad. Sci. (Paris)* **279c**, 813 (1974).
106. S. R. Morrison and J.-P. Bonnelle, *J. Catal.* **25**, 416 (1972).
107. H. T. Spath and K. Torkar, *J. Catal.* **26**, 163 (1972).
108. H. Hattori, M. Itoh, and K. Tanabe, *J. Catal.* **38**, 172 (1975).
109. B. J. Wood, H. Wise, and R. S. Yolles, *J. Catal.* **15**, 355 (1969).
110. M. Nakamura, K. Kawai, and Y. Jujiwara, *J. Catal.* **34**, 345 (1974).
111. H. Clark and D. J. Berets, *Adv. Catal.* **9**, 204 (1957).
112. G. L. Simard, J. F. Steger, R. J. Arnott, and L. A. Siegel, *Ind. Eng. Chem.* **47**, 1424 (1955).
113. A. Kato, K. Tomoda, I. Mochida, and T. Seryama, *Bull. Chem. Soc. Jap.* **45**, 690 (1972).
114. L. le Coustumer, J.-P. Bonnelle, and J.-P. Beaufils, *J. Chim. Phys.* (1976).
115. G. D. Kolovertnov, G. K. Boreskov, V. A. Dzis'ko, B. I. Popov, D. V. Tarasova, and G. G. Belugina, *Kinet. Katal.* **6**, 950, 1052 (1965).
116. W. P. Gomes, *Surface Sci.* **19**, 172 (1970).
117. L. Ya. Margolis, E. Kh. Enikeev, and O. Isaev, *Kinet. Katal.* **3**, 181 (1962).
118. K. Tanaka and G. Blyholder, *Chem. Comm.* **1971**, 736 (1971).
119. K. M. Sancier, *J. Catal.* **9**, 331 (1967).
120. K. Tanaka and K. Miyahara, *JCS Chem. Comm.* **1973**, 877 (1973).
121. P. Amigues and S. J. Teichner, *Disc. Faraday Soc.* **41**, 362 (1966).
122. J. Deren and R. Mania, *Roc. Chem. Ann. Soc. Chem. Pol.* **48**, 1565 (1974).
123. J. Cunningham, J. J. Kelly, and A. L. Penny, *J. Phys. Chem.* **74**, 1992 (1970).
124. G. K. Boreskov and K. I. Matveyev, *Problemy Kinet. Katal.* **8**, 165 (1955).
125. G. Natta, in *Catalysis*, Vol. 3, ed. P. H. Emmett (Reinhold, New York, 1955).
126. B. Viswanathan, M. V. C. Sastri, and V. Srinivisan, *Z. Phys. Chem. NF*, **79**, 216 (1972).
127. D. P. McArthur, H. Bliss, and J. B. Butt, *J. Catal.* **28**, 183 (1973).
128. K. J. Miller and J.-L. Wu, *J. Catal.* **27**, 60 (1972).
129. S. R. Morrison, *Surface Sci.* **13**, 85 (1969).
130. S. R. Morrison, unpublished data.
131. A. C. Herd, T. Onisha, and K. Tamaru, *Bull. Chem. Soc. Jap.* **47**, 575 (1974).
132. A. E. Morgan and G. A. Somorjai, *J. Chem. Phys.* **51**, 3309 (1969).
133. H. P. Bonzel and R. Ku, *Surface Sci.* **33**, 91 (1972); *J. Vac. Sci. Tech.* **9**, 663 (1972).
134. Y. Nishiyama and H. Wise, *J. Catal.* **32**, 50 (1974).
135. E. McCarthy, J. Zahradnik, G. C. Kuczynski, and J. J. Canberry, *J. Catal.* **39**, 29 (1975).
136. J. P. Dauchot and J. van Cakenberghe, *Jap. J. Appl. Phys.*, Suppl. 2, Pt. 2, 533 (1974).
137. G. K. Hori and L. D. Schmidt, *J. Catal.* **38**, 335 (1975).
138. T. Matsushima and J. M. White, *J. Catal.* **39**, 265 (1975).
139. G. Ertl, *Ber. Bunsenges. Phys. Chem.* **75**, 967 (1967).
140. F. P. Netzer and G. Kneringer, *Surface Sci.* **51**, 526 (1975).
141. G. J. K. Acres, *Platinum Met. Rev.* **10**, 60 (1966).
142. J. R. Anderson, R. J. MacDonald, and Y. Shimoyama, *J. Catal.* **20**, 147 (1971).
143. D. R. Kahn, E. E. Petersen, and G. A. Somorjai, *J. Catal.* **34**, 294 (1974).

144. H. C. Egghart, *J. Catal.* **31**, 319 (1973).

145. C. R. Adams and T. J. Jennings, *J. Catal.* **3**, 549 (1964).

146. C. R. Adams, H. H. Voge, C. Z. Morgan, and W. E. Armstrong, *J. Catal.* **3**, 379 (1964).

147. J. Haber and B. Gryzybowska, *J. Catal.* **28**, 489 (1973).

148. G. W. Keulks, *J. Catal.* **19**, 232 (1970).

149. R. D. Wragg, P. G. Ashmore, and J. A. Hockey, *J. Catal.* **22**, 49 (1971).

150. T. Otsubo, H. Miura, Y. Moriwaka, and T. Shirasaki, *J. Catal.* **36**, 240 (1975).

151. J. M. Peacock, A. J. Parker, P. G. Ashmore, and J. A. Hockey, *J. Catal.* **15**, 387 (1969).

152. A. C. A. M. Bleijenberg, B. C. Lippens, and G. C. A. Schuit, *J. Catal.* **4**, 58 (1965).

153. Ph. A. Batist, B. C. Lippens, and G. C. A. Schuit, *J. Catal.* **5**, 55 (1966).

154. Ph. A. Batist, A. H. W. M. der Kinderen, Y. Leeuwenburgh, F. A. M. G. Metz, and G. C. A. Schuit, *J. Catal.* **12**, 45 (1968).

155. K. M. Sancier, T. Dozono, and H. Wise, *J. Catal.* **23**, 270 (1971).

156. J. M. Peacock, A. J. Parker, P. G. Ashmore, and J. A. Hockey, *J. Catal.* **15**, 373 (1969).

157. G. C. A. Schuit, *J. Less Comm. Met.* **36**, 329 (1974).

158. M. Ai and S. Suzuki, *J. Catal.* **26**, 202 (1972); **30**, 362 (1973).

159. C. R. Adams, Proc. 3rd Int. Cong. on Catalysis, Vol. 1, p. 240, Amsterdam, 1965.

160. I. Matsuura and G. C. A. Schuit, *J. Catal.* **20**, 19 (1971); **25**, 314 (1972).

161. N. Pernicone, F. Lazzerin, G. Liberti, and G. Lanzavecchia, *J. Catal.* **14**, 293 (1969).

162. G. Liberti, N. Pernicone, and S. Soattini, *J. Catal.* **27**, 52 (1972).

163. G. Fagherazzi and N. Pernicone, *J. Catal.* **16**, 321 (1970).

* Author Index

The numbers in italics refer to text pages on which the reference appears, the boldface numbers refer to the chapter numbers, and the arabic numbers to the reference numbers.

389

✱ | Subject Index

Entries in italics indicate the subject is continued for more than one page.